T0310127

SLIDING MODE CONTROL OF UNCERTAIN PARAMETER-SWITCHING HYBRID SYSTEMS

SLIDING MODE CONTROL OF UNCERTAIN PARAMETER-SWITCHING HYBRID SYSTEMS

Ligang Wu
Harbin Institute of Technology, China

Peng Shi
The University of Adelaide; and Victoria University, Australia

Xiaojie Su
Chongqing University, China

This edition first published 2014

Registered office
John Wiley & Sons Ltd, The Atrium, Southern Gate, Chichester, West Sussex, PO19 8SQ, United Kingdom

For details of our global editorial offices, for customer services and for information about how to apply for permission to reuse the copyright material in this book please see our website at www.wiley.com.

Library of Congress Cataloging-in-Publication Data applied for.

ISBN 9781118862599

Set in 10/12pt Times by Aptara Inc., New Delhi, India
Printed and bound in Malaysia by Vivar Printing Sdn Bhd

1 2014

To Jingyan and Zhixin
L. Wu

To my family
P. Shi

To my family
X. Su

Contents

Series Preface

Electromechanical systems permeate the engineering and technology fields in aerospace, automotive, mechanical, biomedical, civil/structural, electrical, environmental, and industrial systems. The Wiley Book Series on dynamics and control of electromechanical systems will cover a broad range of engineering and technology within these fields. As demand increases for innovation in these areas, feedback control of these systems is becoming essential for increased productivity, precision operation, load mitigation, and safe operation. Furthermore, new applications in these areas require a reevaluation of existing control methodologies to meet evolving technological requirements, for example the distributed control of energy systems. The basics of distributed control systems are well documented in several textbooks, but the nuances of its use for future applications in the evolving area of energy system applications, such as wind turbines and wind farm operations, solar energy systems, smart grids, and the generation, storage and distribution of energy, require an amelioration of existing distributed control theory to specific energy system needs. The book series serves two main purposes: 1) a delineation and explication of theoretical advancements in electromechanical system dynamics and control, and 2) a presentation of application-driven technologies in evolving electromechanical systems.

This book series will embrace the full spectrum of dynamics and control of electromechanical systems from theoretical foundations to real-world applications. The level of the presentation should be accessible to senior undergraduate and first-year graduate students, and should prove especially well-suited as a self-study guide for practicing professionals in the fields of mechanical, aerospace, automotive, biomedical, and civil/structural engineering. The aim is to provide an interdisciplinary series, ranging from high-level undergraduate/graduate texts, explanation and dissemination of science and technology and good practice, through to important research that is immediately relevant to industrial development and practical applications. It is hoped that this new and unique perspective will be of perennial interest to students, scholars, and employees inthe engineering disciplines mentioned. Suggestions for new topics and authors for the series are always welcome.

This book, *Sliding Mode Control of Uncertain Parameter-Switching Hybrid Systems*, has the objective of providing a theoretical foundation as well as practical insights on the topic at hand. It is broken down into three parts: 1) sliding mode control (SMC) of Markovian jump singular systems, 2) SMC of switched state-delayed hybrid systems, and 3) SMC of switched stochastic hybrid systems. The book provides detailed derivations from first principles to allow the reader to thoroughly understand the particular topic. This is especially useful for Markovian jump singular systems with stochastic perturbations because a comprehensive knowledge of stochastic analysis is not required before understanding the material. Readers can simply dive into the material. It also provides several illustrative examples to bridge the gap between theory and practice. It is a welcome addition to the Wiley Electromechanical Systems Series because no other book is focused on the topic of SMC with a specific emphasis on uncertain parameter-switching hybrid systems.

Mark J. Balas
John L. Crassidis
Florian Holzapfel
Series Editors

Preface

Since the 1950s, sliding mode control (SMC) has been recognized as an effective robust control strategy for nonlinear systems and incompletely modeled systems. In the past two decades, SMC has been successfully applied to a wide variety of real world applications such as robot manipulators, aircraft, underwater vehicles, spacecraft, flexible space structures, electrical motors, power systems, and automotive engines. Basically, the idea of SMC is to utilize a discontinuous control to force the system state trajectories to some predefined sliding surfaces on which the system has desired properties such as stability, disturbance rejection capability, and tracking ability. Many important results have been reported for this kind of control strategy. However, when the controlled plants are uncertain parameter-switching hybrid systems including parameter-switching (Markovian jump or arbitrary switching), state-delay, stochastic perturbation, and singularly perturbed terms, the common SMC methodologies cannot meet the requirements.

It is known that the SMC of uncertain parameter-switching hybrid systems is much more complicated because sliding mode controllers must be designed so that not only is the sliding surface robustly reachable, but also the sliding mode dynamics can converge the system's equilibrium automatically by choosing a suitable switching function. This book aims to present up-to-date research developments and novel methodologies on SMC of uncertain parameter-switching hybrid systems in a unified matrix inequality setting. The considered uncertain parameter-switching hybrid systems include Markovian switching hybrid systems, switched state-delayed hybrid systems, and switched stochastic hybrid systems. These new methodologies provide a framework for stability and performance analysis, SMC design, and state estimation for these classes of systems. Solutions to the design problems are presented in terms of linear matrix inequalities (LMIs). In this book, a large number of references are provided for researchers who wish to explore the area of SMC of uncertain parameter-switching hybrid systems, and the main contents of the book are also suitable for a one-semester graduate course.

In this book, we present new SMC methodologies for uncertain parameter-switching hybrid systems. The systems under consideration include Markovian jump systems, singular systems, switched hybrid systems, stochastic systems, and time-delay systems.

The content of this book are divided into three parts. The first part is focused on SMC of Markovian jump singular systems. Some necessary and sufficient conditions are derived for the stochastic stability, stochastic admissibility, and optimal performances by developing new techniques for the considered Markovian jump singular systems. Then a set of new SMC methodologies are proposed, based on the analysis results. The main contents are as follows:

Chapter 2 is concerned with the state estimation and SMC of singular Markovian switching systems; Chapter 3 studies the optimal SMC problem for singular Markovian switching systems with time delay; and Chapter 4 establishes the integral SMC method for singular Markovian switching stochastic systems.

In the second part, the problem of SMC of switched state-delayed hybrid systems is investigated. A unified approach of the piecewise Lyapunov function combining with the average dwell time technique is developed for analysis and synthesis of the considered systems. By this approach, some sufficient conditions are established for the stability and synthesis of the switched state-delayed hybrid system. More importantly, a set of SMC methodologies under a unique framework are proposed for the considered hybrid systems. The main contents of this part are as follows: Chapter 5 is devoted to the stability analysis and the stabilization problems for switched state-delayed hybrid systems; Chapter 6 investigates the optimal dynamic output feedback (DOF) control of switched state-delayed hybrid systems; and Chapters 7 and 8 study the SMC of continuous- and discrete-time switched state-delayed hybrid systems, respectively.

In the third part, the parallel theories and techniques developed in the second part are extended to deal with switched stochastic hybrid systems. The main contents include the following: Chapters 9 and 10 are concerned with the control of switched stochastic hybrid systems for continuous- and discrete-time cases, respectively; Chapter 11 studies the observer-based SMC of switched stochastic hybrid systems; and Chapter 12 focuses on the dissipativity-based SMC of switched stochastic hybrid systems.

This book is a research monograph whose intended audience is graduate and postgraduate students, academics, scientists and engineers who are working in the field.

Ligang Wu
Harbin, China

Peng Shi
Melbourne, Australia

Xiaojie Su
Chongqing, China
December 2013

Acknowledgments

There are numerous individuals without whose help this book would not have been completed. Special thanks go to Professor James Lam from The University of Hong Kong, Professor Daniel W. C. Ho from City University of Hong Kong, Professor Zidong Wang from Brunel University, Professor Wei Xing Zheng from University of Western Sydney, Professor Yugang Niu from East China University of Science and Technology and Professor Huijun Gao from Harbin Institute of Technology, for their valuable suggestions, constructive comments and support.

Next, our acknowledgements go to many colleagues who have offered support and encouragement throughout this research effort. In particular, we would like to acknowledge the contributions from Jianbin Qiu, Ming Liu, Guanghui Sun, and Hongli Dong. Thanks also go to our students, Rongni Yang, Xiuming Yao, Fanbiao Li, Xiaozhan Yang, Chunsong Han, Yongyang Xiong, and Huiyan Zhang, for their comments. The authors are especially grateful to their families for their encouragement and never-ending support when it was most required. Finally, we would like to thank the editors at Wiley for their professional and efficient handling of this project.

The writing of this book was supported in part by the National Natural Science Foundation of China (61174126, 61222301, 61134001, 61333012, 61174058), the Fok Ying Tung Education Foundation (141059), the Fundamental Research Funds for the Central Universities (HIT.BRETIV.201303), the Australian Research Council (DP140102180), the Engineering and Physical Sciences Research Council, UK (EP/F029195), the Fundamental Research Funds for the Central Universities (2013YJS021), the National Key Basic Research Program, China (2011CB710706, 2012CB215202), the 111 Project (B12018), and the Key Laboratory of Integrated Automation for the Process Industry, Northeast University.

Abbreviations and Notations

Abbreviations

CCL	cone complementary linearization
CQLF	common quadratic Lyapunov function
DOF	dynamic output feedback
LMI	linear matrix inequality
LQR	linear-quadratic regulator
LTI	linear time-invariant
MIMO	multiple-input multiple-output
MJLS	Markovian jump linear system
MLF	multiple Lyapunov function
SISO	single-input single-output
SMC	sliding mode control
SOF	static output feedback
SQLF	switched quadratic Lyapunov functions

Notations

$\blacksquare$	end of proof
$\blacklozenge$	end of remark
$\triangleq$	is defined as
$\in$	belongs to
$\forall$	for all
$\sum$	sum
$\mathbf{C}$	field of complex numbers
$\mathbf{R}$	field of real numbers
$\mathbf{Z}$	field of integral numbers
$\mathbf{R}^n$	space of n-dimensional real vectors
$\mathbf{R}^{n\times m}$	space of $n \times m$ real matrices
$\mathbf{C}_{n,d}$	set of $\mathbf{R}^n$-valued continuous functions on $[-d, 0]$
$\mathbf{E}\{\cdot\}$	mathematical expectation operator
lim	limit
max	maximum
min	minimum

sup	supremum
inf	infimum
rank$(\cdot)$	rank of a matrix
trace$(\cdot)$	trace of a matrix
$\lambda_{\min}(\cdot)$	minimum eigenvalue of a real symmetric matrix
$\lambda_{\max}(\cdot)$	maximum eigenvalue of a real symmetric matrix
diag	block diagonal matrix with blocks $\{X_1, \ldots, X_m\}$
$\sigma_{\min}(\cdot)$	minimum singular value of a real symmetric matrix
$\sigma_{\max}(\cdot)$	maximum singular value of a real symmetric matrix
I	identity matrix with appropriate dimension
I_n	$n \times n$ identity matrix
0	zero matrix with appropriate dimension
$0_{n\times m}$	zero matrix of dimension $n \times m$
X^T	transpose of matrix X
X^{-1}	inverse of matrix X
$X^{\perp}$	full row rank matrix satisfying $X^{\perp}X = 0$ and $X^{\perp}X^{\perp T} > 0$
$X > (<)0$	X is real symmetric positive (negative) definite
$X \geq (\leq)0$	X is real symmetric positive (negative) semi-definite
$\mathcal{L}_2[0, \infty)$	space of square integrable functions on $[0, \infty)$ (continuous case)
$\ell_2[0, \infty)$	space of square summable infinite vector sequences over $[0, \infty)$ (discrete case)
$\lvert\cdot\rvert$	Euclidean vector norm
$\lVert\cdot\rVert$	Euclidean matrix norm (spectral norm)
$\lVert\cdot\rVert_2$	$\mathcal{L}_2$-norm: $\sqrt{\int_0^\infty \lvert\cdot\rvert^2\, dt}$ (continuous case) ℓ_2-norm: $\sqrt{\sum_0^\infty \lvert\cdot\rvert^2}$ (discrete case)
$\star$	symmetric terms in a symmetric matrix

1

Introduction

1.1 Sliding Mode Control

Sliding mode control (SMC) has proven to be an effective robust control strategy for incompletely modeled or nonlinear systems since its first appearance in the 1950s [70, 103, 197]. One of the most distinguished properties of SMC is that it utilizes a discontinuous control action which switches between two distinctively different system structures such that a new type of system motion, called sliding mode, exists in a specified manifold. The peculiar characteristic of the motion in the manifold is its insensitivity to parameter variations, and its complete rejection of external disturbances [260]. SMC has been developed as a new control design method for a wide spectrum of systems including nonlinear, time-varying, discrete, large-scale, infinite-dimensional, stochastic, and distributed systems [101]. Also, in the past two decades, SMC has successfully been applied to a wide variety of practical systems such as robot manipulators, aircraft, underwater vehicles, spacecraft, flexible space structures, electrical motors, power systems, and automotive engines [60, 77, 199, 259].

In this section, we will first present some preliminary background and fundamental theory of SMC, which will be helpful to some readers who have little or no knowledge on SMC, and then we will give an overview of recent development of SMC methodologies.

1.1.1 Fundamental Theory of SMC

We first formulate the SMC problem as follows. For a general nonlinear system of the form

$$\dot{x}(t) = f(x, u, t), \tag{1.1}$$

where $x(t) \in \mathbf{R}^n$ is the system state vector, $u(t) \in \mathbf{R}^m$ is the control input. We need to design a sliding surface

$$s(x) = 0,$$

where $s(x)$ is called the switching function, and the order of $s(x)$ is usually the same as that of the control input, i.e. $s(x) \in \mathbf{R}^m$, and

$$s(x) = \begin{bmatrix} s_1(x) & s_2(x) & \cdots & s_m(x) \end{bmatrix}^T.$$

Sliding Mode Control of Uncertain Parameter-Switching Hybrid Systems, First Edition. Ligang Wu, Peng Shi and Xiaojie Su.

Then a sliding mode controller $u(t) = [\, u_1(t) \quad u_2(t) \quad \cdots \quad u_m(t)\,]^T$ is designed in the form of

$$u_i(t) = \begin{cases} u_i^+(t), & \text{when } s_i(x) > 0, \\ u_i^-(t), & \text{when } s_i(x) < 0, \end{cases} \qquad i = 1, 2, \ldots, m,$$

where $u_i^+(t) \neq u_i^+(t)$, such that the following two conditions hold:

Condition 1. The sliding mode is reached in a finite time and subsequently maintained, that is, the system state trajectories can be driven onto the specified sliding surface $s(x) = 0$ by the sliding mode controller in a finite time and maintained there for all subsequent time;
Condition 2. The dynamics in sliding surface $s(x) = 0$, that is, the sliding mode dynamics, is stable with some specified performances.

Further consider (1.1) with single input, that is, $u(t) \in \mathbf{R}$ and $s(x) \in \mathbf{R}$, and suppose that the sliding mode can be reached in a finite time, then the solutions of the equation

$$\dot{x}(t) = f(x, u^+(t), t), \quad s(x) > 0,$$

will approach $s(x) = 0$ and reach there in a finite time. During the approaching phase, $\dot{s}(x) < 0$. Similarly, the solutions of the equation

$$\dot{x}(t) = f(x, u^-(t), t), \quad s(x) < 0,$$

will also approach $s(x) = 0$ and reach there in a finite time, thus we have $\dot{s}(x) > 0$. To summarize the above analysis, we have

$$\begin{cases} \dot{s}(x) < 0, & \text{when } s(x) > 0, \\ \dot{s}(x) > 0, & \text{when } s(x) < 0, \end{cases}$$

or, equivalently,

$$s(x)\dot{s}(x) < 0.$$

which is the so-called 'reaching condition'. This is the condition under which the state will move toward and reach a sliding surface. The system state trajectories under the reaching condition is called the reaching phase [77, 101].

In summary, Condition 1 requires the reachability of a sliding mode, which is guaranteed through designing a sliding mode controller, while Condition 2 requires the sliding mode dynamics to be stable with some specified performances, which is assured by designing an appropriate sliding mode surface. Therefore, a conventional SMC design consists of two steps:

Step 1. Design a sliding surface $s(x) = 0$ such that the dynamics restricted to the sliding surface has the desired properties such as stability, disturbance rejection capability, and tracking;
Step 2. Design a discontinuous feedback control $u(t)$ such that the system state trajectories can be attracted to the designed sliding surface in a finite time and maintained on the surface for all subsequent time.

In the following, we will briefly introduce some commonly used methods in the design of sliding surfaces and sliding mode controllers, and in the elimination/reduction of chattering. Readers can refer to various books on SMC theory for more details, for example, [60, 77, 197, 199].

Sliding Surface Design

In this section, three kinds of sliding surfaces, namely, linear sliding surface, integral sliding surface, and terminal sliding surface, are introduced.

Linear Sliding Surface

The linear sliding surface, due to its simplicity of implementation, is commonly used in SMC design. There are two approaches to designing linear sliding surface. First, we introduce the 'regular form' model transformation approach. Consider the following nonlinear system:

$$\dot{x}(t) = f(x,t) + B(x,t)u(t), \tag{1.2}$$

where $x(t) \in \mathbf{R}^n$ and $u(t) \in \mathbf{R}^m$ are the system states and control inputs, respectively. $f(x,t) \in \mathbf{R}^n$ and $B(x,t) \in \mathbf{R}^{n\times m}$ are assumed to be continuous with bounded continuous derivatives with respect to x. $B(x,t)$ is bounded away from zero at any time.

By applying an appropriate diffeomorphic transformation $z(t) = \begin{bmatrix} z_1(t) \\ z_2(t) \end{bmatrix} = Tx(t)$, system (1.2) can be written in the following regular form [120]:

$$\begin{bmatrix} \dot{z}_1(t) \\ \dot{z}_2(t) \end{bmatrix} = \begin{bmatrix} \hat{f}_1(z,t) \\ \hat{f}_2(z,t) \end{bmatrix} + \begin{bmatrix} 0 \\ \hat{B}_1(z,t) \end{bmatrix} u(t),$$

where $z_1(t) \in \mathbf{R}^{n-m}$ and $z_2(t) \in \mathbf{R}^m$ are the transformed system states. $\hat{B}_1(z,t) \in \mathbf{R}^{m\times m}$ is nonsingular (to ensure this, the matrix $B(x,t)$ should be of full column rank for all t for the existence of such a transformation).

Design a switching function as

$$s(z) = z_2(t) + \hbar(z_1(t)),$$

where $\hbar(\cdot)$ is a function to be defined. When the system state trajectories reach onto the sliding surface, we have $s(z) = 0$, thus $z_2(t) = -\hbar(z_1(t))$. Substituting this into the first equation of the regular form yields

$$\dot{z}_1(t) = \hat{f}_1\left(z_1, z_2, t\right) = \hat{f}_1\left(z_1, -\hbar(z_1(t)), t\right).$$

which is a reduced-order system representing the sliding mode dynamics. The remaining work of the sliding surface design is to choose a function $\hbar(\cdot)$ such that the above nonlinear sliding mode dynamics is stable and/or satisfies a specified performance.

For a linear time-invariant (LTI) system of the form

$$\dot{x}(t) = Ax(t) + Bu(t), \tag{1.3}$$

where $x(t) \in \mathbf{R}^n$ is the system state vector, $u(t) \in \mathbf{R}^m$ is the control input, and the matrices $A \in \mathbf{R}^{n \times n}$ and $B \in \mathbf{R}^{n \times m}$. The matrix B is assumed to have full column rank and the pair (A, B) is assumed to be controllable.

It is well known that for the controllable system (1.3) there exists a nonsingular transformation, defined by

$$\begin{bmatrix} z_1(t) \\ z_2(t) \end{bmatrix} = z(t) = Tx(t),$$

such that

$$TAT^{-1} = \begin{bmatrix} A_{11} & A_{12} \\ A_{21} & A_{22} \end{bmatrix}, \quad TB = \begin{bmatrix} 0 \\ B_1 \end{bmatrix}.$$

Thus, by $z(t) = Tx(t)$ system (1.3) can be transformed into the following regular form:

$$\begin{cases} \dot{z}_1(t) = A_{11}z_1(t) + A_{12}z_2(t), \\ \dot{z}_2(t) = A_{21}z_1(t) + A_{22}z_2(t) + B_1 u(t), \end{cases} \tag{1.4}$$

where $z_1(t) \in \mathbf{R}^{n-m}$ and $z_2(t) \in \mathbf{R}^m$ are the transformed system states. $A_{11} \in \mathbf{R}^{(n-m)\times(n-m)}$, $A_{12} \in \mathbf{R}^{(n-m)\times m}$, $A_{21} \in \mathbf{R}^{m\times(n-m)}$, $A_{22} \in \mathbf{R}^{m\times m}$, $B_1 \in \mathbf{R}^{m\times m}$, and B_1 is nonsingular.

Now, a sliding surface can be designed under the model of (1.4). For example, we can choose the following linear one:

$$s(z) = z_2(t) + Cz_1(t), \tag{1.5}$$

where C is the design parameter to be designed. Similarly, when the system state trajectories reach onto the sliding surface, that is, $s(z) = 0$, it follows that

$$z_2(t) = -Cz_1(t). \tag{1.6}$$

Substituting (1.6) into the first equation of (1.4) yields

$$\dot{z}_1 = \left(A_{11} - A_{12}C\right) z_1(t). \tag{1.7}$$

The above reduced-order system is the so-called sliding mode dynamics (that is, the motion equation in the sliding surface), which is an autonomous system. Therefore, the design of sliding surfaces becomes choosing the matrix parameter C such that the sliding mode dynamics is stable. Furthermore, since it can be shown that, if the pair (A, B) is controllable, then the pair (A_{11}, A_{12}) is controllable as well, the problem of finding the design matrix C is in fact a classical state feedback problem with matrix C as a feedback gain and A_{12} as an input matrix. Therefore, all existing linear state feedback control design methods can be used to solve this problem, for example, the conventional eigenvalue allocation method and linear-quadratic regulator (LQR) design method.

There is another approach to linear surface design, named the Lyapunov approach [186]. Let $V(x)$ be a Lyapunov function for system (1.2), that is, $V(x) > 0$ and $\dot{V}(x) < 0$. The sliding surface can be chosen as

$$s(x) = B^T(x,t)\left[\frac{\partial V(x)}{\partial x}\right]^T = 0, \tag{1.8}$$

where

$$\frac{\partial V(x)}{\partial x} = \left[\frac{\partial V(x)}{\partial x_1} \quad \frac{\partial V(x)}{\partial x_2} \quad \cdots \quad \frac{\partial V(x)}{\partial x_n}\right].$$

Lemma 1.1.1 [186] *System (1.2) with sliding mode on the sliding surface (1.8) is asymptotically stable.*

For linear system (1.3), since (A, B) is controllable we know that there exists a feedback matrix K such that $\bar{A} = A + BK$ is stabilizable. Thus, there exist matrices $P > 0$ and $Q > 0$ such that the following Lyapunov equation holds:

$$P\bar{A} + \bar{A}^T P = -Q.$$

Now, design the sliding surface as

$$s(x) = B^T Px(t) = 0, \tag{1.9}$$

and rewrite system (1.3) as

$$\dot{x}(t) = \bar{A}x(t) + B\bar{u}(t), \tag{1.10}$$

where $\bar{u}(t) = u(t) - Kx(t)$, and Kx is a fictitious feedback to system (1.3).

Let $V(x) = x^T(t)Px(t) > 0$, and we have

$$\dot{V}(x) = x^T(t)\left(P\bar{A} + \bar{A}^T P\right)x(t) + 2x^T(t)PB\bar{u}(t).$$

When the system state trajectories are driven onto the sliding surface, that is, $s(x) = B^T Px(t) = 0$, it follows that

$$\dot{V}(x) = -x^T(t)Qx(t) < 0,$$

for $x(t) \neq 0$. Therefore, the system states are asymptotically stable on the sliding surface. Therefore, we have the following lemma.

Lemma 1.1.2 [186] *System (1.10) with sliding mode on the sliding surface (1.9) is asymptotically stable.*

We can see from Lemmas 1.1.1–1.1.2 that for the Lyapunov approach, the design of sliding surface is given by the positive definite matrix P.

Integral Sliding Surface

In the above-mentioned linear sliding surface design, the order of the resulting sliding mode dynamics is $(n - m)$, with n being the dimension of the state space and m being the dimension of the control input. Unlike in linear sliding surface design, in the integral sliding surface, the order of the sliding motion equation is the same as that of the original system, rather than being reduced by the number of the dimension of the control input. As the result, the robustness of the system can be guaranteed throughout an entire response of the system starting from the initial time instance [198].

Consider system (1.3) with a nonlinear perturbation included in the input channel (called matched perturbation), that is,

$$\dot{x}(t) = Ax(t) + B\left(u(t) + d(x,t)\right).$$

where $d(x,t)$ is a nonlinear perturbation with known upper bound $d_0(x,t)$, that is, $|d(x,t)| < d_0(x,t)$. Design control $u(t) = u_0(t) + u_1(t)$ for the above system, and suppose that there exists a feedback control law $u(t) = u_0(t)$ such that the perturbation-free system, that is, $\dot{x}(t) = Ax(t) + Bu(t)$ can be stabilized in a desired way. That is, the state trajectories of the closed-loop system $\dot{x}(t) = Ax(t) + Bu_0(t)$ follow pre-specified reference trajectories with a desired accuracy. Here, $u_0(t)$ may be designed through linear static feedback control, such as $u_0(t) = Kx(t)$ in which the feedback gain K can be determined by eigenvalue allocation or LQR methods.

Design the integral switching function as

$$s(x) = Cx(t) - Cx(t_0) - C\int_{t_0}^{t} (A + BK)\, x(\tau)d\tau, \tag{1.11}$$

and C is the parameter matrix to be designed such that CB is nonsingular. Notice that, at $t = t_0$, the switching function $s(x)|_{t=t_0} = 0$, and hence the reaching phase is eliminated. By (1.11), the resulting sliding mode dynamics coincides with that of the ideal system $\dot{x}(t) = Ax(t) + Bu_0(t)$, which means that the integral sliding surface is robust to the perturbation throughout the entire response of the system starting from the initial time instance.

The approaches to integral sliding surface design were then developed for uncertain systems with mismatched uncertainties/perturbations [27, 30, 32, 43, 170], higher order SMC systems [114, 118], stochastic systems [12, 155–158], singular systems [219, 221, 223, 225], and switched hybrid systems [125]. For discrete-time SMC systems, the integral sliding surface design approaches were developed in [1, 107, 233].

Terminal Sliding Surface

The terminal SMC technique was first proposed in [201]. Compared to the conventional SMC, the terminal SMC has some superior properties such as fast and finite-time convergence and high steady-state tracking precision.

Consider the second-order linear system

$$\begin{cases} \dot{x}_1(t) = x_2(t), \\ \dot{x}_2(t) = a_1x_1(t) + a_2x_2(t) + bu(t), \end{cases}$$

where $x_1(t)$ and $x_2(t)$ are the system states and $u(t)$ is the control input.

The following terminal switching function is designed:

$$s(x_1, x_2) = x_2(t) + \beta x_1^{q/p}, \quad \beta > 0,$$

where p and q are positive odd integers that satisfy $p > q$.

Similar to the conventional SMC technique, if the system state trajectories are driven onto the sliding surface, that is, $s(x_1, x_2) = 0$, then

$$\dot{x}_1(t) = -\beta x_1^{q/p}.$$

Let the initial condition of $x_1(t)$ at $t = 0$ be $x_1(0)(\neq 0)$, then the relaxation time t_1 for a solution of above equation is

$$t_1 = -\beta^{-1} \int_{x_1(0)}^{0} \frac{dx_1(\tau)}{x_1^{q/p}(\tau)} = \frac{|x_i(0)|^{(1-q/p)}}{\beta(1 - q/p)},$$

which means that on the terminal sliding surface, the system state trajectories converge to zero in a finite time.

For a high-order single-input single-output (SISO) linear system

$$\begin{cases} \dot{x}_i(t) = x_{i+1}(t), \quad i = 1, 2, \ldots, n-1, \\ \dot{x}_2(t) = \sum_{j=1}^{n} a_j x_j(t) + u(t), \end{cases}$$

the following terminal switching functions are designed:

$$\begin{cases} s_0(x) = x_1(t), \\ s_1(x) = \dot{s}_0(x) + \beta_1 s_0^{q_1/p_1}(x), \\ s_2(x) = \dot{s}_1(x) + \beta_2 s_1^{q_2/p_2}(x), \\ \vdots \\ s_{n-1}(x) = \dot{s}_{n-2}(x) + \beta_{n-1} s_{n-2}^{q_{n-1}/p_{n-1}}(x), \end{cases}$$

where $\beta_i > 0$ are constants and p_i and q_i are positive odd integers satisfying $p_i > q_i$, $i = 1, 2, \ldots, n-1$.

The terminal SMC technique for multiple-input multiple-output (MIMO) systems was proposed in [141], and then developed in [39, 72, 142].

Sliding Mode Controller Design

Having designed the sliding mode via the design of switching functions, the next step is to design a sliding mode controller such that the system state trajectories can be driven onto the specified sliding surface in a finite time and maintained there for all subsequent time. The main requirement in this step is that the control should be designed to satisfy the reaching condition, thus guaranteeing the existence of a sliding mode on the sliding surface. Additional requirements in this reaching phase include some desired properties such as fast reaching and low chattering. In the following, we will introduce some commonly used methods to the sliding mode controller design.

Equivalent Control Design

Equivalent control is designed in the reaching phase, which can satisfy the reachability of the system state trajectories to the sliding surfaces if the system is free of parameter uncertainties and external disturbances. Consider system (1.2) with switching function being $s(x)$. Suppose that the system state trajectories reach onto the sliding surface at time instant t_1 and remain there in the subsequent time. We then have $s(x) = 0$ for all $t > t_1$. Along sliding mode trajectories, $s(x)$ is constant, and so sliding mode trajectories are described by the differential equation $\dot{s}(x) = 0$. Differentiating $s(x)$ yields

$$\dot{s}(x) = \frac{\partial s(x)}{\partial x}\dot{x}(t) = \frac{\partial s(x)}{\partial x}\left[f(x,t) + B(x,t)u_{eq}(t)\right] = 0,$$

where

$$\frac{\partial s(x)}{\partial x} = \begin{bmatrix} \frac{\partial s(x)}{\partial x_1} & \frac{\partial s(x)}{\partial x_2} & \cdots & \frac{\partial s(x)}{\partial x_n} \end{bmatrix}$$

is called the gradient of $s(x)$. Here, we suppose that $\frac{\partial s(x)}{\partial x}B(x,t)$ is nonsingular for all x and t, thus the equivalent control can be solved as follows:

$$u_{eq}(t) = -\left(\frac{\partial s(x)}{\partial x}B(x,t)\right)^{-1}\frac{\partial s(x)}{\partial x}f(x,t). \tag{1.12}$$

Substituting the above equivalent control into the original system, it follows that on the sliding surface $s(x) = 0$ the system dynamics satisfy

$$\dot{x}(t) = \left[I - B(x,t)\left(\frac{\partial s(x)}{\partial x}B(x,t)\right)^{-1}\frac{\partial s(x)}{\partial x}\right]f(x,t).$$

The above differential equation represents the sliding mode dynamics, which is actually is a reduced-order model of order $n - m$. (Considering $s(x) = 0$, thus m of the system states can be eliminated from the equation.)

Reaching Condition Approach

A straightforward method of sliding mode controller design is based on the reaching condition, that is, for $i = 1, 2, \ldots, m$,

$$\begin{cases} \dot{s}_i(x) < 0, & \text{when } s_i(x) > 0, \\ \dot{s}_i(x) > 0, & \text{when } s_i(x) < 0, \end{cases}$$

or, equivalently,

$$s_i(x)\dot{s}_i(x) < 0, \quad i = 1, 2, \dots, m.$$

With the designed controller satisfying the above reachability condition, the tangent vectors of the state trajectories are guaranteed to point toward the sliding surface, hence, the reachability of the system state trajectories to the sliding surface can be guaranteed. Some more discussions of this approach can be found in [101].

Lyapunov Function Approach

The Lyapunov function approach is commonly used in sliding mode controller design. Choosing a Lyapunov function of the form

$$V(s) = s^T(x)s(x),$$

a sufficient condition for the sliding surface to be globally attractive is that the control $u(t)$ is designed such that

$$\dot{V}(s) < 0, \quad \text{when } s(x) \neq 0.$$

Finite reaching time can be guaranteed by [103]

$$\dot{V}(s) < -\epsilon, \quad \text{when } s(x) \neq 0,$$

where $\epsilon > 0$ is a constant.

For system (1.2), design the sliding mode controller as

$$u(t) = u_{eq}(t) + u_N(t),$$

where $u_{eq}(t)$ is the equivalent control which is designed in (1.12), and the discontinuous control $u_N(t)$ is to be chosen such that

$$\begin{aligned}
\dot{V}(s) &= 2s^T(x)\dot{s}(x) \\
&= 2s^T(x)\frac{\partial s(x)}{\partial x}\left[f(x,t) + B(x,t)\left(u_{eq}(t) + u_N(t)\right)\right] \\
&= 2s^T(x)\frac{\partial s(x)}{\partial x}B(x,t)u_N(t) < 0.
\end{aligned}$$

Clearly, this approach leads to the global attraction of the system state trajectories to the sliding surface.

Reaching Law Approach

The reaching law is a differential equation which specifies the dynamics of a switching function, and by the choice of the parameters in the reaching law, the dynamic quality of SMC system in the reaching mode can be controlled [78]. A general form of the reaching law can be described by the differential equation

$$\dot{s}(x) = -\Upsilon \text{sign}(s(x)) - Kg(s(x)), \tag{1.13}$$

where

$$\Upsilon = \text{diag}\{\varepsilon_1, \varepsilon_2, \dots, \varepsilon_m\}, \quad \varepsilon_i > 0,$$
$$K = \text{diag}\{k_1, k_2, \dots, k_m\}, \quad k_i > 0,$$
$$\text{sign}(s(x)) = \begin{bmatrix} \text{sign}(s_1(x)) \\ \text{sign}(s_2(x)) \\ \vdots \\ \text{sign}(s_m(x)) \end{bmatrix}, \quad g(s(x)) = \begin{bmatrix} g_1(s_1(x)) \\ g_2(s_2(x)) \\ \vdots \\ g_m(s_m(x)) \end{bmatrix}.$$

The functions $g_i(s_i(x))$ satisfy $g_i(0) = 0$ and

$$s_i(x)g_i(s_i(x)) > 0, \quad \text{when } s_i(x) \neq 0, i = 1, 2, \dots, m.$$

Therefore, using reaching law (1.13) directly to system (1.2) with

$$\begin{aligned} \dot{s}(x) &= \frac{\partial s(x)}{\partial x}\Big(f(x,t) + B(x,t)u(t)\Big) \\ &= -\Upsilon \text{sign}(s(x)) - Kg(s(x)), \end{aligned}$$

the sliding mode controller can be obtained as

$$u(t) = -\left(\frac{\partial s(x)}{\partial x}B(x,t)\right)^{-1}\left(\frac{\partial s(x)}{\partial x}f(x,t) + \Upsilon \text{sign}(s(x)) + Kg(s(x))\right).$$

Equation (1.13) is a general form of the reaching law, and some special cases are

1. The constant rate reaching law:

$$\dot{s}(t) = -\Upsilon \text{sign}(s(x))$$

2. The constant plus proportional rate reaching law:

$$\dot{s}(t) = -\Upsilon \text{sign}(s(x)) - Ks(t)$$

3. The power rate reaching law:

$$\dot{s}_i(t) = -\varepsilon_i |s_i(x)|^{\alpha} \text{sign}(s_i(x)), \quad 0 < \alpha < 1; i = 1, 2, \dots, m.$$

The reaching law approach not only guarantees the reaching condition but also specifies the dynamic characteristics of the motion during the reaching phase [78, 101].

Chattering Problem

The chattering problem is one of the most common handicaps for applying SMC to real applications. Chattering in SMC systems is usually caused by 1) the unmodeled dynamics with small time constants, which are often neglected in the ideal model; and 2) utilization of digital controllers with finite sampling rate, which causes so called 'discretization chattering'. Theoretically, the ideal sliding mode implies infinite switching frequency. Since the control is constant within a sampling interval, switching frequency can not exceed that of sampling, which also leads to chattering. From the control engineer's point of view, chattering is undesirable because it often causes control inaccuracy, high heat loss in electric circuitry, and high wear of moving mechanical parts. In addition, the chattering action may excite the unmodeled high-order dynamics, which probably leads to unforeseen instability. Therefore, a good deal of research work has been reported in literature on the chattering elimination/reduction problem; see for example, [2, 9–11, 19, 36, 44, 83, 116, 117, 183, 200, 209, 258] and references therein. In the following, we will review some chattering elimination/reduction approaches.

Boundary Layer Approach

Roughly speaking, an SMC law consists of two parts, that is, $u(t) = u_{eq}(t) + u_N(t)$. The continuous control $u_{eq}(t)$, known as the equivalent control, controls the system when its states are on the sliding surface, and the discontinuous control $u_N(t)$ handles the system uncertainties. Since the discontinuous control $u_N(t)$ will switch between two structures during operation, the SMC system will undergo oscillation near the sliding surface. A commonly used method to alleviate chattering is to insert a boundary layer near the sliding surface so that a continuous control replaces the discontinuous one when the system is inside the boundary layer [52, 183, 196]. For this purpose, the discontinuous controller of

$$u_N(t) = -K_s \text{sign}(s(x)),$$

is often replaced by the saturation control of

$$u_N(t) \approx -K_s \text{sat}\left(\frac{s(x)}{\delta}\right) = \begin{cases} -K_s \dfrac{s(x)}{\|s(x)\|}, & \text{when } \|s(x)\| \geq \delta, \\ -K_s \dfrac{s(x)}{\delta}, & \text{when } \|s(x)\| < \delta, \end{cases}$$

or

$$u_N(t) \approx -K_s \frac{s(x)}{\|s(x)\| + \delta},$$

for some, preferably small, $\delta > 0$.

The boundary layer approach has been utilized extensively in practical applications. However, this method has some disadvantages such as: 1) it may give a chattering-free system but a finite steady-state error must exist; 2) the boundary layer thickness has a trade-off relation between control performance of SMC and chattering migration; and 3) within the boundary layer, the characteristics of robustness and accuracy of the system are no longer assured.

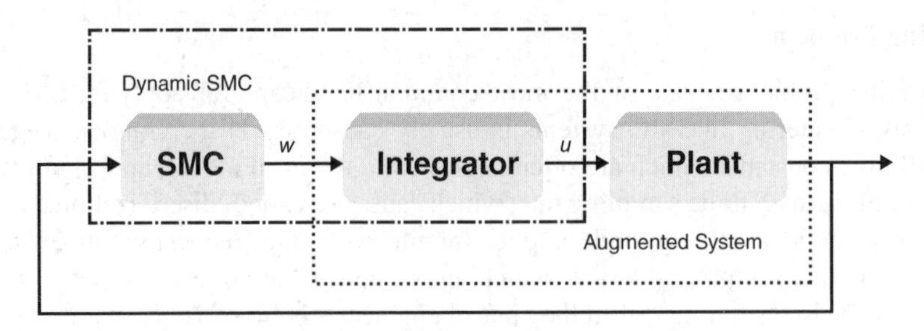

Figure 1.1 Dynamic sliding mode control

Dynamic SMC Approach

The second way to eliminate chatter is the dynamic SMC approach [9, 10, 36]. The main idea of this method is to insert an integrator (or any other strictly proper low-pass filter) between the SMC and the controlled plant, see Figure 1.1. The time derivative of the control input, $w = \dot{u}$, is treated as the new control input for the augmented system (including the original system and the integrator). Since the low-pass integrator in Figure 1.1 filters out the high frequency chattering in w, the control input to the real plant $u = \int w dt$ becomes chattering free [36].

Chattering reduction using the dynamic SMC approach is achieved by using an integrator, and the property of perfect disturbance rejection is guaranteed (no boundary layer is used in the controller). Such a method can eliminate chattering and ensure zero steady-state error; however, the system order is increased by one and the transient responses will be degraded [209].

Reaching Law Approach

Another way of reducing chattering is to decrease the amplitude of the discontinuous control. However, the robustness property of the controller is affected, and the transient performance of the system will also be degraded. There is a trade-off between the chattering reduction and the robustness property. A compromise approach is to decrease the amplitude of the discontinuous control when the system state trajectories are near to the sliding surface (to reduce the chattering), and to increase the amplitude when the system states are not near to the sliding surface (to guarantee the robustness to system uncertainties and unmodeled dynamics). This can be implemented by tuning the parameters of the reaching law

$$\dot{s}_i(x) = -\varepsilon_i \text{sign}(s_i(x)) - k_i s_i(x),$$

where ε_i and k_i, $i = 1, 2, \ldots, m$, are positive parameters to be tuned. When the system state trajectories are closed to the sliding surface, we have $s_i(x, t) \approx 0$ and $\dot{s}_i(x, t) \approx -\varepsilon_i \text{sign}(s_i(x, t))$. Here, the parameter ε_i represents the reaching velocity. By choosing ε_i small, the momentum of the motion will be reduced as the system state trajectories approach the sliding surface. As a result, the amplitude of the chattering will be reduced. However, in this case, the transient performance of the system is also degraded. To guarantee the transient performance, a large value for the parameter k_i should be chosen to increase the reaching rate when the state is not near the sliding surface.

Apart from the above-mentioned chattering elimination/reduction approaches, there have been some others, which can be found in [2, 11, 19, 44, 83, 116, 117, 200, 258].

1.1.2 Overview of SMC Methodologies

Due to its simplicity and robustness against parameter variations and disturbances, SMC has been studied extensively for many kinds of systems such as uncertain systems, time-delay systems, stochastic systems, parameter-switching systems, and singular systems. Many important SMC methodologies have been reported in literature. Here, we review some recently developed results in this area.

SMC of Uncertain Systems

Uncertainties exist in all practical physical systems, and the robust control, as a branch of control theory, is invented to explicitly deal with system uncertainties and to achieve robust performance and/or stability for controlled systems. SMC, as one of the robust control strategies, is well known for its strong robustness to system uncertainties in sliding motion. However, the uncertainties should satisfy the so-called 'matching' condition, that is, the uncertainties act within channels implicit in the control input. If a system has mismatched uncertainties in the state matrix or/and the input matrix, the conventional SMC approaches are not directly applicable. Therefore, in the past two decades, many researchers have investigated the SMC of uncertain systems with mismatched uncertainties/disturbances – see for example [31, 40, 41, 43, 112, 193] and references therein. To mention a few, in [112], the SMC of uncertain second-order single-input systems with mismatched uncertainties, was considered; in [40, 193], the authors investigated the SMC design for uncertain systems, in which the uncertainties are mismatched and exist only in state matrix. The related approaches were then developed in [41,43] to deal with a more complicated case that the mismatched uncertainties are involved in not only the state matrix but also the input matrix. In addition, the integral SMC techniques were extensively used to deal with uncertain systems with mismatched uncertainties – see for example, [27,30,43,170,233] – and some other SMC approaches to deal with uncertain systems can be found in [65,108,172,194,229].

SMC of Time-Delay Systems

It is well known that time delays appear commonly in various practical systems, such as communication, electronic, hydraulic, and chemical processes. Their existence can introduce instability, oscillation, and poor performance [168]. Time-delay systems have continuously been receiving considerable attention over the past decades. The main reason is that many processes include after-effect phenomena in their inner dynamics, and engineers need their models to approximate the real processes more accurately due to the ever-increasing expectations of dynamic performance. Stability analysis is a fundamental and vital issue in studying time-delay systems, and the conservativeness of a stability condition is an important index to evaluate a stability result. Several methods have been proposed to develop delay-dependent stability conditions (which have less conservativeness compared to delay-independent ones), such as the model transformation approach (based on Newton–Leibniz formula) [110,121], the descriptor system approach [74], the slack matrix approach [228, 243], the delay partitioning approach [86], and the input-output method (based on the small gain theorem) [88]. There have been a number of excellent survey papers on the stability analysis of time-delay systems – see for example, [168, 246]. SMC of time-delay systems have also been receiving considerable attention over the past decades – see for example, [69, 75, 85, 91, 120, 123, 160, 162, 212, 234, 250] and the references therein. To mention a few, El-Khazali in [69] proposed an output feedback robust SMC for uncertain time-delay systems, and the delay variables were considered as external perturbation when designing the sliding surface; Fridman *et al.* in [75] presented a descriptor approach to SMC of systems with time-varying delays; Xia and Jia in [234] considered the SMC of time-delay systems with mismatched parametric uncertainties by using a delay-independent approach and the LMI technique; Yan in [250] studied the SMC of uncertain time-delay systems with a class of nonlinear inputs by using a delay-dependent approach; Wu *et al.* in [212] investigated a sliding mode observer design and an observer-based SMC for a class of uncertain nonlinear neutral delay systems; Han *et al.* in [91] addressed the SMC design for time-varying input-delayed systems by using a singular perturbation approach.

SMC of Stochastic Systems

Stochastic systems and processes have come to play an important role in many fields of science, engineering, and economics. Thus, stochastic systems have received considerable attention, in which the stochastic differential equations are the most useful stochastic models with extensive applications in aeronautics, astronautics, chemical or process control system, and economic systems. A great number of methods and techniques have been developed for stochastic systems governed by Itô stochastic differential equations – see for example, [144, 145, 240, 241]. SMC design scheme for stochastic systems has also been developed – see for example, [8, 12, 13, 33, 98, 99, 155, 156, 158] and references therein. In [33], based on the concept of SMC, the steady-state covariance assignment problem was investigated for perturbed stochastic multivariable systems. The robust integral SMC and the robust sliding mode observer were designed for uncertain stochastic systems with time-varying delay in [155, 156], respectively. In [98], SMC of nonlinear stochastic systems was addressed by using a fuzzy approach. In [158], by utilizing the $\mathcal{H}_\infty$ disturbance attenuation technique, a novel SMC method was proposed for nonlinear stochastic systems. In [8], a covariance control scheme was proposed for stochastic uncertain multivariable systems via SMC strategy. In [99], a robust SMC design scheme was developed for discrete-time stochastic systems with mixed time delays, randomly occurring uncertainties, and randomly occurring nonlinearities.

SMC of Parameter-Switching Hybrid Systems

The parameter-switching hybrid system, which is the main plant considered in this book, consists of two types: Markovian jump systems and switched hybrid systems. Parameter-switching systems have received considerable research attention in the past two decades – see for example, [53, 131, 136] – since such systems are capable of modeling a wide range of practical systems that are subject to abrupt variations in their structures, owing to random failures or repairs of components, sudden environmental disturbances, changing subsystem interconnections, abrupt variations, and so on. An overview of the development of uncertain parameter-switching hybrid systems is presented in Section 1.2, from which we can see that the study on such systems, including the problems of stability analysis, stabilization, optimal control, filtering and model reduction, have been fully developed. However, SMC of parameter-switching hybrid systems, as a relatively new problem, has had only limited attention, and further research in this area is needed. There have been some results reported in the literature – see for example, [34,35,125,135,157,178,210,224,227] and references therein. More recently, for Markovian jump systems, Shi *et al.* in [178] presented an SMC design scheme by designing a linear mode-dependent sliding surface; Niu *et al.* in [157] investigated the SMC of Markovian stochastic systems by designing an integral mode-dependent sliding surface; Ma and Boukas in [135] proposed a singular system approach to robust sliding mode control for uncertain Markovian jump systems; Chen *et al.* in [35] developed an adaptive SMC for stochastic Markovian jump systems with actuator degradation. For switched hybrid systems, Wu and Lam in [210] proposed a linear mode-independent sliding surface in designing SMC for switched hybrid systems with time delay, and then the results were developed to deal with the SMC design problem for switched stochastic systems in [224]. Wu *et al.* in [227] investigated the dissipativity-based SMC design for switched stochastic systems, in which an integral mode-dependent sliding surface was designed such that the sliding motion is strictly dissipative.

Output Feedback SMC

The conventional implementation of SMC schemes is usually based on state feedback, which requires the assumption that all the state variables of the controlled systems are completely accessible for feedback. Such an assumption, however, is not always valid in practice since some state components cannot be measured. Roughly, there are two commonly used methods to deal with the controller design in the case that the system state components are not fully accessible. One approach is first to design an observer or a filter to estimate the immeasurable state components, and then synthesize an observer-based sliding mode controller – see for example, [154, 184, 212, 252]. However, the observer-based SMC scheme will require more hardware and will increase system dimension. The other approach is to design a feedback controller by using the measurable output information, which is called the output feedback SMC approach.

During the past two decades, output feedback SMC approaches have been intensively studied, and many important results have been reported in the literature – see for example, [32, 42, 48, 61–64, 68, 90, 113, 163, 253, 263] and references therein. To mention a few, output feedback SMC design for uncertain dynamic systems was investigated in [263], and an algorithm for output-dependent hyperplane design was proposed based upon eigenvector methods. The eigenvalue assignment approach was proposed in [68] to design the sliding surface of the output feedback SMC scheme. The LMI technique was applied to output feedback SMC design in [62, 63]. Output feedback SMC design for state-delayed systems was investigated in [90, 253]. The above-mentioned results are all for static output feedback (SOF) SMC problems. In fact, output feedback control has two different forms: the SOF control and dynamic output feedback (DOF) control. Generally speaking, DOF control is more flexible than SOF control since the additional dynamics of the controller is introduced. Although DOF control involves more design parameters, for linear systems the closed-loop system can usually be written in a more compact form where certain parameters can be embedded into augmented matrix variables. Compared to the SOF SMC design, the DOF SMC design problem has received less attention, and only a few results have been reported, for example, in [32, 163] the DOF SMC was studied for MIMO linear systems with mismatched norm-bounded uncertainties and matched nonlinear disturbances.

1.2 Uncertain Parameter-Switching Hybrid Systems

1.2.1 Analysis and Synthesis of Switched Hybrid Systems

Switched systems form a class of hybrid systems consisting of a family of subsystems described by continuous- or discrete-time dynamics, and a rule specifying the switching among them [129, 191]. The switching rule in such systems is usually considered to be arbitrary. Switched systems have received increasing attention in the past few years, since many real-world systems such as, chemical processes, transportation systems, computer-controlled systems, and communication industries can be modeled as switched systems [131]. More importantly, many intelligent control strategies are designed based on the idea of controllers switching to overcome the shortcomings of the traditionally used single controller and to improve their performance, thus making the corresponding closed-loop systems into switched systems.

Switched hybrid systems with all subsystems described by linear differential or difference equations are called switched linear hybrid systems. A continuous-time switched linear system can be modeled as

$$\dot{x}(t) = A(\alpha(t))x(t) + B(\alpha(t))u(t),$$

where $x(t) \in \mathbf{R}^n$ is the state vector; $u(t) \in \mathbf{R}^m$ is the control input; $\{(A(\alpha(t)), B(\alpha(t))) : \alpha(t) \in \mathcal{N}\}$ is a family of matrices parameterized by an index set $\mathcal{N} = \{1, 2, \dots, N\}$ and $\alpha(t) : \mathbf{R} \rightarrow \mathcal{N}$ is a piecewise constant function of time t called a switching signal. At a given time t, the value of $\alpha(t)$, denoted by α for simplicity, might depend on t or $x(t)$, or both, or may be generated by any other hybrid scheme. Therefore, the switched hybrid system effectively switches among N subsystems with the switching sequence controlled by $\alpha(t)$. It is assumed that the value of $\alpha(t)$ is unknown *a priori*, but its instantaneous value is available in real time.

Similarly, a discrete-time switched linear hybrid system can be described by

$$x(k+1) = A(\alpha(k))x(k) + B_u(\alpha(k))u(k),$$

where $x(k) \in \mathbf{R}^n$ is the state vector; $u(k) \in \mathbf{R}^m$ is the control input; $\{(A(\alpha(k)), B(\alpha(k))) : \alpha(k) \in \mathcal{N}\}$ is a family of matrices parameterized by an index set $\mathcal{N} = \{1, 2, \dots, N\}$, and $\alpha(k) : \mathbf{Z}^+ \rightarrow \mathcal{N}$ is a piecewise constant function of time, called a switching signal, which takes its values in the finite set $\mathcal{N}$. At an arbitrary discrete time k, the value of $\alpha(k)$, denoted by α for simplicity, might depend on k or $x(k)$, or both, or may be generated by any other hybrid scheme.

Stability of Switched Hybrid Systems

The stability analysis of switched hybrid systems is a fundamental issue for the synthesis of such systems. Note that there are two facts related with the stability of switched hybrid systems: 1) a switched hybrid system may have divergent trajectories even when all the subsystems are stable; and 2) a switched hybrid system may have convergent trajectories even when some of the subsystems are unstable. These two facts show that the stability of a switched hybrid system depends not only on the dynamics of each subsystem but also on the properties of the switching signals.

When focusing on stability analysis of switched hybrid systems, there are many valuable results that have appeared in the past two decades, and interested readers may refer to survey papers, such as [53, 128, 131, 148], and books, such as [129, 191]. In the following, we will briefly overview some recently developed results.

Arbitrary Switching

We first consider the stability analysis of the switched hybrid systems without any restrictions on switching signal, that is, the switching is arbitrary. Several approaches have been reported on the stability analysis of switched hybrid systems with arbitrary switching, for example:

1. Common Quadratic Lyapunov Functions. For the stability analysis problem of switched hybrid systems under arbitrary switching, it is necessary to require that all the subsystems are asymptotically stable. However, even when all the subsystems of a switched system are exponentially stable, the stability of the switched hybrid system still can not be guaranteed [129]. Therefore, in general, all subsystems' stability assumptions are not sufficient to ensure stability for the switched systems under arbitrary switching. On the other hand, if there exists a common quadratic Lyapunov function (CQLF) for all the subsystems, then the stability of the switched system is guaranteed under arbitrary switching. Generally speaking, the existence of a CQLF is only sufficient for the asymptotic stability of linear switched hybrid systems under arbitrary switching, and could be rather conservative. For the switched linear system $\dot{x}(t) = A(\alpha(t))x(t)$ with the parameter matrices $A(\alpha(t))$ replaced by $A(i)$ denoting that the ith subsystem is activated, by constructing a CQLF as $V(x) = x^T(t)Px(t)$ where $P > 0$, it can be shown that the switched linear system is asymptotically stable if there exists a positive definite symmetric matrix P such that

$$PA(i) + A^T(i)P < 0, \quad i \in \mathcal{N}.$$

For discrete-time switched linear system $x(k+1) = A(\alpha(k))x(k)$, it is asymptotically stable if there exists a positive definite symmetric matrix P such that

$$A^T(i)PA(i) - P < 0, \quad i \in \mathcal{N}.$$

The above stability results are both expressed in the form of LMIs, which can be tested easily by using standard software such as the LMI Toolbox in Matlab [25]. In [127], a sufficient condition was presented for asymptotic stability of a switched linear system in terms of Lie algebra generated by the individual matrices. Namely, if this Lie algebra is solvable, then the switched system is exponentially stable for arbitrary switching. In [146], a stability criterion was proposed for switched nonlinear systems which involves Lie brackets of the individual vector fields but does not require that these vector fields commute. However, the stability conditions are both only sufficient conditions, not necessary and sufficient ones. In [180], some necessary and sufficient conditions were proposed for the existence of a CQLF for two stable second-order LTI systems, and then the related results were extended for a set of stable LTI systems in [181, 182]. In [109], the authors studied a singularity test for the existence of CQLF for pairs of stable LTI systems, and some necessary and sufficient algebraic conditions were given. A necessary and sufficient condition for the existence of a common Lyapunov function for all subsystems was proposed in [128] for a switched hybrid

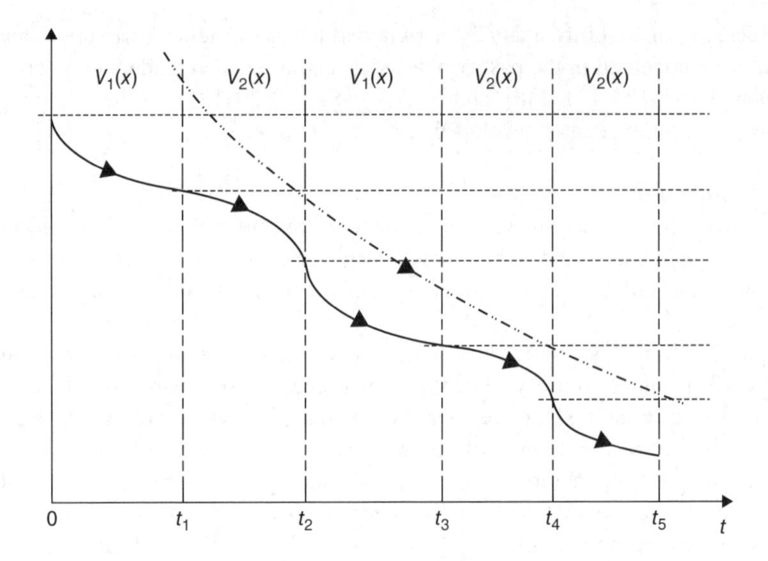

Figure 1.2 Switched quadratic Lyapunov functions

system under arbitrary switching. A considerable number of approaches to construct such a CQLF were presented in [159].

2. Switched Quadratic Lyapunov Functions. Since the existence conditions of a CQLF are conservative for all subsystems of a switched hybrid system with arbitrary switching, some attention has been paid to a less conservative class of Lyapunov functions, namely switched quadratic Lyapunov functions (SQLF). By using the SQLF, the values of such a Lyapunov function still decrease at the switching instants – see Figure 1.2. Compared with the CQLF, the SQLF contains the switching information (mode-dependent), and a typical form of such Lyapunov function can be constructed as $V(x) = x^T(t)P(\alpha(t))x(t)$ for continuous-time switched systems or $V(x) = x^T(k)P(\alpha(k))x(k)$ for discrete-time switched systems, where $P(\cdot) > 0, i \in \mathcal{N}$ are mode-dependent. Using an SQLF approach, the stability analysis condition for the discrete-time switched linear system $x(k+1) = A(\alpha(k))x(k)$ can be formulated as: it is asymptotically stable if there exist positive definite symmetric matrices $P(i), i \in \mathcal{N}$ such that

$$\begin{bmatrix} -P(i) & A^T(i)P(j) \\ \star & -P(j) \end{bmatrix} < 0, \quad i,j \in \mathcal{N}.$$

The above stability analysis result based on the SQLF approach will turn out to be the above-mentioned one with the CQLF approach if $P(i) = P(j), i,j \in \mathcal{N}$. Obviously, the SQLF approach is less conservative than the CQLF approach. Some results on the SQLF approach to stability analysis and control synthesis for switched hybrid systems can be found in [46, 71].

Restricted Switching

Stability analysis approaches for arbitrary switching have been developed, but a natural question may still be raised, that is, can switched hybrid systems be stable under some restricted switchings in spite of the fact that they fail to preserve stability under arbitrary switching? If so, what kinds of restrictions should be put on the switching signals to guarantee the stability of switched hybrid systems? To answer such questions, there have been some stability analysis approaches for switched hybrid systems under restricted switching, for example,

1. Dwell Time Approach. Recently, there has been enormous growth of interest in using the dwell time approach to deal with stability analysis of switched hybrid systems – see for example, [92, 93, 102, 149, 151, 165, 188, 215–217, 219, 220, 264, 265]. A positive constant $T_d \in \mathbf{R}$ is called the dwell time of a switching signal if the time interval between any two consecutive switchings is no smaller than T_d. The basic idea of the dwell time approach can be formulated as follows: given a dwell time, and let $\mathcal{S}(T_d)$ denote the set of all switching signals with interval between consecutive discontinuities not smaller than T_d, it has been shown that one can pick T_d sufficiently large such that the switched system considered is exponentially stable for any switching signal belonging to $\mathcal{S}(T_d)$. The dwell time approach was used to analyze the local asymptotic stability of nonlinear switched systems. Subsequently, this concept was extended and the average dwell time approach was developed [92], which means that the average time interval between consecutive switchings is no less than a specified constant T_a. Specifically, a positive constant T_a is called an average dwell time for a switching signal $\alpha(t)$ if

$$N_\alpha(T_1, T_2) \leq N_0 + \frac{T_2 - T_1}{T_a}.$$

   For any $T_2 > T_1 \geq 0$, let $N_\alpha(T_1, T_2)$ denote the number of switching of $\alpha(t)$ over (T_1, T_2). Here, T_a is called an average dwell time and N_0 is the chatter bound. It has been proved in [92] that if all the subsystems are exponentially stable then the switched hybrid system remains exponentially stable provided that the average dwell time T_d is sufficiently large. By using the average dwell time approach, Zhai *et al.* in [264] investigated the disturbance attenuation properties of continuous-time switched hybrid systems, and then the exponential stability and ℓ_2 gain properties for discrete-time switched hybrid systems was investigated in [265]; Sun *et al.* in [188] studied the exponential stability and weighted $\mathcal{L}_2$-gain for switched delay systems; Wu and Lam in [215] considered the filtering problem of switched hybrid systems with time-varying delay. As well as the above-mentioned results, the model reduction problem for switched hybrid systems with time-varying delay was addressed in [219] by using the average dwell time approach incorporated with a piecewise Lyapunov function; and the DOF controller design problem was considered in [216, 217].
2. Multiple Lyapunov Functions Approach. By using the Lyapunov function approach to the stability analysis of switched hybrid systems, the above-mentioned CQLF and SQLF approaches require that the Lyapunov functions are globally monotonically decreasing as with the state trajectories. This is, however, conservative since such Lyapunov functions may not exist for all subsystems of switched hybrid systems. For such cases, one can construct a set of Lyapunov-like functions, which only require non-positive Lie-derivatives for certain subsystems in certain regions of the state space, instead of being negative globally. Multiple Lyapunov functions (MLF), is a non-traditional Lyapunov stability approach, and the key point of the method is the non-increasing requirement on any Lyapunov function over the exiting (switch from) or starting (switch to) time sequences of the corresponding subsystem [94, 128, 148, 164]. Specifically, the Lyapunov-like function is selected for each subsystem,

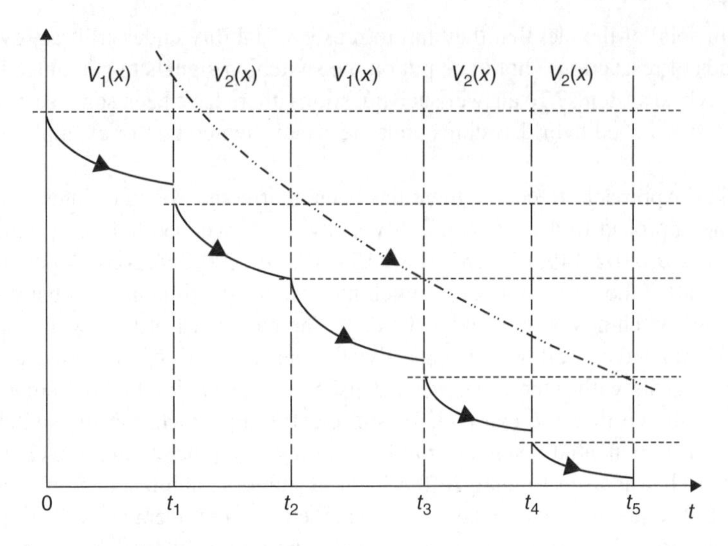

Figure 1.3 Multiple Lyapunov stability (Case 1: the values of Lyapunov-like functions at the switching instants form a monotonically decreasing sequence)

and the values of the Lyapunov-like function at the exiting (the starting) instant of the next running interval are smaller than that of the current running interval, then the energy of the Lyapunov-like functions are decreasing globally. There are several versions of MLF results in the literature, for example, Case 1: the Lyapunov-like function is decreasing when the corresponding mode is active and does not increase its value at each switching instant [53] – see Figure 1.3 – and in this case, the switched hybrid system is asymptotically stable; Case 2: the value of the Lyapunov-like function at every exiting instant is smaller than its value at the previous exiting time, then the switched system is asymptotically stable [26] – see Figure 1.4. Case 3: the Lyapunov-like function may increase its value during a time interval, only if the increment is bounded by certain kind of continuous functions [257] – see Figure 1.5 – and in this case, the switched system can remain stable.

Synthesis of Switched Hybrid Systems

Over the past several decades, considerable interest has been devoted to synthesis problems of switched hybrid systems, including stabilization, robust/optimal control, state estimation/ filering, fault detection, model approximation, and so on. Here, we will review some relevant literature on the synthesis of switched hybrid systems. First, we introduce two important properties of switched hybrid systems, namely, the controllability and the observability. Roughly speaking, the concept of controllability denotes the ability to move a system around in its entire configuration space using only certain admissible manipulations. Observability is a measure for how well internal states of a system can be inferred by knowledge of its external outputs. Observability and controllability are dual aspects of the same problem. Some results on the controllability and the observability analysis for switched hybrid systems were reported in [16, 37, 105, 132, 166, 185, 189, 206, 235, 236, 270] and references therein.

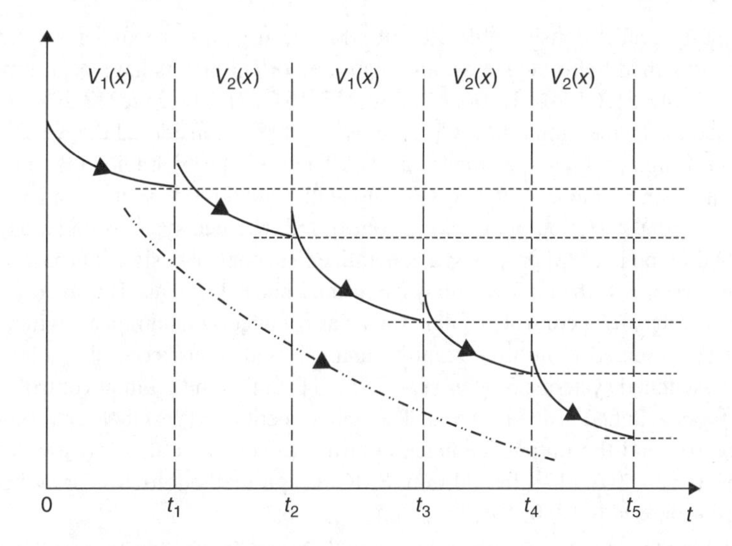

Figure 1.4 Multiple Lyapunov stability (Case 2: the values of Lyapunov-like function for each subsystem at every exiting instant form a monotonically decreasing sequence)

In the previous section, we discussed the stability properties of switched hybrid systems. As mentioned earlier, the stability of a switched hybrid system depends not only on the dynamics of each subsystem but also on the properties of the switching signals, thus the synthesis problems include two strategies for implementation. The first is based on the subsystems' dynamics with given switching signals, and the second is based on the switching signals. The stabilization problem for switched hybrid systems was investigated in [4, 6, 14, 38, 45, 46, 79, 84, 95, 100, 102, 124, 130, 140, 143, 190, 192, 222, 237, 247, 249, 267] and references therein.

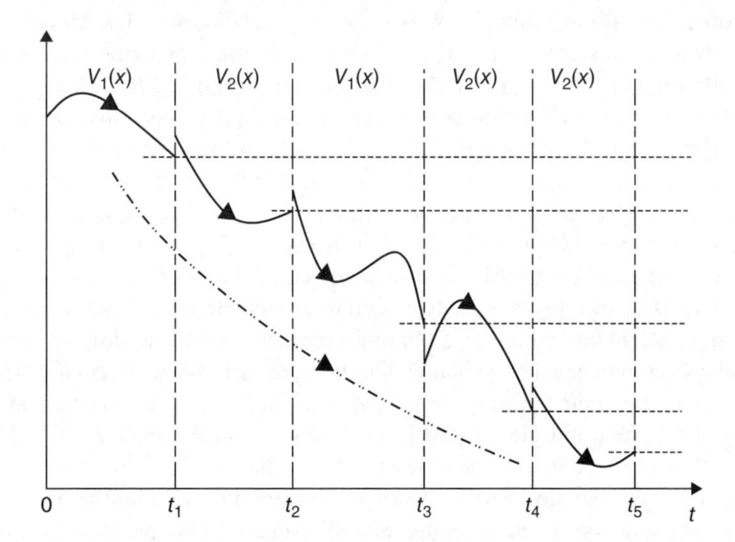

Figure 1.5 Multiple Lyapunov stability (Case 3: the Lyapunov-like function for each subsystem increases its value during a certain period)

Over the past decade, considerable attention has been paid to robust and optimal control problems for switched hybrid systems, and many important results have been reported – see for example, [17,47,51,67,80,81,106,126,138,153,161,171,216,217,232,248,264,268] and references therein. To mention a few, Geromel *et al.* in [81] considered the passivity analysis and controller design problems; Kamgarpour and Tomlin in [106] studied the optimal control problem for non-autonomous switched systems with a fixed mode sequence; Lian and Ge in [126] addressed robust $\mathcal{H}_\infty$ output tracking control for switched systems under asynchronous switching; Mahmoud in [138] proposed a generalized $\mathcal{H}_2$ control design approach for discrete-time switched systems with unknown delays; Niu and Zhao in [153] used the average dwell time approach to the robust $\mathcal{H}_\infty$ control problem for a class of uncertain nonlinear switched systems; Orlov in [161] presented finite time stability analysis and robust control synthesis methods for uncertain switched systems; Seatzu *et al.* in [171] studied the optimal control problem for continuous-time switched affine systems. The above-mentioned results are all based on state feedback control, and the output feedback control problem was also investigated – see for example, [51, 67, 80, 216, 217]. In addition, SMC design methodologies for switched hybrid systems were proposed in [210, 224, 227, 262].

It is well known that one of the fundamental problems in control systems and signal processing is the estimation of the state variables of a dynamical system through available noisy measurements, which is referred to as the filtering problem. The celebrated Kalman filter has been considered as the best possible (optimal) estimator for a large class of systems; it is an algorithm that uses a series of measurements observed over time, containing noise (random variations) and other inaccuracies, and produces estimates of unknown variables that tend to be more precise than those based on a single measurement alone. The Kalman filter for switched discrete-time linear systems was designed in [3]. However, the application of the Kalman filter is subject to two initial assumptions: the underlying system is linear with complete knowledge of the dynamical model, and the noise concerned is white/colored with known spectral density. Thus the Kalman filtering scheme is no longer applicable when *a priori* information on the external noises is not precisely known. Therefore, the past two decades have witnessed significant progress on robust filtering involving various approaches such as $\mathcal{H}_2$ filtering, $\mathcal{H}_\infty$ filtering, $\mathcal{L}_2$-$\mathcal{L}_\infty$ filtering, and mixed $\mathcal{H}_2/\mathcal{H}_\infty$ filtering. The robust filtering problem for switched hybrid systems has also been developed over the past decade – see for example, [50, 137, 139, 167, 202, 215, 216, 272] and references therein. To mention a few important robust filtering results for switched hybrid systems, Deaecto *et al.* in [50] developed a trajectory-dependent filter design approach for discrete-time switched linear systems; Mahmoud in [137] presented a delay-dependent $\mathcal{H}_\infty$ filter design approach for a class of discrete-time switched systems with state delay; Qiu *et al.* in [167] investigated the robust mixed $\mathcal{H}_2/\mathcal{H}_\infty$ filtering design for discrete-time switched polytopic linear systems; Wang *et al.* in [202] addressed the $\mathcal{H}_\infty$ filtering problem for discrete-time switched systems with state delays via switched the Lyapunov function approach; Wu and Lam in [215] proposed an average dwell time approach to the weighted $\mathcal{H}_\infty$ filter design for switched systems with time-varying delay; Wu and Ho in [216] developed a reduced-order $\mathcal{L}_2$-$\mathcal{L}_\infty$ filter design scheme for a class of nonlinear switched stochastic systems.

The issues of fault detection and isolation are increasingly required in various kinds of practical complex systems for guaranteeing reliability and pursuing performance. Hence, how to develop effective methods for timely and accurate diagnosis of faults becomes a crucial problem. To combat this, many significant schemes have been introduced, such as model-based approaches and knowledge-based methods. Among them, the model-based approach is the most favored. The basic idea of model-based fault detection is to construct a residual signal and, based on this, determine a residual evaluation function to compare with a predefined threshold. When the residual evaluation function has a value larger than the threshold, an fault alarm is generated. Since accurate mathematical models are not always available, the unavoidable modeling errors and external disturbances may seriously affect the performance of model-based fault detection systems. Thus, the designed fault detection systems should be both sensitive to faults and suppressive to external disturbances. Fortunately, the $\mathcal{H}_\infty$ fault detection filter or observer is known to be able to do a good job of achieving the above-mentioned requirements. The $\mathcal{H}_\infty$ fault detection problem for switched hybrid systems was studied in [203, 261], and some other approaches can be found in [15, 49, 119]. Fault-tolerant control is a related issue that makes it possible to develop a control feedback that allows the required system performance to be maintained in the case of faults. The fault-tolerant control problem for switched hybrid systems has also been investigated: for example, Du *et al.* in [59] proposed an active fault-tolerant controller design scheme for switched systems with time delay; Li and Yang in [119] developed a simultaneous fault detection and control technique for switched systems under asynchronous switching; Wang *et al.* in [205] designed a robust fault-tolerant controller for a class of switched nonlinear systems in lower triangular form.

Mathematical modeling of physical systems often results in complex high-order models, which bring serious difficulties to analysis and synthesis of the systems concerned. Therefore, in practical applications it is desirable to replace high-order models by reduced ones with respect to some given criterion, which is the model reduction problem. Over the past decades, the model reduction problem has been the concern of many researchers. Many important results have been reported, which involve various efficient model reduction approaches, such as the balanced truncation approach [89], the Hankel-norm approach [82], Krylov projection approach [87], the Padé reduction approach [7], the $\mathcal{H}_2$ approach [251], and the $\mathcal{L}_2$ approach [217, 219]. Readers can refer to [5] for a detailed survey of model reduction. The model reduction problem for switched hybrid systems has also received considerable attention – see for example, [18, 133, 150, 173, 179, 204, 217, 219] and references therein. To mention a few important results, Birouche *et al.* in [18] investigated the model order-reduction for discrete-time switched linear systems by the balanced truncation approach; Monshizadeh *et al.* in [150] developed a simultaneous balanced truncation approach to model reduction of switched linear systems; Shi *et al.* in [179] studied the model reduction problem for discrete-time switched linear systems over finite frequency ranges; Wang *et al.* in [204] developed a delay-dependent model reduction approach for continuous-time switched state-delayed systems; and Wu and Zheng in [219] proposed a weighted $\mathcal{H}_\infty$ model reduction approach for linear switched systems with time-varying delay.

1.2.2 Analysis and Synthesis of Markovian Jump Linear Systems

Markovian jump linear systems (MJLSs) are another typical class of parameter-switching systems, and they are modeled by a set of linear systems with the transitions between the models determined by a Markov chain, taking values in a finite set [136]. MJLSs can also be considered as a special case of switched hybrid systems with the switching signals governed by a Markov chain. Applications of MJLSs may be found in many processes, such as target tracking problems, manufactory processes, solar thermal receivers, fault-tolerant systems, and economic problems. From a mathematical point of view, MJLSs can be regarded as a special class of stochastic system with system matrices changed randomly at discrete time points governed by a Markov process and remaining LTI between random jumps. Over the past decades, owing to a large number of applications in control engineering, MJLSs have received increasing interest. Many results in this field can be found in the literature, and in the following, we will review some recently published results on MJLSs.

The stability analysis and stabilization problems for MJLSs were addressed in [20, 22, 28, 56, 76, 104, 144, 174, 176, 187, 207, 238, 239]. Specifically, Cao and Lam in [28] investigated the stochastic stabilizability and $\mathcal{H}_\infty$ control for discrete-time jump linear systems with time delay; de Souza in [56] studied the robust stability and stabilization problems for uncertain discrete-time MJLSs; Gao *et al.* in [76] considered the stabilization and $\mathcal{H}_\infty$ control problems for two-dimensional MJLSs; Sun *et al.* in [187] dealt with the robust exponential stabilization of MJLSs with mode-dependent input delay; Xiong *et al.* in [238] studied the robust stabilization problem for MJLSs with uncertain switching probabilities. In addition, there have been some results on the stability and stabilization for Markovian jump stochastic systems. For example, Boukas and Yang in [20] proposed an exponential stabilizability condition for stochastic systems with Markovian jump parameters; Wang *et al.* in [207] solved the stabilization problem for bilinear uncertain time-delay stochastic systems with Markovian jump parameters; and some other results on Markovian jump stochastic systems can be found in [144].

The $\mathcal{H}_\infty$ control for MJLSs was investigated in [21, 24, 28, 29, 76, 115, 230, 245]; robust $\mathcal{H}_\infty$ control of MJLSs with unknown nonlinearities was studied in [21]. $\mathcal{H}_\infty$ control was addressed in [24] for discrete-time MJLSs with bounded transition probabilities; the robust $\mathcal{H}_\infty$ control problem was considered in [29] for uncertain MJLSs with time delay; the robust $\mathcal{H}_\infty$ control of descriptor discrete-time Markovian jump systems is covered in [115]; delay-dependent $\mathcal{H}_\infty$ control for singular Markovian jump systems with time delay appears in [230]; delay-dependent $\mathcal{H}_\infty$ control and filtering for uncertain Markovian jump systems with time-varying delays are found in [245].

The filtering problem for MJLSs was considered in [54, 55, 57, 134, 175, 177, 208, 211, 231, 242, 254, 256]. To mention a few, de Souza and Fragoso studied the $\mathcal{H}_\infty$ filter design problem for continuous- and discrete-time MJLSs in [54, 55], respectively; Ma and Boukas in [134] investigated robust $\mathcal{H}_\infty$ filtering for uncertain discrete Markovian jump singular systems with mode-dependent time delay; Shi *et al.* in [175] considered Kalman filtering for continuous-time uncertain MJLSs; Wu *et al.* in [211] addressed the $\mathcal{H}_\infty$ filtering problem for Markovian jump two-dimensional systems; Yao *et al.* in [254] dealt with robust $\mathcal{H}_\infty$ filtering of Markovian jump stochastic systems with uncertain transition probabilities; and then they studied quantized $\mathcal{H}_\infty$ filtering for Markovian jump LPV systems with intermittent measurements in [256].

The fault detection problem for MJLSs was investigated in [147, 152, 226, 255, 273, 274]. Specifically, Meskin and Khorasani in [147] investigated fault detection and isolation problems for discrete-time MJLSs with application to a network of multi-agent systems having imperfect communication channels; Nader and Khashayar proposed a geometric approach to fault detection and isolation of continuous-time MJLSs in [152]; Wu *et al.* in [226] studied generalized $\mathcal{H}_2$ fault detection for Markovian jump two-dimensional systems; Yao *et al.* in [255] considered fault detection filter design for Markovian jump singular systems with intermittent measurements; and Zhong *et al.* in [273, 274] addressed robust fault detection problem for continuous- and discrete-time MJLSs, respectively.

The SMC design problem was also addressed for MJLSs in [34, 35, 135, 157, 178, 221, 223, 225]. Chen *et al.* studied SMC of MJLSs with actuator nonlinearities in [34], and adaptive SMC design for Markovian jump stochastic systems with actuator degradation in [35]; Ma and Boukas in [135] proposed a singular system approach to robust SMC for uncertain MJLSs; Shi *et al.* in [178] considered the SMC design problem for MJLSs; Wu and Ho in [221] solved the SMC problem for Markovian jump singular stochastic hybrid systems; Wu *et al.* in [223] investigated state estimation and SMC of Markovian jump singular systems; and then they considered SMC design with bounded $\mathcal{L}_2$ gain performance for Markovian jump singular time-delay systems in [225].

Apart from the above-mentioned synthesis problems for MJLSs, the model reduction problem for such systems was also investigated – see for example, [111, 266]. Kotsalis *et al.* in [111] studied the model reduction problem for discrete-time MJLSs; and Zhang *et al.* in [266] considered $\mathcal{H}_\infty$ model reduction for both continuous- and discrete-time MJLSs.

1.3 Contribution of the Book

This book represents the first of a number of attempts to reflect the state-of-the-art of the research area for handling the SMC problem for uncertain parameter-switching hybrid systems (including Markovian jump systems, switched hybrid systems, singular systems, stochastic systems, and time-delay systems). The content of this book can be divided into three parts. The first part is focused on SMC of Markovian jump singular systems. Some necessary and sufficient conditions are derived for the stochastic stability, stochastic admissibility, and optimal performances by developing new techniques for the considered Markovian jump singular systems. Then, a set of new SMC methodologies are proposed, based on the analysis results. In the second part, the problem of SMC of switched delayed hybrid systems is investigated. A unified framework under 'average dwell time' is established for analyzing the considered switched delayed hybrid systems. Then some sufficient conditions are derived for the stability, stabilizability, existence of the desired DOF controllers, and existence of the sliding mode dynamics in the SMC issue. More importantly, a set of SMC methodologies under a unique framework are proposed for the considered hybrid systems. In the third part, the parallel theories and techniques developed in the previous part are extended to deal with switched stochastic hybrid systems. Specifically, in this third part, the main attention will be focused on stochastic stability analysis, stabilization, $\mathcal{H}_\infty$ control, and SMC of switched stochastic hybrid systems. Sufficient conditions are established first for the stochastic exponential stability and optimal performances (such as $\mathcal{H}_\infty$ and dissipativity) of the continuous- and discrete-time switched stochastic systems. Based on the obtained analysis results, the synthesis issues, including $\mathcal{H}_\infty$ control and SMC design, are solved.

The features of this book can be highlighted as follows. 1) A unified framework is established for SMC of Markovian jump singular systems, where the parameters are jumping from one mode to another stochastically, and at the same time there are time delays in existing system states. 2) A series of problems are solved with new approaches for analysis and synthesis of continuous- and discrete-time switched hybrid systems, including stability analysis and stabilization, DOF control, and SMC. 3) Three correlated problems, $\mathcal{H}_\infty$ control (state feedback control and DOF control), SMC, and state estimation problems, are dealt with for switched stochastic systems. 4) A set of newly developed techniques (e.g. average dwell time method, piecewise Lyapunov function approach, parameter-dependent Lyapunov function approach, cone complementary linearization (CCL) approach, slack matrix approach, and sums of squares technique) are exploited to handle the emerging mathematical/computational challenges.

This book is a timely reflection on the developing area of system analysis and SMC theories for systems with uncertain switching parameters, typically resulting from varying operation environments. It is a collection of a series of latest research results and therefore serves as a useful textbook for senior and/or graduate students who are interested in knowing: 1) the state of the art of the SMC area; 2) recent advances in Markovian jump systems; 3) recent advances in switched hybrid systems; and 4) recent advances in singular systems, stochastic systems and time-delay systems. Readers will also benefit from new concepts, models and methodologies with theoretical significance in system analysis and control synthesis. The book can also be used as a practical research reference for engineers dealing with SMC, optimal control, and state estimation problems for uncertain parameter-switching hybrid systems. The aim of this book is to close the gap in literature by providing a unified, neat framework for SMC of uncertain parameter-switching hybrid systems.

In general, this book aims at third- or fourth-year undergraduates, postgraduates and academic researchers. Prerequisite knowledge includes linear algebra, matrix analysis, linear control system theory, and stochastic systems. It should be described as an advanced book.

More specifically, the readers should include: 1) control engineers working on nonlinear control, switching control, and optimal control; 2) system engineers working on switched hybrid systems and stochastic systems; 3) mathematicians and physicists working on hybrid systems and singular systems; and 4) postgraduate students majoring on control engineering, system sciences, and applied mathematics. This book could also serve as a useful reference to: 1) mathematicians and physicists working on complex dynamic systems; 2) computer scientists working on algorithms and computational complexity; and 3) third- or fourth-year students who are interested in knowing about advances in control theory and applications.

1.4 Outline of the Book

The organization structure of this book is shown in Figure 1.6. The general layout of this book is divided into three parts: Part One: SMC of Markovian jump singular systems; Part Two: SMC of switched hybrid systems with time-varying delay; and Part Three: SMC of switched stochastic hybrid systems. The main contents of this book are shown in Figure 1.7.

Chapter 1 first presents the research background, motivations and research problems of this book which mainly involve SMC methodologies and the uncertain parameter-switching hybrid systems. A survey is provided on the fundamental theory of the SMC methodologies, which include some basic concepts (SMC problem, reaching condition, and two SMC design steps), sliding surface design (linear sliding surface, integral sliding surface, and terminal sliding surface), sliding mode controller design (equivalent control design, reaching condition approach, Lyapunov function approach, and reaching law approach), chattering problem (boundary layer approach, dynamic SMC approach, and reaching law approach). Then, an overview of recent developments of SMC Methodologies is also presented, which includes SMC of uncertain systems, SMC of time-delay systems, SMC of stochastic systems, SMC of parameter-switching hybrid systems, and the output feedback

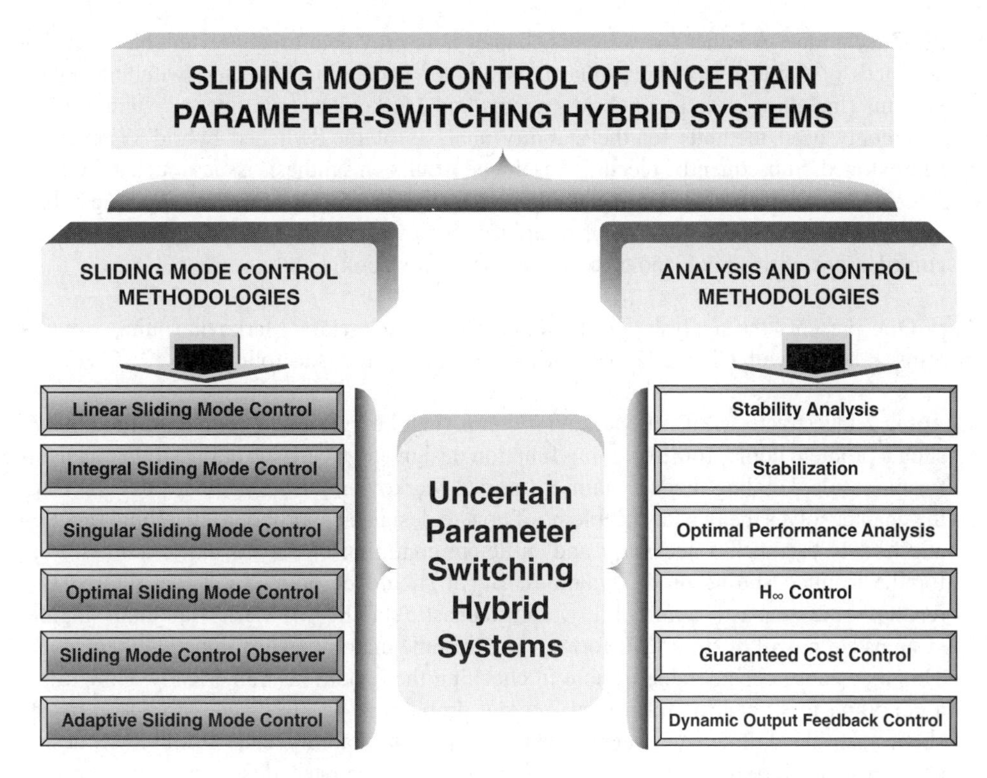

Figure 1.6 The organization structure of the book

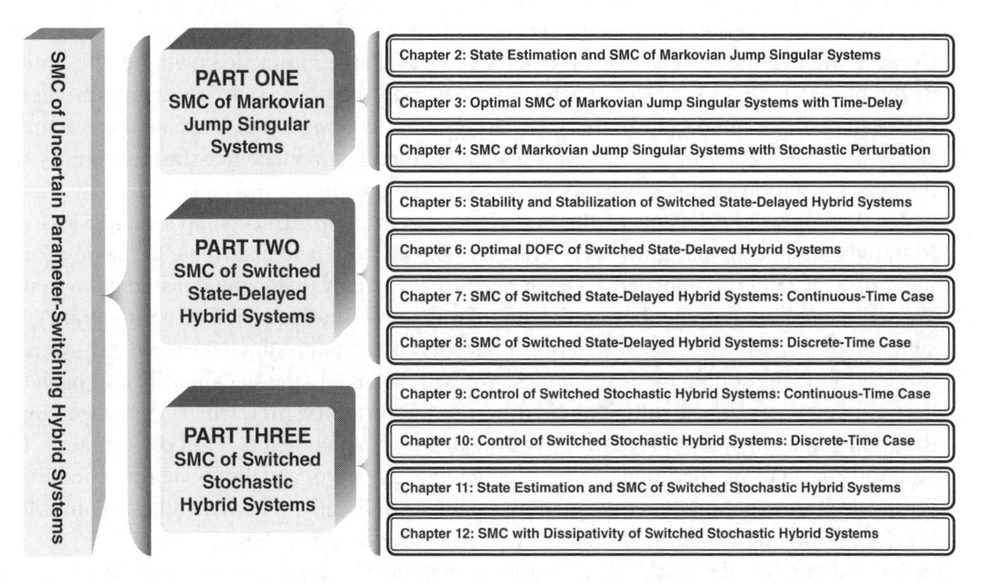

Figure 1.7 The main contents of the book

SMC technique. Another focus in this chapter is to provide a timely review on the recent advances of the analysis and synthesis issues for uncertain parameter-switching hybrid systems (including switched hybrid systems and Markovian jump linear systems). Most commonly used methods for the stability analysis of the switched hybrid systems are summarized. Subsequently, recently developed results on synthesis issues (such as control, filtering, fault detection, and model reduction) for the uncertain parameter-switching hybrid systems are reviewed with a lot of references involved. Finally, we summarize the main contributions of this book and give the outline of this book.

Part One presents the analysis and SMC design procedure for Markovian jump singular systems. It begins with Chapter 2, and consists of three chapters as follows.

Chapter 2 investigates SMC of Markovian jump singular systems. The main difficulties of such a problem come from switching function design and stochastic admissibility analysis for the resulted sliding mode dynamics (termed Markovian jump singular systems). Thus, the chapter solves the two key problems of how to design an appropriate switching function and how to establish a necessary and sufficient condition of the stochastic admissibility for the resulted sliding mode dynamics. But it should be pointed out that the existing results on the stochastic admissibility (and stochastic stability) of Markovian jump singular systems are not all of strict LMI form owing to some matrix equality constraints involved. This may cause considerable trouble in checking the conditions numerically. How to get a condition in strict LMI form is also a key problem to be discussed. Motivated by the above-mentioned three key issues, we will consider the stability analysis and SMC design problems for Markovian jump singular systems in this chapter. First, a new integral-type switching function is designed by taking the singular matrix E into account, by which the sliding mode dynamics can be derived. Then, a necessary and sufficient condition for the existence of such a sliding mode (the stochastic stability of the sliding mode dynamics) is established in terms of strict LMIs, by which the sliding surface can be designed. Considering that the system states are not always available in practice owing to the limits of the physical situation or the expense of measuring them, the state estimation problem has become more important. In this case, an observer is designed to estimate the system's states, and an observer-based SMC law is then synthesized to guarantee the reachability of the state trajectories of the closed-loop system to the predefined sliding surface.

Chapter 3 studies the problems of the bounded $\mathcal{L}_2$ gain performance analysis and SMC of Markovian jump singular time-delay systems. The purpose is to contribute to the development of SMC and the bounded $\mathcal{L}_2$ gain performance analysis for the considered system. We will pay particular attention to the singular matrix E in the design of an integral-type switching function, which leads to a full-order Markovian jump singular time-delay system for describing the sliding mode dynamics. We will then apply the slack matrix approach to derive a delay-dependent sufficient condition in the form of LMIs, which guarantees that the sliding mode dynamics is stochastically stable with a bounded $\mathcal{L}_2$ gain performance. In addition, the analysis conditions of the sliding mode dynamics and the solvability condition for the desired switching function are both established. Finally, we will synthesize an SMC law for driving the system state trajectories onto the predefined sliding surface.

Chapter 4 addresses the SMC of a nonlinear singular stochastic system with Markovian switching. An integral switching function is designed, and the resulting sliding mode dynamics is expressed by a full-order Markovian jump singular stochastic system. By introducing some specified matrices, a sufficient condition is proposed in terms of strict LMIs, which guarantees the stochastic stability of the sliding mode dynamics (thus the existence of such a sliding mode can be guaranteed). A sliding mode controller is then synthesized for reaching motion. Moreover, when there is an external disturbance in the considered control system, the $\mathcal{L}_2$ disturbance attenuation performance ($\mathcal{H}_\infty$ performance) is analyzed. Some corresponding sufficient conditions are also established for the existence of the sliding mode dynamics, and some algorithms (including the CCL algorithm) are presented to cast the SMC design problem into a nonlinear minimization problem involving LMI conditions instead of the original nonconvex feasibility problem.

Part Two presents the analysis and SMC design procedure for switched state-delayed hybrid systems. It begins with Chapter 5, and consists of four chapters as follows.

Chapter 5 deals with the stability analysis and stabilization problems for continuous- and discrete-time switched hybrid systems with time-varying delays. For a continuous-time system, the time-varying delay $d(t)$ is assumed to satisfy either (A1) $0 \leq d(t) \leq d$ and $\dot{d}(t) \leq \tau$ or (A2) $0 \leq d(t) \leq d$. By using the average dwell time approach and the piecewise Lyapunov function technique, two delay-dependent sufficient conditions are established for the exponential stability of the considered hybrid system with (A1) and (A2), respectively. Here, the slack matrix approach is applied to further reduce the conservativeness of the stability conditions caused by the time delay. For the discrete-time system, the stability conditions are also derived by the average dwell time approach, and the results are all delay-dependent, and thus less conservative. The stabilization problem is then solved by designing a memoryless state feedback controller, and then an explicit expression for the desired controller is given. The research in this chapter is an important foundation for the development of the SMC methodologies for switched hybrid systems in subsequent chapters.

Chapter 6 is concerned with the DOF control problem for continuous-time switched hybrid systems with time-varying delays. Specifically, two issues are investigated: 1) the $\mathcal{L}_2$-$\mathcal{L}_\infty$ control problem for continuous-time switched hybrid systems with time-varying delay. A DOF controller is designed, which is assumed to be switching with the same switching signal as in the original system. A delay-dependent sufficient condition is proposed, to guarantee the exponential stability and a weighted $\mathcal{L}_2$-$\mathcal{L}_\infty$ performance for the closed-loop system with the decay estimate is explicitly given. The corresponding solvability condition for a desired DOF controller is established, and an explicit parametrization of all desired DOF controllers is also given; 2) the guaranteed cost DOF controller design for continuous-time linear switched hybrid system with both discrete and neutral delays. A sufficient condition is first proposed, in terms of a set of LMIs, to guarantee the exponential stability and a certain bound for the cost function of the closed-loop system, where the decay estimate is explicitly given to quantify the convergence rate. Then, the corresponding solvability conditions for a desired DOF controller under guaranteed cost are established by using the approach of linearizing variable transforms. Since these obtained conditions are not all expressed by strict LMIs, the CCL algorithm is exploited to cast them into sequential minimization problems subject to LMI constraints, which can be easily solved numerically.

Chapter 7 studies the SMC design problem for continuous-time switched hybrid systems with time-varying delay. First, the original system is transformed into a regular form through model transformation, and then, by designing a linear sliding surface, the dynamical equation for the sliding mode dynamics is derived. A delay-dependent sufficient condition for the existence of a desired sliding mode is proposed, and an explicit parametrization of the desired sliding surface is also given. Since the obtained conditions are not all expressed in terms of strict LMIs (some matrix equality constraints are involved), the CCL method is exploited to cast them into a sequential minimization problem subject to LMI constraints, which can be easily solved numerically. Then, a discontinuous SMC law is synthesized, by which the system state trajectories can be driven onto the prescribed sliding surface in a finite time and maintained there for all subsequent time. Since the designed SMC law contains state-delay terms, it requires the time-varying delay to be explicitly known *a priori* in the practical implementation of the controller. However, in some practical situations, the information for time delay is usually unavailable, or difficult to measure. In such a case, the designed SMC law is not applicable. To overcome this, we suppose that the state-delay terms in the controller are norm-bounded with an unknown upper bound. We will design an adaptive law to estimate the unknown upper bound, and thus an adaptive SMC law is synthesized, which can also guarantee the system state trajectories reach the prescribed sliding surface.

Chapter 8 is concerned with the problem of SMC of discrete-time switched delayed hybrid systems with time-varying delay. First, we transform the original system into a new one with regular form, and then by designing a linear switching function, a reduced-order sliding mode dynamics, described by a switched state-delayed hybrid system, is developed. By utilizing the average dwell time approach and the piecewise Lyapunov function technique, a delay-dependent sufficient condition for the existence of the desired sliding mode is proposed in terms of LMIs, and an explicit parametrization of the desired switching surface is also given. Here, to reduce the conservativeness induced by the time delay in the system, both the slack matrix technique and also the delay partitioning method are employed, which make the proposed existence condition less conservative. In this chapter, the time delay considered is a time-varying one with a known lower bound. In this case, combined with construction of an appropriate Lyapunov–Krasovskii function, the delay partitioning method is employed by partitioning the lower bound evenly into several components. We then show that the conservativeness of the obtained existence condition becomes less and less with the partitioning getting thinner. Finally, a discontinuous SMC law is designed to drive the state trajectories of the closed-loop system onto a prescribed sliding surface in a finite time and maintained there for all subsequent time.

Part Three presents the analysis and SMC design procedure for switched stochastic hybrid systems. It begins with Chapter 9, and consists of four chapters as follows.

Chapter 9 investigates the problems of stability and performance analysis, stabilization and $\mathcal{H}_\infty$ control (including state feedback control and DOF control) for continuous-time switched stochastic hybrid systems. The average dwell time approach combined with the piecewise Lyapunov function technique is applied to derive the main results. There are two main advantages to using this approach to the switched system. First, this approach uses a mode-dependent Lyapunov function, which avoids some conservativeness caused by using a common Lyapunov function for all the subsystems. Then the other advantage is that the obtained result is not just an asymptotic stability condition, but an exponential one. Therefore, by this approach, a sufficient condition is first proposed, which guarantees the mean-square exponential stability of the unforced switched stochastic hybrid system. When system states are available, a state feedback controller is designed such that the closed-loop system is mean-square exponentially stable with an $\mathcal{H}_\infty$ performance. However, when system states are not all available, a DOF controller is designed and the mean-square exponential stability with an $\mathcal{H}_\infty$ performance is also guaranteed.

Sufficient solvability conditions for the desired controllers are proposed in terms of LMIs.

Chapter 10 considers the stability and performance analysis, stabilization and $\mathcal{H}_\infty$ control problems for discrete-time switched stochastic systems with time-variant delays. By applying the average dwell time method and the piecewise Lyapunov function technique, a sufficient condition is first proposed to guarantee the mean-square exponential stability for the considered system. A condition on a weighted $\mathcal{H}_\infty$ performance is also proposed. Then, the stabilization and $\mathcal{H}_\infty$ state feedback control problems are solved with some sufficient conditions presented in terms of LMI.

Chapter 11 is concerned with the SMC of a continuous-time switched stochastic hybrid system, and some results developed in Chapter 9 are used in the research. Firstly, by designing an integral switching function, we obtain the sliding mode dynamics, which is expressed by a switched stochastic hybrid system with the same order as the original systems. Based on the stability analysis result in Chapter 9, a sufficient condition for the existence of the sliding mode is proposed in terms of LMIs, and an explicit parametrization of the desired switching function is also given. Then, a discontinuous SMC law for reaching motion is synthesized such that the state trajectories of the SMC system can be driven onto a prescribed sliding surface and maintained there for all subsequent time. Moreover, considering that some system state components may not be available in practical applications, we further study the state estimation problem by designing an observer. Sufficient conditions are also established for the existence of the sliding mode and the solvability of the desired observer, and then the observer-based SMC law is synthesized.

Chapter 12 shows the dissipativity analysis and the SMC design for switched stochastic hybrid systems. A more general supply rate is proposed, and a strict $(\mathcal{Z}, \mathcal{Y}, \mathcal{X})$-dissipativity is defined, which includes $\mathcal{H}_\infty$, positive realness, and passivity as its special cases. The main idea is to introduce the strict $(\mathcal{Z}, \mathcal{Y}, \mathcal{X})$-dissipativity concept into the analysis of the sliding mode dynamics so as to improve the transient performance of the SMC system. The objective is to conduct dissipativity analysis and investigate the dissipativity-based SMC design scheme, with a view to contributing to the development of SMC design and the dissipativity analysis methods for the switched stochastic hybrid system. Specifically, an integral sliding surface is designed such that the sliding mode dynamics exists with the same order as the original system. Then, a sufficient condition, which guarantees the sliding mode dynamics mean-square exponentially stable with a strict dissipativity, is then established in terms of LMIs by using the average dwell time approach and the piecewise Lyapunov function technique. In addition, a solution to the dissipativity synthesis is provided by designing a discontinuous SMC law such that the system state trajectories can be driven onto the predefined sliding surface in a finite time.

Part One

SMC of Markovian Jump Singular Systems

2

State Estimation and SMC of Markovian Jump Singular Systems

2.1 Introduction

In this chapter, we are aiming at the investigation of state estimation and SMC problems for Markovian jump singular systems. Although there has been some existing work on the stability analysis of such systems based on the LMI technique, the results are not all of strict LMI form since there are usually some matrix equality constraints. This may cause a lot of trouble in checking the analysis results numerically. Therefore, a natural question is immediately raised: are there any techniques that can release the matrix equality constraints? In this chapter, we extend the approach proposed in [195] to the stability analysis of Markovian jump singular systems, and a new necessary and sufficient stability condition is established in terms of strict LMI. Also, the analysis and synthesis of singular systems have been extensively investigated in the past decades, but little progress has been made toward solving the SMC problem of singular systems. This problem may become difficult and complicated due to the singular matrix E in the systems. Since the rank of E may not be equal to that of B in a simple singular LTI system of $E\dot{x}(t) = Ax(t) + Bu(t)$, it is difficult to obtain the so-called 'regular form' through conventional model transformation approach. As a result, the linear sliding surface is not suitable for singular systems. Therefore, a key issue in the study of this problem is how to design a suitable sliding surface such that the resulting sliding mode dynamics exists.

In this chapter, a new integral-type sliding surface is designed by taking the singular matrix E into account. Then, by using the integral SMC technique, the sliding mode dynamics described by a Markovian jump singular differential equation can be derived. A necessary and sufficient condition for the stochastic stability of the sliding mode dynamics is presented in terms of strict LMI, by which the sliding surface can be designed. In practice, the system states are not always available owing to the limit of physical conditions or the expense of measuring it. Thus, the estimation problem has become more important in this case. In this chapter, we investigate the state estimation and SMC problems for Markovian jump singular systems with unmeasured states. An observer is first designed to estimate the system states, and then a

Sliding Mode Control of Uncertain Parameter-Switching Hybrid Systems, First Edition. Ligang Wu, Peng Shi and Xiaojie Su.

discontinuous SMC law is synthesized based on feedback of the estimated states, which forces the system state trajectories onto the sliding surface in a finite time.

2.2 System Description and Preliminaries

Consider the continuous-time Markovian jump singular system described by

$$E\dot{x}(t) = A(\mathrm{r}_t)x(t) + B(\mathrm{r}_t)u(t), \tag{2.1a}$$

$$y(t) = C(\mathrm{r}_t)x(t), \tag{2.1b}$$

where $x(t) \in \mathbf{R}^n$ is the state vector; $u(t) \in \mathbf{R}^m$ is the control input; $y(t) \in \mathbf{R}^p$ is the measured output. Matrix $E \in \mathbf{R}^{n\times n}$ may be singular, and we assume that rank$(E) = r \leq n$. $A(\cdot)$, $B(\cdot)$, and $C(\cdot)$ are known real matrices with appropriate dimensions. These matrices are functions of r_t. Here, let $\{\mathrm{r}_t, t \geq 0\}$ be a continuous-time Markov process which takes values in a finite state space $\mathcal{S} = \{1, 2, \ldots, N\}$, and the generator matrix $\Pi = \pi_{ij}$, $i, j \in \mathcal{S}$ with transition probability from mode i at time t to mode j at time $t + \Delta$ is given by

$$\mathbf{P}_{ij} = \mathbf{P}\left\{\mathrm{r}_{t+\Delta} = j|\mathrm{r}_t = i\right\} = \begin{cases} \pi_{ij}\Delta + o(\Delta), & \text{if } i \neq j, \\ 1 + \pi_{ii}\Delta + o(\Delta), & \text{if } i = j, \end{cases} \tag{2.2}$$

where $\Delta > 0$ and $\lim_{\Delta\to 0} o(\Delta)/\Delta = 0$; $\pi_{ij} > 0, i \neq j$, and $\pi_{ii} = -\sum_{j\neq i} \pi_{ij}$ for each $i \in \mathcal{S}$.

For each possible value $\mathrm{r}_t = i \in \mathcal{S}$, $A(\mathrm{r}_t) = A_i$, $B(\mathrm{r}_t) = B_i$, and $C(\mathrm{r}_t) = C_i$. Then, the system (2.1a)–(2.1b) can be described by

$$E\dot{x}(t) = A_i x(t) + B_i u(t), \tag{2.3a}$$

$$y(t) = C_i x(t). \tag{2.3b}$$

The following preliminary assumption is made for system (2.3a)–(2.3b).

Assumption 2.1 *For each $i \in \mathcal{S}$, the pair (A_i, B_i) in (2.3a)–(2.3b) is controllable, the pair (A_i, C_i) is observable, and matrix B_i is full column rank.*

Before proceeding, we first consider the unforced system of (2.3a), that is,

$$E\dot{x}(t) = A_i x(t). \tag{2.4}$$

Definition 2.2.1

I. The Markovian jump singular system in (2.4) is said to be regular if det $(sE - A_i)$ *is not identically zero for each $i \in \mathcal{S}$.*

II. The Markovian jump singular system in (2.4) is said to be impulse free if deg $(\det(sE - A_i)) = \mathrm{rank}(E)$ *for each $i \in \mathcal{S}$.*

III. The Markovian jump singular system in (2.4) is said to be stochastically stable if, for any $x_0 \in \mathbf{R}^n$ and $\mathrm{r}_0 \in \mathcal{S}$, there exists a positive scalar $T(x_0, \mathrm{r}_0)$ such that

$$\min_{t\to\infty} \mathbf{E}\left\{ \int_0^t \|x(s, x_0, \mathrm{r}_0)\|^2 \, ds | (x_0, \mathrm{r}_0) \right\} \leq T(x_0, \mathrm{r}_0).$$

IV. The Markovian jump singular system in (2.4) is said to be stochastically admissible if it is regular, impulse free and stochastically stable.

The following lemma provides a necessary and sufficient condition for the stochastic admissibility of the Markovian jump singular system in (2.4).

Lemma 2.2.2 *[244] The Markovian jump singular system in (2.4) is stochastically admissible if and only if there exist nonsingular matrices P_i such that for $i \in \mathcal{S}$,*

$$E^T P_i = P_i^T E \geq 0, \tag{2.5a}$$

$$A_i^T P_i + P_i^T A_i + \sum_{j=1}^{N} \pi_{ij} E^T P_j < 0. \tag{2.5b}$$

Remark 2.1 *Notice that the conditions in Lemma 2.2.2 are not all of strict LMI form due to the matrix equality constraint of (2.5a). This may cause major problems in checking the conditions numerically, since the matrix equality constraint is fragile and is not usually perfectly satisfied. Therefore, the strict LMI conditions are more desirable than non-strict ones from the numerical point of view.* ♦

2.3 Stochastic Stability Analysis

In the section, we propose a strict LMI condition (easy to check by using standard software) of the stochastic admissibility for the Markovian jump singular system in (2.4), and present the following result.

Theorem 2.3.1 *The Markovian jump singular system in (2.4) is stochastically admissible if and only if there exist matrices $X_i > 0$, Y_i, U, and W such that for $i \in \mathcal{S}$,*

$$A_i^T \left(X_i E + U^T Y_i W^T\right) + \left(E^T X_i + W Y_i^T U\right) A_i + \sum_{j=1}^{N} \pi_{ij} E^T X_j E < 0, \tag{2.6}$$

where $U \in \mathbf{R}^{(n-r)\times n}$ and $W \in \mathbf{R}^{n\times(n-r)}$ are matrices satisfying $UE = 0$ and $EW = 0$.

Proof. (Sufficiency) Letting $P_i \triangleq X_i E + U^T Y_i W^T$, $i \in \mathcal{S}$ in (2.6), we can satisfy (2.5a) and (2.5b). Thus, according to Lemma 2.2.2 we know that the continuous Markovian jump singular system in (2.4) is stochastically admissible.

(Necessity) Suppose that the system in (2.4) is stochastically admissible, then there exist nonsingular matrices M and N such that, for each $i \in \mathcal{S}$,

$$MEN = \begin{bmatrix} I & 0 \\ 0 & 0 \end{bmatrix}, \quad MA_iN = \begin{bmatrix} A_{1i} & A_{2i} \\ A_{3i} & A_{4i} \end{bmatrix}. \tag{2.7}$$

Since the system in (2.4) is regular and impulse free we have that matrices A_{4i} are nonsingular for $i \in \mathcal{S}$. Thus, we can set

$$\tilde{M}_i \triangleq \begin{bmatrix} I & -A_{2i}A_{4i}^{-1} \\ 0 & I \end{bmatrix} M.$$

Then, it follows that

$$\begin{aligned} \tilde{E} = \tilde{M}_iEN &= \begin{bmatrix} I & -A_{2i}A_{4i}^{-1} \\ 0 & I \end{bmatrix} MEN \\ &= \begin{bmatrix} I & -A_{2i}A_{4i}^{-1} \\ 0 & I \end{bmatrix} \begin{bmatrix} I & 0 \\ 0 & 0 \end{bmatrix} = \begin{bmatrix} I & 0 \\ 0 & 0 \end{bmatrix}, \end{aligned} \tag{2.8}$$

$$\begin{aligned} \tilde{A}_i = \tilde{M}_iA_iN &= \begin{bmatrix} I & -A_{2i}A_{4i}^{-1} \\ 0 & I \end{bmatrix} MA_iN \\ &= \begin{bmatrix} I & -A_{2i}A_{4i}^{-1} \\ 0 & I \end{bmatrix} \begin{bmatrix} A_{1i} & A_{2i} \\ A_{3i} & A_{4i} \end{bmatrix} = \begin{bmatrix} \tilde{A}_{1i} & 0 \\ A_{3i} & A_{4i} \end{bmatrix}, \end{aligned} \tag{2.9}$$

where $\tilde{A}_{1i} \triangleq A_{1i} - A_{2i}A_{4i}^{-1}A_{3i}$. Therefore, it is easy to see that the stochastic stability of system (2.4) implies that the following continuous Markovian jump system is stochastically stable:

$$\dot{\xi}(t) = \tilde{A}_{1i}\xi(t)$$

It follows that there exist matrices $\tilde{X}_i > 0$ such that, for $i \in \mathcal{S}$,

$$\tilde{A}_{1i}^T\tilde{X}_i + \tilde{X}_i\tilde{A}_{1i} + \sum_{j=1}^{N} \pi_{ij}\tilde{X}_j < 0.$$

Now, let $N \triangleq \begin{bmatrix} N_1 & N_2 \end{bmatrix}$ and $\tilde{M}_i \triangleq \begin{bmatrix} \tilde{M}_{1i}^T & \tilde{M}_{2i}^T \end{bmatrix}^T$, thus by (2.9) we have $\tilde{M}_{2i}A_iN = \begin{bmatrix} A_{3i} & A_{4i} \end{bmatrix}$, where the partitions of N and $\tilde{M}_i$ are compatible for algebraic operations. Therefore, for a sufficient small $\alpha > 0$, we have

$$\begin{aligned} &\tilde{A}_{1i}^T\tilde{X}_i + \tilde{X}_i\tilde{A}_{1i} + \sum_{j=1}^{N} \pi_{ij}\tilde{X}_j - \alpha\left(A_{3i}^TN_2^TN_1 + N_1^TN_2A_{3i}\right) + \alpha\left(A_{3i}^TN_2^TN_2 + N_1^TN_2A_{4i}\right) \\ &\times \left(A_{4i}^TN_2^TN_2 + N_2^TN_2A_{4i}\right)^{-1}\left(A_{4i}^TN_2^TN_1 + N_2^TN_2A_{3i}\right) < 0. \end{aligned} \tag{2.10}$$

By Schur complement, (2.10) is equivalent to

$$\begin{bmatrix} \Psi_{11i} & -\alpha A_{3i}^T N_2^T N_2 - \alpha N_1^T N_2 A_{4i} \\ \star - & \alpha A_{4i}^T N_2^T N_2 - \alpha N_2^T N_2 A_{4i} \end{bmatrix} < 0, \tag{2.11}$$

where

$$\Psi_{11i} \triangleq \tilde{A}_{1i}^T \tilde{X}_i + \tilde{X}_i \tilde{A}_{1i} + \sum_{j=1}^{N} \pi_{ij} \tilde{X}_j - \alpha A_{3i}^T N_2^T N_1 - \alpha N_1^T N_2 A_{3i}.$$

Furthermore, (2.11) can be rewritten as

$$\begin{aligned} &\begin{bmatrix} \tilde{A}_{1i} & 0 \\ A_{3i} & A_{4i} \end{bmatrix}^T \begin{bmatrix} \tilde{X}_i & 0 \\ 0 & I \end{bmatrix} \begin{bmatrix} I & 0 \\ 0 & 0 \end{bmatrix} + \begin{bmatrix} I & 0 \\ 0 & 0 \end{bmatrix} \begin{bmatrix} \tilde{X}_i & 0 \\ 0 & I \end{bmatrix} \begin{bmatrix} \tilde{A}_{1i} & 0 \\ A_{3i} & A_{4i} \end{bmatrix} \\ &+ \sum_{j=1}^{N} \pi_{ij} \begin{bmatrix} I & 0 \\ 0 & 0 \end{bmatrix} \begin{bmatrix} \tilde{X}_j & 0 \\ 0 & I \end{bmatrix} \begin{bmatrix} I & 0 \\ 0 & 0 \end{bmatrix} \\ &+ \begin{bmatrix} A_{3i}^T \\ A_{4i}^T \end{bmatrix} (-\alpha I_{n-r}) N_2^T \begin{bmatrix} N_1 & N_2 \end{bmatrix} \\ &+ \begin{bmatrix} N_1^T \\ N_2^T \end{bmatrix} N_2 (-\alpha I_{n-r}) \begin{bmatrix} A_{3i} & A_{4i} \end{bmatrix} < 0. \end{aligned} \tag{2.12}$$

Considering (2.7), it follows from (2.12) that

$$\begin{aligned} &N^T A_i^T \tilde{M}_i^T \begin{bmatrix} \tilde{X}_i & 0 \\ 0 & I \end{bmatrix} \tilde{M}_i E N + N^T E^T \tilde{M}_i^T \begin{bmatrix} \tilde{X}_i & 0 \\ 0 & I \end{bmatrix} \tilde{M}_i A_i N \\ &+ \sum_{j=1}^{N} \pi_{ij} N^T E^T \tilde{M}_j^T \begin{bmatrix} \tilde{X}_j & 0 \\ 0 & I \end{bmatrix} \tilde{M}_j E N \\ &- \alpha N^T \left(N_2 \tilde{M}_{2i} A_i + A_i^T \tilde{M}_{2i}^T N_2^T \right) N < 0. \end{aligned} \tag{2.13}$$

Let $X_i \triangleq \tilde{M}_i^T \begin{bmatrix} \tilde{X}_i & 0 \\ 0 & I \end{bmatrix} \tilde{M}_i$ in (2.13) (obviously, $X_i > 0$), we have

$$N^T \left[A_i^T \left(X_i E - \alpha \tilde{M}_{2i}^T N_2^T \right) + \left(E^T X_i - \alpha N_2 \tilde{M}_{2i} \right) A_i + \sum_{j=1}^{N} \pi_{ij} E^T X_j E \right] N < 0.$$

Since $\tilde{M}_{2i}$ is of full row rank, it can be written as $\tilde{M}_{2i} = \tilde{M}_{3i}U$, where $\tilde{M}_{3i} \in \mathbf{R}^{(n-r)\times(n-r)}$ is nonsingular (thus, by (2.8) $\tilde{M}_{2i}EN = \tilde{M}_{3i}UEN = 0$ implies $UE = 0$). Then, defining $Y_i \triangleq -\alpha\tilde{M}_{3i}^T$ and $W \triangleq N_2$ (it is easily seen from (2.7) that $EW = EN_2 = 0$), we have

$$N^T\left[A_i^T\left(X_iE + U^TY_iW^T\right) + \left(E^TX_i + WY_i^TU\right)A_i + \sum_{j=1}^{N}\pi_{ij}E^TX_jE\right]N < 0,$$

which is equivalent to (2.6). This completes the proof. ■

Remark 2.2 *Note that Theorem 2.3.1 presents a new necessary and sufficient condition of stochastic admissibility in terms of strict LMI for the Markovian jump singular system in (2.4), which is less conservative and more useful than Lemma 2.2.2.* ♦

2.4 Main Results

In this section, we consider the state estimation and SMC problems for the Markovian jump singular systems with unmeasured states in (2.1a)–(2.1b). First, we design an observer to estimate unmeasured states, and then we design a sliding surface and an SMC law based on the state estimates. The designed observer-based SMC law can drive the system state trajectories onto the predefined sliding surface in a finite time.

2.4.1 Observer and SMC Law Design

The following observer is employed to provide the estimates of the unmeasured states for the system in (2.3a)–(2.3b):

$$E\dot{\hat{x}}(t) = A_i\hat{x}(t) + B_iu(t) + L_i\left(y(t) - \hat{y}(t)\right), \tag{2.14a}$$

$$\hat{y}(t) = C_i\hat{x}(t), \tag{2.14b}$$

where $\hat{x}(t) \in \mathbf{R}^n$ represents the estimate of $x(t)$, and $L_i \in \mathbf{R}^{n\times p}$ are the observer gains to be designed.

Let $e(t) \triangleq x(t) - \hat{x}(t)$ denote the estimation error. Considering (2.3a)–(2.3b) and (2.14a)–(2.14b), the estimation error dynamics is obtained as

$$E\dot{e}(t) = \left(A_i - L_iC_i\right)e(t). \tag{2.15}$$

Design the following integral switching function:

$$s(t) = G_iE\hat{x}(t) - \int_0^t G_i\left(A_i + B_iK_i\right)\hat{x}(\theta)d\theta, \tag{2.16}$$

where $K_i \in \mathbf{R}^{m\times n}$ are real matrices to be designed such that

$$E\dot{\hat{x}}(t) = \left(A_i + B_iK_i\right)\hat{x}(t) \tag{2.17}$$

is stochastically admissible. The matrices $G_i \in \mathbf{R}^{m\times n}$ are to be chosen so that G_iB_i are nonsingular.

Design the following state estimate-based SMC law:

$$u(t) = K_i\hat{x}(t) - (\varepsilon + \rho(t))\,\mathrm{sign}\,(s(t)), \tag{2.18}$$

where $\varepsilon > 0$ is a real constant and

$$\rho(t) \triangleq \max_{i\in S} \|G_iB_i\| \left(\|G_iL_iy(t)\| + \|G_iL_iC_i\hat{x}(t)\|\right).$$

The following theorem shows that the sliding motion in the specified sliding surface $s(t) = 0$ is attained in a finite time.

Theorem 2.4.1 *Under the SMC law (2.18), the state trajectories of the observer dynamics (2.14a)–(2.14b) can be driven onto the sliding surface $s(t) = 0$ in a finite time and remain there in subsequent time.*

Proof. Choose the following Lyapunov function:

$$V(t) = s^T(t)\left(B_i^TZ_iB_i\right)^{-1} s(t),$$

where $Z_i > 0$ are matrices to be specified such that $B_i^TZ_iB_i > 0$. Thus, we choose $G_i = B_i^TZ_i$ in (2.16), and then $G_iB_i = B_i^TZ_iB_i > 0$ are nonsingular. According to (2.14a)–(2.14b) and (2.16), we have

$$\begin{aligned}\dot{s}(t) &= G_iE\dot{\hat{x}}(t) - G_i\left(A_i + B_iK_i\right)\hat{x}(t)\\ &= B_i^TZ_iB_i\left(u(t) - K_i\hat{x}(t)\right) + G_iL_i\left(y(t) - \hat{y}(t)\right).\end{aligned} \tag{2.19}$$

Substituting (2.18) into (2.19) yields

$$\dot{s}(t) = -B_i^TZ_iB_i\left(\varepsilon + \rho(t)\right)\mathrm{sign}\,(s(t)) + G_iL_i\left(y(t) - \hat{y}(t)\right).$$

Thus taking the derivation of $V(t)$ and considering $|s(t)| \geq \|s(t)\|$, we have

$$\begin{aligned}\dot{V}(t) &= 2s^T(t)\left(B_i^TZ_iB_i\right)^{-1}\dot{s}(t)\\ &= 2s^T(t)\left(B_i^TZ_iB_i\right)^{-1}\left[-B_i^TZ_iB_i\left(\varepsilon + \rho(t)\right)\mathrm{sign}\,(s(t)) + G_iL_i\left(y(t) - \hat{y}(t)\right)\right]\\ &\leq -2\left(\varepsilon + \rho(t)\right)\|s(t)\| + 2\left\|B_i^TZ_iB_i\right\|\left(\|G_iL_iy(t)\| + \|G_iL_iC_i\hat{x}(t)\|\right)\|s(t)\|\\ &\leq -2\varepsilon\,\|s(t)\| \leq -\tilde{\varepsilon}V^{\frac{1}{2}}(t).\end{aligned} \tag{2.20}$$

where $\tilde{\varepsilon} \triangleq 2\varepsilon/\sqrt{\lambda_{\max}\left(B_i^T Z_i B_i\right)}$. It can be seen from (2.20) that there exists a time $t^* = 2\sqrt{V(0)}/\tilde{\varepsilon}$ such that $V(t) = 0$, and consequently $s(t) = 0$, for $t \geq t^*$, which means that the system state trajectories can reach onto the predefined sliding surface in a finite time. This completes the proof. ■

2.4.2 *Sliding Mode Dynamics Analysis*

When the system operates in the sliding mode, it follows that $s(t) = 0$ and $\dot{s}(t) = 0$. Thus, by $\dot{s}(t) = 0$ in (2.19), we can obtain the equivalent control $u_{eq}(t)$ as

$$u_{eq}(t) = K_i\hat{x}(t) - G_i L_i C_i e(t). \tag{2.21}$$

Substituting (2.21) into (2.14a), the sliding mode dynamics can be obtained as

$$E\dot{\hat{x}}(t) = \left(A_i + B_i K_i\right)\hat{x}(t) + \left(I - B_i G_i\right) L_i C_i e(t). \tag{2.22}$$

In the following, we will analyze the stochastic admissibility of the estimation error dynamics in (2.15). By Theorem 2.3.1, we give the following result.

Theorem 2.4.2 *The estimation error dynamics in (2.15) is stochastically admissible if and only if there exist symmetric positive definite matrices $X_i \in \mathbf{R}^{n\times n}$, nonsingular matrices $Y_i \in \mathbf{R}^{(n-r)\times(n-r)}$, and matrices $\mathcal{L}_i \in \mathbf{R}^{n\times p}$, $\mathcal{Q}_i \in \mathbf{R}^{(n-r)\times p}$, $U \in \mathbf{R}^{(n-r)\times n}$, $W \in \mathbf{R}^{n\times(n-r)}$ such that for $i \in \mathcal{S}$,*

$$\begin{aligned}&\left(E^T X_i + W Y_i^T U\right) A_i + A_i^T \left(X_i E + U^T Y_i W^T\right)\\ &- \left(E^T \mathcal{L}_i + W \mathcal{Q}_i\right) C_i - C_i^T \left(\mathcal{L}_i^T E + \mathcal{Q}_i^T W^T\right) + \sum_{j=1}^{N} \pi_{ij} E^T X_j E < 0,\end{aligned} \tag{2.23}$$

where U and W are matrices satisfying $UE = 0$ and $EW = 0$. Moreover, the parametric matrices L_i can be computed by

$$L_i = \left(E^T X_i + W Y_i^T U\right)^{-1} \left(E^T \mathcal{L}_i + W \mathcal{Q}_i\right). \tag{2.24}$$

Proof. According to Theorem 2.3.1, we know that the estimation error dynamics in (2.15) is stochastically admissible if and only if there exist matrices $X_i > 0, Y_i$, U and W such that for $i \in \mathcal{S}$,

$$\left(E^T X_i + W Y_i^T U\right)\left(A_i - L_i C_i\right) + \left(A_i^T - C_i^T L_i^T\right)\left(X_i E + U^T Y_i W^T\right) + \sum_{j=1}^{N} \pi_{ij} E^T X_j E < 0. \tag{2.25}$$

Letting $\mathcal{L}_i \triangleq X_i L_i$ and $\mathcal{Q}_i \triangleq Y_i^T U L_i$ in (2.25) yields (2.23), thus the proof is completed. ■

Next, we shall analyze the stochastic admissibility of the dynamics in (2.17), and give a solution to parameter K_i. Before proceeding, we give the following lemma.

Lemma 2.4.3 *Let X_i be symmetric such that $E_L^T X_i E_L > 0$ and matrices Y_i are nonsingular, then $X_i E + U^T Y_i W^T$ are nonsingular and their inverse are expressed as*

$$\left(X_i E + U^T Y_i W^T\right)^{-1} = \mathcal{X}_i E^T + W \mathcal{Y}_i U, \tag{2.26}$$

where $\mathcal{X}_i$ are symmetric matrices and $\mathcal{Y}_i$ are nonsingular matrices with

$$\mathcal{Y}_i = \left(W^T W\right)^{-1} Y_i^{-1} \left(U U^T\right)^{-1}, \quad E_R^T \mathcal{X}_i E_R = \left(E_L^T X_i E_L\right)^{-1}.$$

Proof. Decompose E as $E = E_L E_R^T$, where $E_L \in \mathbf{R}^{n\times r}$ and $E_R \in \mathbf{R}^{n\times r}$ are of full column rank. Since $UE = 0$ and $EW = 0$, thus we have that $UE_L = 0$, $E_R^T W = 0$ and $\begin{bmatrix} E_R & W \end{bmatrix}$ is nonsingular. Then,

$$\begin{cases} \begin{bmatrix} E_R & W \end{bmatrix}^{-1} = \begin{bmatrix} \left(E_R^T E_R\right)^{-1} E_R^T \\ \left(W^T W\right)^{-1} W^T \end{bmatrix}, \\ E_R \left(E_R^T E_R\right)^{-1} E_R^T + W \left(W^T W\right)^{-1} W^T = I_n. \end{cases} \tag{2.27}$$

According to (2.27), we have

$$\begin{aligned} &\Big[E_R \left(E_R^T E_R\right)^{-1} \left(E_L^T X_i E_L\right)^{-1} E_L^T + W \left(W^T W\right)^{-1} Y_i^{-1} \left(U U^T\right)^{-1} U \\ &\quad - W \left(W^T W\right)^{-1} Y_i^{-1} \left(U U^T\right)^{-1} U X_i E_L \left(E_L^T X_i E_L\right)^{-1} E_L^T\Big] \\ &\quad \times \left(X_i E + U^T Y_i W^T\right) = I_n, \end{aligned}$$

which implies that $X_i E + U^T Y_i W^T$ are nonsingular and

$$\begin{aligned} \left(X_i E + U^T Y_i W^T\right)^{-1} &= E_R \left(E_R^T E_R\right)^{-1} \left(E_L^T X_i E_L\right)^{-1} E_L^T \\ &\quad + W \left(W^T W\right)^{-1} Y_i^{-1} \left(U U^T\right)^{-1} U \\ &\quad - W \left(W^T W\right)^{-1} Y_i^{-1} \left(U U^T\right)^{-1} U X_i E_L \left(E_L^T X_i E_L\right)^{-1} E_L^T \\ &= \Phi^T \Omega_i \Phi E^T + W \left(W^T W\right)^{-1} Y_i^{-1} \left(U U^T\right)^{-1} U, \end{aligned} \tag{2.28}$$

where

$$\begin{cases} \Phi \triangleq \begin{bmatrix} \left(E_R^T E_R\right)^{-1} E_R^T \\ \left(W^T W\right)^{-1} W^T \end{bmatrix}, \\ \Omega_i \triangleq \begin{bmatrix} \left(E_L^T X_i E_L\right)^{-1} & -\left(E_L^T X_i E_L\right)^{-1} E_L^T X_i U^T \left(UU^T\right)^{-1} Y_i^{-T} \\ \star & 0 \end{bmatrix}. \end{cases}$$

Define $\mathcal{X}_i \triangleq \Phi^T \Omega_i \Phi$ and $\mathcal{Y}_i \triangleq \left(W^T W\right)^{-1} Y_i^{-1} \left(UU^T\right)^{-1}$ in (2.28), and we have (2.26)–(2.27). This completes the proof. ∎

Now, according to Theorem 2.3.1, we present the following result without proof.

Theorem 2.4.4 *The dynamics in (2.17) is stochastically admissible if and only if there exist matrices $X_i > 0$, Y_i, U and W such that for $i \in \mathcal{S}$,*

$$\left(E^T X_i + W Y_i^T U\right)\left(A_i + B_i K_i\right) + \left(A_i^T + K_i^T B_i^T\right)\left(X_i E + U^T Y_i W^T\right) + \sum_{j=1}^{N} \pi_{ij} E^T X_j E < 0, \quad (2.29)$$

where $U \in \mathbf{R}^{(n-r)\times n}$ and $W \in \mathbf{R}^{n\times(n-r)}$ are matrices satisfying $UE = 0$ and $EW = 0$.

The following sufficient condition is proposed for the stochastic admissibility of the dynamics in (2.17), by which the parametric matrices K_i can be solved.

Theorem 2.4.5 *The dynamics in (2.17) is stochastically admissible if there exist symmetric positive definite matrices $\mathcal{X}_i \in \mathbf{R}^{n\times n}$, nonsingular matrices $\mathcal{Y}_i \in \mathbf{R}^{(n-r)\times(n-r)}$, and matrices $\mathcal{K}_i \in \mathbf{R}^{m\times n}$, $\mathcal{R}_i \in \mathbf{R}^{m\times(n-r)}$, $U \in \mathbf{R}^{(n-r)\times n}$, and $W \in \mathbf{R}^{n\times(n-r)}$ such that for $i \in \mathcal{S}$,*

$$\begin{bmatrix} \Psi_{11i} + \Psi_{11i}^T - \pi_{ii} E \mathcal{X}_i E^T & \Psi_{12i} \\ \star & \Psi_{22i} \end{bmatrix} < 0, \quad (2.30)$$

where $\mathcal{Z}_i \triangleq \mathcal{X}_i E^T + W \mathcal{Y}_i U$ and

$$\begin{cases} \Psi_{11i} \triangleq A_i \mathcal{Z}_i + B_i\left(\mathcal{K}_i E^T + \mathcal{R}_i U\right) + \pi_{ii} E \mathcal{Z}_i, \\ \Psi_{22i} \triangleq -\mathrm{diag}\left\{E_R^T \mathcal{X}_1 E_R,\ E_R^T \mathcal{X}_2 E_R,\ \ldots,\ E_R^T \mathcal{X}_{i-1} E_R,\right. \\ \qquad \left. E_R^T \mathcal{X}_{i+1} E_R,\ \ldots,\ E_R^T \mathcal{X}_{N-1} E_R,\ E_R^T \mathcal{X}_N E_R\right\}, \\ \Psi_{12i} \triangleq \left[\sqrt{\pi_{i1}} \mathcal{Z}_i^T E_R \quad \sqrt{\pi_{i2}} \mathcal{Z}_i^T E_R \quad \cdots \quad \sqrt{\pi_{i(i-1)}} \mathcal{Z}_i^T E_R \right. \\ \qquad \left. \sqrt{\pi_{i(i+1)}} \mathcal{Z}_i^T E_R \quad \cdots \quad \sqrt{\pi_{i(N-1)}} \mathcal{Z}_i^T E_R \quad \sqrt{\pi_{iN}} \mathcal{Z}_i^T E_R\right], \end{cases}$$

where U and W are matrices satisfying $UE = 0$ and $EW = 0$. Moreover, the parametric matrices K_i are given by

$$K_i = \left(\mathcal{K}_i E^T + \mathcal{R}_i U\right) \mathcal{Z}_i^{-1}. \quad (2.31)$$

Proof. By Theorem 2.4.4 we know that the dynamics in (2.17) is stochastically admissible if there exist matrices $X_i > 0$ and nonsingular matrices Y_i such that (2.29) holds for $i \in \mathcal{S}$. However, by Lemma 2.4.3, $X_iE + U^TY_iW^T$ are nonsingular and their inverse matrices are $\mathcal{X}_iE^T + W\mathcal{Y}_iU$. Now, performing a congruence transformation to (2.29) by $\mathcal{Z}_i \triangleq \mathcal{X}_iE^T + W\mathcal{Y}_iU$, we have

$$\left(A_i + B_iK_i\right)\mathcal{Z}_i + \mathcal{Z}_i^T\left(A_i^T + K_i^TB_i^T\right) + \sum_{j=1}^{N}\pi_{ij}\mathcal{Z}_i^TE^TX_jE\mathcal{Z}_i < 0. \tag{2.32}$$

Letting $\mathcal{K}_i \triangleq K_i\mathcal{X}_i$ and $\mathcal{R}_i \triangleq K_iW\mathcal{Y}_i$ in (2.32), we have

$$\begin{aligned}&A_i\mathcal{Z}_i + \mathcal{Z}_i^TA_i^T + B_i\left(\mathcal{K}_iE^T + \mathcal{R}_iU\right) + \left(E\mathcal{K}_i^T + U^T\mathcal{R}_i^T\right)B_i^T + \pi_{ii}\mathcal{Z}_i^TE_R\left(E_R^T\mathcal{X}_iE_R\right)^{-1}E_R^T\mathcal{Z}_i\\ &+ \sum_{j=1,j\neq i}^{N}\pi_{ij}\mathcal{Z}_i^TE_R\left(E_R^T\mathcal{X}_jE_R\right)^{-1}E_R^T\mathcal{Z}_i < 0.\end{aligned} \tag{2.33}$$

Also, the following fact is true:

$$\begin{aligned}0 &\le \left[E_R^T\mathcal{Z}_i - \left(E_R^T\mathcal{X}_iE_R\right)E_L^T\right]^T\left(E_R^T\mathcal{X}_iE_R\right)^{-1}\left[E_R^T\mathcal{Z}_i^T - \left(E_R^T\mathcal{X}_iE_R\right)E_L^T\right]\\ &= -E\mathcal{Z}_i - \mathcal{Z}_i^TE^T + E\mathcal{X}_iE^T + \mathcal{Z}_i^TE_R\left(E_R^T\mathcal{X}_iE_R\right)^{-1}E_R^T\mathcal{Z}_i.\end{aligned}$$

Considering $\pi_{ii} < 0$ in (2.2), we have

$$\pi_{ii}\mathcal{Z}_i^TE_R\left(E_R^T\mathcal{X}_iE_R\right)^{-1}E_R^T\mathcal{Z}_i \le \pi_{ii}E\mathcal{Z}_i + \pi_{ii}\mathcal{Z}_i^TE^T - \pi_{ii}E\mathcal{X}_iE^T.$$

Therefore, (2.33) holds if the following inequality holds:

$$\Psi_{11i} + \Psi_{11i}^T - \pi_{ii}E\mathcal{X}_iE^T + \sum_{j=1,j\neq i}^{N}\pi_{ij}\mathcal{Z}_i^TE_R\left(E_R^T\mathcal{X}_jE_R\right)^{-1}E_R^T\mathcal{Z}_i < 0, \tag{2.34}$$

where Ψ_{11i} is defined in (2.30). By Schur complement, LMI (2.30) implies inequality (2.34). This completes the proof. ■

Remark 2.3 *Notice from Definition 2.2.1 that the stochastic admissability implies the stochastic stability of a Markovian jump singular system. Thus, we know that the estimation error dynamics in (2.15) is stochastically stable if LMI (2.23) in Theorem 2.4.2 holds. Also, the dynamics in (2.17) is stochastically stable if LMI (2.30) in Theorem 2.4.5 holds. It is not difficult to show from stochastic stability of dynamics (2.15) and (2.17) that the sliding mode dynamics (2.22) is stochastically stable.* ♦

2.5 Illustrative Example

Example 2.5.1 Consider the Markovian jump singular system in (2.1a)–(2.1b) with two operating modes, that is, $N = 2$, and the following parameters:

$$A_1 = \begin{bmatrix} 1.3 & 0.8 & 1.0 \\ 0.7 & 0.8 & 0.9 \\ 0.4 & 0.2 & -0.7 \end{bmatrix}, \ B_1 = \begin{bmatrix} 1.5 \\ 0.9 \\ 1.1 \end{bmatrix}, \ E = \begin{bmatrix} 1.0 & 0.0 & 0.0 \\ 0.0 & 1.0 & 0.0 \\ 0.0 & 0.0 & 0.0 \end{bmatrix}, \ S = \begin{bmatrix} 0.0 \\ 0.0 \\ 1.0 \end{bmatrix},$$

$$A_2 = \begin{bmatrix} 0.7 & 0.9 & 0.3 \\ 1.1 & 1.4 & -0.4 \\ 0.5 & 0.3 & 1.6 \end{bmatrix}, \ B_2 = \begin{bmatrix} 0.9 \\ 1.8 \\ 1.4 \end{bmatrix}, \ E_L = E_R = \begin{bmatrix} 1.0 & 0.0 \\ 0.0 & 1.0 \\ 0.0 & 0.0 \end{bmatrix},$$

$$\Pi = \begin{bmatrix} -0.6 & 0.6 \\ 0.8 & -0.8 \end{bmatrix}, \ C_1 = \begin{bmatrix} 1.1 & 1.6 & 0.9 \end{bmatrix}, \ C_2 = \begin{bmatrix} 1.5 & 1.3 & 0.7 \end{bmatrix}, \ R = S^T.$$

Our aim is to design an observer in the form of (2.14a)–(2.14b) to estimate the states of system (2.1a)–(2.1b), and then synthesize an SMC law $u(t)$ as (2.18) (based on the state estimate) such that the closed-loop system is stochastically admissible.

Solving the LMI condition (2.23) in Theorem 2.4.2 by using LMI Toolbox in the Matlab environment and then by (2.24), we have

$$L_1 = \begin{bmatrix} 3.6359 \\ 0.1377 \\ 0.1525 \end{bmatrix}, \ L_2 = \begin{bmatrix} 0.6149 \\ 1.8942 \\ 1.0853 \end{bmatrix}.$$

However, solving the LMI condition (2.30) in Theorem 2.4.5, and then by (2.31) we have

$$K_1 = \begin{bmatrix} -2.3459 & -0.0756 & -0.1750 \end{bmatrix},$$
$$K_2 = \begin{bmatrix} -1.6968 & -1.0136 & -0.7452 \end{bmatrix}.$$

Here, parameters G_1 and G_2 in (2.16) are chosen as

$$G_1 = \begin{bmatrix} 0.3513 & 0.2108 & 0.2576 \end{bmatrix},$$
$$G_2 = \begin{bmatrix} 0.1498 & 0.2995 & 0.2329 \end{bmatrix}.$$

Thus, the switching function in (2.16) can be computed as

$$s(t) = \begin{cases} s_1(t) = \begin{bmatrix} 0.3513 & 0.2108 & 0.0000 \end{bmatrix} \hat{x}(t) \\ \quad - \int_0^t \begin{bmatrix} -1.6387 & 0.4256 & 0.1857 \end{bmatrix} \hat{x}(s) ds, & i = 1, \\ s_2(t) = \begin{bmatrix} 0.1498 & 0.2995 & 0.0000 \end{bmatrix} \hat{x}(t) \\ \quad - \int_0^t \begin{bmatrix} -1.1460 & -0.3897 & -0.4474 \end{bmatrix} \hat{x}(s) ds, & i = 2. \end{cases}$$

Let the adjustable parameter ε be $\varepsilon = 0.5$, then the observer-based SMC law designed in (2.18) can be obtained as

$$u(t) = \begin{cases} u_1(t) = \begin{bmatrix} -2.3459 & -0.0756 & -0.1750 \end{bmatrix} \hat{x}(t) \\ \qquad - (0.5 + \rho(t)) \operatorname{sign}\left(s_1(t)\right), & i = 1, \\ u_2(t) = \begin{bmatrix} -1.6968 & -1.0136 & -0.7452 \end{bmatrix} \hat{x}(t) \\ \qquad - (0.5 + \rho(t)) \operatorname{sign}\left(s_2(t)\right), & i = 2, \end{cases}$$

where $\rho(t) = 1.3456\,(\|y(t)\| + \|\hat{x}(t)\|)$.

To prevent the control signals from chattering, we replace $\operatorname{sign}\left(s_i(t)\right)$ with

$$\frac{s_i(t)}{0.01 + \left\|s_i(t)\right\|}, \qquad i \in \{1, 2\}.$$

For given initial condition of $x(0) = \begin{bmatrix} -0.8 & -1.2 & -0.6 \end{bmatrix}^T$, the simulation results are given in Figures 2.1–2.2. Specifically, in Figure 2.1 shows the states of the closed-loop system, while Figure 2.2 depicts the switching function $s(t)$.

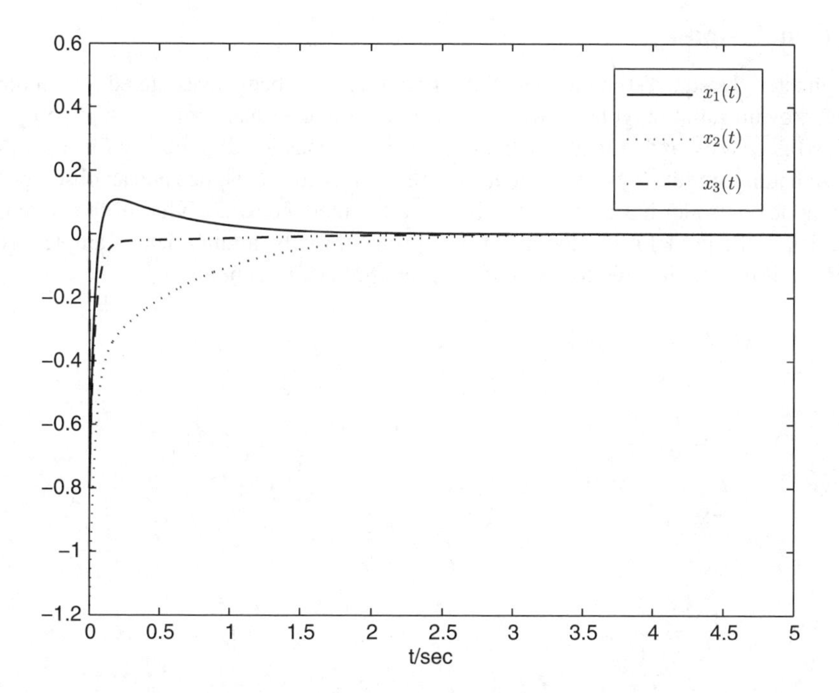

Figure 2.1 States of the closed-loop system

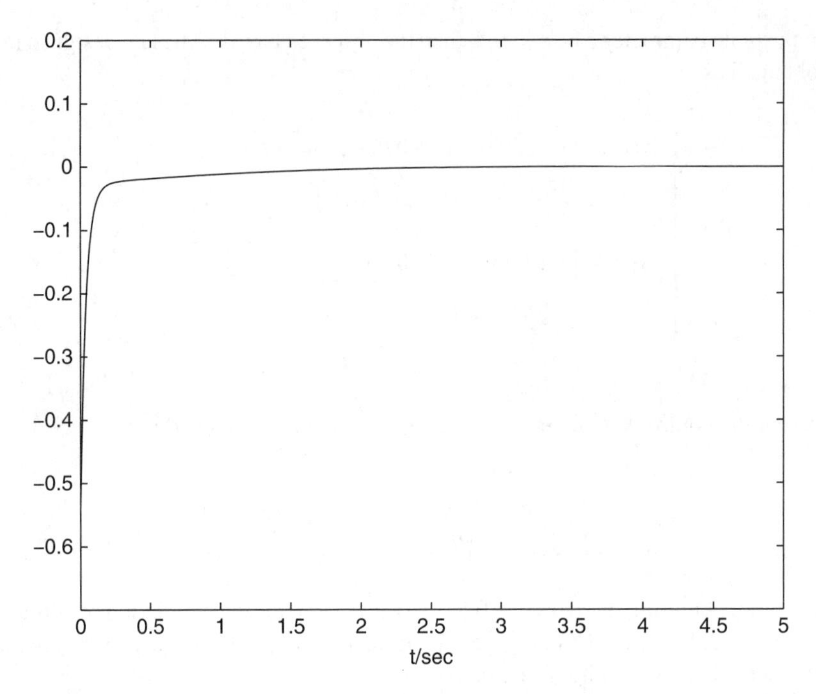

Figure 2.2 Switching function

2.6 Conclusion

In this chapter, the state estimation and SMC problems have been investigated for continuous-time Markovian jump singular systems with unmeasured states. First, we have proposed a strict LMI necessary and sufficient condition of the stochastic admissibility for the unforced Markovian jump singular systems. Then, an observer has been designed and an observer-based sliding mode controller has been synthesized to guarantee the reachability of the system state trajectories to the predefined integral sliding surface. Finally, a numerical example has been provided to illustrate the effectiveness of the proposed design scheme.

3

Optimal SMC of Markovian Jump Singular Systems with Time Delay

3.1 Introduction

It is recognized that the sliding mode of an SMC system is invariant to system perturbations and external disturbances, only if the perturbations/disturbances satisfy the so-called 'matching condition'. Although many researchers have paid considerable attention to the mismatched uncertainties in SMC design, the obtained results are very conservative. If the undesired uncertainties/disturbances can not be eliminated in the sliding mode, it is possible to attenuate its effect on the system performance. In this chapter, we will consider the disturbance attenuation problem in sliding mode with $\mathcal{H}_\infty$ performance. For this purpose, we design an integral switching function. The plant considered in this chapter is the Markovian jump singular time-delay system, which is a typical kind of hybrid systems of high complexity (including system parameter jumping, time delay in states, and singularity). How to establish a less conservative stability condition is a key issue in SMC design. As is well known, the slack matrix technique [228, 243], usually combined with the Lyapunov–Krasovskii approach, has been proved to be an effective tool to establish less conservative stability conditions for time-delay systems. Unfortunately, little progress has been made in dealing with singular time-delay systems by this technique, probably due to the particularity and complexity caused by the singular matrix E, thus it is difficult to choose a suitable Lyapunov–Krasovskii function.

In this chapter, we will pay particular attention to the singular matrix E in the design of an integral-type switching function, which leads to a full-order Markovian jump singular time-delay system for describing the sliding mode dynamics. We will then apply the slack matrix technique combining with the Lyapunov–Krasovskii approach to derive a delay-dependent sufficient condition, which guarantees that the sliding mode dynamics is stochastically stable with a bounded $\mathcal{L}_2$ gain performance. In addition, the analysis result and the solvability condition for the desired switching function are both established. All the obtained results are in terms of strict LMI, which can be solved by efficient interior-point algorithms [25]. Finally, a discontinuous SMC law is designed to drive the system state trajectories onto the predefined sliding surface in a finite time.

Sliding Mode Control of Uncertain Parameter-Switching Hybrid Systems, First Edition. Ligang Wu, Peng Shi and Xiaojie Su.

3.2 System Description and Preliminaries

Consider a Markovian jump singular time-delay system described by

$$E\dot{x}(t) = A(\mathrm{r}_t)x(t) + A_d(\mathrm{r}_t)x(t-d) + B_w(\mathrm{r}_t)\omega(t) + B(\mathrm{r}_t)\left(u(t) + f(x(t),t)\right), \tag{3.1a}$$

$$z(t) = C(\mathrm{r}_t)x(t) + C_d(\mathrm{r}_t)x(t-d) + D_w(\mathrm{r}_t)\omega(t), \tag{3.1b}$$

$$x(t) = \phi(t), \quad t \in [-d, 0], \tag{3.1c}$$

where $\{\mathrm{r}_t, t \geq 0\}$ is a continuous-time Markov process on the probability space which has been defined in (2.2) of Chapter 2, and $x(t) \in \mathbf{R}^n$ is the system state vector; $u(t) \in \mathbf{R}^m$ is the control input; $z(t) \in \mathbf{R}^p$ is the controlled output; $\omega(t) \in \mathbf{R}^q$ is the exogenous input (which represents either the exogenous disturbance input or the exogenous reference input) belonging to $\mathcal{L}_2[0,\infty)$. Matrix $E \in \mathbf{R}^{n\times n}$ may be singular, and it is assumed that $\mathrm{rank}(E) = r \leq n$. $A(\cdot)$, $B(\cdot)$, $C(\cdot)$, $A_d(\cdot)$, $C_d(\cdot)$, $B_w(\cdot)$ and $D_w(\cdot)$ are known real matrices with appropriate dimensions. d represents the constant time-delay and $\phi(t) \in \mathcal{C}_{n,d}$ is a compatible vector-valued initial function. In addition, $f(x(t),t) \in \mathbf{R}^m$ is an unknown nonlinear function (which represents the unmodeled dynamics of a physical plant), and there exists a known constant $\eta > 0$ such that

$$\|f(x(t),t)\| \leq \eta \|x(t)\|.$$

For each $\mathrm{r}_t = i \in \mathcal{S}$, $A(\mathrm{r}_t) = A_i$, $B(\mathrm{r}_t) = B_i$, $C(\mathrm{r}_t) = C_i$, $A_d(\mathrm{r}_t) = A_{di}$, $C_d(\mathrm{r}_t) = C_{di}$, $B_w(\mathrm{r}_t) = B_{wi}$, and $D_w(\mathrm{r}_t) = D_{wi}$. Then, system (3.1a)–(3.1c) can be described by

$$E\dot{x}(t) = A_i x(t) + A_{di}x(t-d) + B_{wi}\omega(t) + B_i\left(u(t) + f(x(t),t)\right), \tag{3.2a}$$

$$z(t) = C_i x(t) + C_{di}x(t-d) + D_{wi}\omega(t), \tag{3.2b}$$

$$x(t) = \phi(t), \quad t \in [-d, 0], \tag{3.2c}$$

Assumption 3.1 *For each $i \in \mathcal{S}$, the pair (A_i, B_i) in (3.2a) is controllable, and matrix B_i is full column rank.*

Before proceeding, we first consider the unforced system of (3.2a)–(3.2c), that is,

$$E\dot{x}(t) = A_i x(t) + A_{di}x(t-d), \tag{3.3a}$$

$$x(t) = \phi(t), \quad t \in [-d, 0]. \tag{3.3b}$$

We introduce the following definition for the Markovian jump singular time-delay system in (3.3a)–(3.3b).

Definition 3.2.1

I. *The Markovian jump singular time-delay system in (3.3a)–(3.3b) is said to be regular and impulse free if the pairs* (E, A_i) *and* $(E, A_i + A_{di})$ *are regular and impulse free for each* $i \in \mathcal{S}$.

II. *The Markovian jump singular time-delay system in (3.3a)–(3.3b) is said to be stochastically stable if for any* $x_0 \in \mathbf{R}^n$ *and* $\mathrm{r}_0 \in S$, *there exists a positive scalar* $T(x_0, \phi(\cdot))$ *such that*

$$\min_{t\to\infty} \mathbf{E}\left\{\int_0^t \|x(t)\|^2\, dt|\mathrm{r}_0, x(s) = \phi(s), s \in [-d, 0]\right\} \le T(x_0, \phi(\cdot)).$$

III. *The Markovian jump singular time-delay system in (3.3a)–(3.3b) is said to be stochastically admissible if it is regular, impulse free and stochastically stable.*

In addition, we introduce the following definition for the Markovian jump singular time-delay system of

$$E\dot{x}(t) = A_i x(t) + A_{di} x(t-d) + B_{wi}\omega(t), \tag{3.4a}$$

$$z(t) = C_i x(t) + C_{di} x(t-d) + D_{wi}\omega(t), \tag{3.4b}$$

$$x(t) = \phi(t), \quad t \in [-d, 0], \tag{3.4c}$$

Definition 3.2.2 *Given a scalar* $\gamma > 0$, *the Markovian jump singular time-delay system in (3.4a)–(3.4c) is said to be stochastically admissible with a bounded* $\mathcal{L}_2$ *gain performance* γ, *if the system (3.4a)–(3.4c) with* $\omega(t) \equiv 0$ *is stochastically admissible, and under zero condition, for nonzero* $\omega(t) \in \mathcal{L}_2[0, \infty)$, *it holds that*

$$\mathbf{E}\left\{\int_0^\infty z^T(t)z(t)dt\right\} < \gamma^2 \int_0^\infty \omega^T(t)\omega(t)dt. \tag{3.5}$$

3.3 Bounded $\mathcal{L}_2$ Gain Performance Analysis

This section is concerned with the bounded $\mathcal{L}_2$ gain performance analysis for the Markovian jump singular time-delay system in (3.4a)–(3.4c) in the sense of Definition 3.2.2, and we give the following theorem.

Theorem 3.3.1 *Given a scalar* $\gamma > 0$, *the Markovian jump singular time-delay system in (3.4a)–(3.4c) is stochastically admissible with bounded* $\mathcal{L}_2$ *gain performance* γ, *if there exist* *matrices* $Q > 0$, $R > 0$, $P_i \triangleq \begin{bmatrix} P_{11i} & P_{12i} \\ 0 & P_{22i} \end{bmatrix}$, $W_i \triangleq \begin{bmatrix} W_{1i} & 0_{n\times(n-r)} \end{bmatrix}$ *(with* $0 < P_{11i} \in \mathbf{R}^{r\times r}$ *and* $W_{1i} \in \mathbf{R}^{n\times r}$*) such that for each* $i \in \mathcal{S}$,

$$\begin{bmatrix} \Psi_{11i} & P_iA_{di} - W_i & P_iB_{\omega i} & dW_i & dA_i^TR & C_i^T \\ \star & -Q & 0 & 0 & dA_{di}^TR & C_{di}^T \\ \star & \star & -\gamma^2 I & 0 & dB_{wi}^TR & D_{wi}^T \\ \star & \star & \star & -dR & 0 & 0 \\ \star & \star & \star & \star & -dR & 0 \\ \star & \star & \star & \star & \star & -I \end{bmatrix} < 0, \tag{3.6}$$

where

$$\Psi_{11i} \triangleq P_iA_i + A_i^TP_i^T + Q + \sum_{j=1}^{N} \pi_{ij}P_jE + W_i + W_i^T.$$

Proof. First, we consider the nominal case of (3.4a)–(3.4c), that is, $\omega(t) = 0$ in (3.4a)–(3.4c). Without loss of generality, we assume that the matrix E and the state vector $x(t)$ in (3.4a)–(3.4c) have the form of

$$E = \begin{bmatrix} I_{r\times r} & 0_{r\times(n-r)} \\ 0_{(n-r)\times r} & 0_{(n-r)\times(n-r)} \end{bmatrix}, \quad x(t) = \begin{bmatrix} x_1(t) \\ x_2(t) \end{bmatrix},$$

where $x_1(t) \in \mathbf{R}^r$ and $x_2(t) \in \mathbf{R}^{n-r}$.

In the following, we will consider the stochastic stability of the system in (3.4a)–(3.4c) with $\omega(t) \equiv 0$. To this end, we choose a Lyapunov function as

$$V(x_t, \mathrm{r}_t, t) \triangleq x^T(t)P(\mathrm{r}_t)Ex(t) + \int_{t-d}^{t} x^T(\tau)Qx(\tau)d\tau + \int_{-d}^{0}\int_{t+\theta}^{t} \dot{x}^T(\tau)E^TRE\dot{x}(\tau)d\tau d\theta, \tag{3.7}$$

where $x_t \triangleq x(\theta)$, $\theta \in [t-2d, t]$, thus $\{(x_t, \mathrm{r}_t), t \geq d\}$ is a Markov process with initial condition $(\phi(\cdot), \mathrm{r}_0)$. Matrices Q and R are positive definite, and

$$P(\mathrm{r}_t) \triangleq \begin{bmatrix} P_{11}(\mathrm{r}_t) & P_{12}(\mathrm{r}_t) \\ P_{21}(\mathrm{r}_t) & P_{22}(\mathrm{r}_t) \end{bmatrix}, \text{ or equivalently, } P_i \triangleq \begin{bmatrix} P_{11i} & P_{12i} \\ P_{21i} & P_{22i} \end{bmatrix},$$

with $P_{11}(\mathrm{r}_t) > 0$ and $P_{21}(\mathrm{r}_t) = 0$ (which can be found from $P(\mathrm{r}_t)E = E^T P^T(\mathrm{r}_t) \geq 0$). Let $\mathcal{A}$ be the weak infinitesimal generator of the random process $\{x_t, \mathrm{r}_t\}$. Thus, for each possible value $\mathrm{r}_t = i \in \mathcal{S}$ and $t \geq d$, we have

$$\begin{aligned}\mathcal{A}V(x_t, i, t) &= 2x^T(t)P_i[A_i x(t) + A_{di}x(t-d)] + x^T(t)Qx(t) - x^T(t-d)Qx(t-d) \\ &\quad + d[A_i x(t) + A_{di}x(t-d)]^T R[A_i x(t) + A_{di}x(t-d)] \\ &\quad + x^T(t)\left(\sum_{j=1}^{N} \pi_{ij}P_j E\right)x(t) - \int_{t-d}^{t} \dot{x}^T(\tau)E^T RE\dot{x}(\tau)d\tau. \end{aligned} \tag{3.8}$$

On the other hand, Newton–Leibniz formula gives

$$x(t) - x(t-d) = \int_{t-d}^{t} \dot{x}(\tau)d\tau.$$

Thus, for $W_i \triangleq \begin{bmatrix} W_{1i} & 0_{n\times(n-r)} \end{bmatrix}$ with $W_{1i} \in \mathbf{R}^{n\times r}$, it holds that

$$2x^T(t)W_i\left[x(t) - x(t-d) - \int_{t-d}^{t} \dot{x}(\tau)d\tau\right] = 0. \tag{3.9}$$

By (3.8)–(3.9) and noting

$$W_i E = \begin{bmatrix} W_{1i} & 0_{n\times(n-r)} \end{bmatrix}\begin{bmatrix} I_{r\times r} & 0_{r\times(n-r)} \\ 0_{(n-r)\times r} & 0_{(n-r)\times(n-r)} \end{bmatrix} = W_i,$$

we have

$$\begin{aligned}\mathcal{A}V(x_t, i, t) &= 2x^T(t)P_i[A_i x(t) + A_{di}x(t-d)] + x^T(t)Qx(t) - x^T(t-d)Qx(t-d) \\ &\quad + x^T(t)\left(\sum_{j=1}^{N} \pi_{ij}P_j E\right)x(t) + 2x^T(t)W_i[x(t) - x(t-d)] \\ &\quad + d[A_i x(t) + A_{di}x(t-d)]^T R[A_i x(t) + A_{di}x(t-d)] \\ &\quad + \int_{t-d}^{t} x^T(t)W_i R^{-1}W_i^T x(t)d\tau - \int_{t-d}^{t} x^T(t)W_i R^{-1}W_i^T x(t)d\tau \\ &\quad - \int_{t-d}^{t} 2x^T(t)W_i E\dot{x}(\tau)d\tau - \int_{t-d}^{t} \dot{x}^T(\tau)E^T RE\dot{x}(\tau)d\tau \\ &= \psi^T(t)\Phi_i\psi(t) \\ &\quad - \int_{t-d}^{t} \left[W_i^T x(t) + RE\dot{x}(\tau)\right]^T R^{-1}\left[W_i^T x(t) + RE\dot{x}(\tau)\right] d\tau, \end{aligned} \tag{3.10}$$

where $\psi(t) \triangleq \begin{bmatrix} x(t) \\ x(t-d) \end{bmatrix}$ and

$$\Phi_i \triangleq \begin{bmatrix} \Phi_{11i} & P_i A_{di} - W_i \\ \star & -Q \end{bmatrix} + d \begin{bmatrix} A_i^T \\ A_{di}^T \end{bmatrix} R \begin{bmatrix} A_i & A_{di} \end{bmatrix},$$

with

$$\Phi_{11i} \triangleq P_i A_i + A_i^T P_i^T + Q + \sum_{j=1}^{N} \pi_{ij} P_j E + W_i + W_i^T + d W_i R^{-1} W_i^T.$$

By Schur complement, LMI (3.6) implies $\Phi_i < 0$. Moreover, noting that the last integral term in (3.10) is semi-positive, thus (3.10) implies that there exists a scalar $\varepsilon > 0$ such that for each $i \in \mathcal{S}$,

$$\mathcal{A}V(x_t, i, t) \leq -\varepsilon \left\| x(t) \right\|^2.$$

The rest of the proof on stochastic stability can be found in [230, 231], and so we omit it here.

Moreover, (3.6) implies $\Psi_{11i} < 0$. Now partition matrices A_i and Q as

$$A_i = \begin{bmatrix} A_{11i} & A_{12i} \\ A_{21i} & A_{22i} \end{bmatrix}, \quad Q = \begin{bmatrix} Q_{11} & Q_{12} \\ \star & Q_{22} \end{bmatrix} > 0,$$

and then substituting them and P_i into $\Psi_{11i} < 0$ yields

$$P_{22i} A_{22i} + A_{22i}^T P_{22i}^T + Q_{22} < 0,$$

which implies that A_{22i} are nonsingular for $i \in \mathcal{S}$, thus the pairs (E, A_i) are regular and impulse free for $i \in \mathcal{S}$. Therefore, the system in (3.4a)–(3.4c) with $\omega(t) = 0$ is regular and impulse free.

Now, we establish the bounded $\mathcal{L}_2$ gain performance of system (3.4a)–(3.4c). Consider the Lyapunov function in (3.7) again and the following index:

$$J = \mathbf{E}\left\{ \int_0^t \left[z^T(\tau) z(\tau) - \gamma^2 \omega^T(\tau) \omega(\tau) \right] d\tau \right\}.$$

Then, under zero initial condition, it can be shown that for any nonzero $\omega(t) \in \mathcal{L}_2[0, \infty)$,

$$\begin{aligned} J &\leq \mathbf{E}\left\{ \int_0^t \left[z^T(\tau) z(\tau) - \gamma^2 \omega^T(\tau) \omega(\tau) + \mathcal{A}V(x_\tau, i, \tau) \right] d\tau \right\} \\ &\leq \mathbf{E}\left\{ \int_0^t \bar{\psi}^T(\tau) \bar{\Phi}_i \bar{\psi}(\tau) d\tau \right\}, \end{aligned}$$

where $\bar{\psi}(t) \triangleq \begin{bmatrix} x(t) \\ x(t-d) \\ \omega(t) \end{bmatrix}$ and

$$\bar{\Phi}_i \triangleq \begin{bmatrix} \Phi_{11i} & P_iA_{di} - W_i & P_iB_{\omega i} \\ \star & -Q & 0 \\ \star & \star & -\gamma^2 I \end{bmatrix} + d \begin{bmatrix} A_i^T \\ A_{di}^T \\ B_{\omega i}^T \end{bmatrix} R \begin{bmatrix} A_i & A_{di} & B_{\omega i} \end{bmatrix}$$

$$+ \begin{bmatrix} C_i^T \\ C_{di}^T \\ D_{\omega i}^T \end{bmatrix} \begin{bmatrix} C_i & C_{di} & D_{\omega i} \end{bmatrix}.$$

By Schur complement, LMI (3.6) implies $\bar{\Phi}_i < 0$, thus $J \leq 0$, and hence (3.5) is true for any nonzero $\omega(t) \in \mathcal{L}_2[0, \infty)$. This completes the proof. ■

Remark 3.1 *It should be noted that Theorem 3.3.1 presents a delay-dependent sufficient condition of the stochastic admissibility with the bounded $\mathcal{L}_2$ gain performance defined in Definition 3.2.2 for the Markovian jump singular time-delay system in (3.4a)–(3.4c). Notice that the slack matrix variables W_i are introduced in the derivation of the delay-dependent result in Theorem 3.3.1, which avoids some conservativeness caused by the commonly used model transformation approach when dealing with time-delay systems.* ♦

3.4 Main Results

3.4.1 Sliding Mode Dynamics Analysis

We design the following integral-type switching function:

$$s(t) = G_iEx(t) - \int_0^t G_i \left(A_i + B_iK_i\right) x(\theta)d\theta, \tag{3.11}$$

where $G_i \in \mathbf{R}^{m\times n}$ and $K_i \in \mathbf{R}^{m\times n}$ are real matrices. In particular, the matrices G_i are to be chosen such that G_iB_i are nonsingular for $i \in \mathcal{S}$. The solution of $Ex(t)$ can be given by

$$Ex(t) = Ex(0) + \int_0^t [A_ix(\theta) + A_{di}x(\theta - d) + B_{wi}\omega(\theta) + B_i\left(u(\theta) + f(x(\theta), \theta)\right)]d\theta. \tag{3.12}$$

It follows from (3.11) and (3.12) that

$$s(t) = G_iEx(0) + G_i \int_0^t [-B_iK_ix(\theta) + A_{di}x(\theta - d) + B_{wi}\omega(\theta) + B_iu(\theta) + B_if(x(\theta), \theta)]d\theta. \tag{3.13}$$

When the system state trajectories reach onto the sliding surface, it follows that $s(t) = 0$ and $\dot{s}(t) = 0$. Therefore, by $\dot{s}(t) = 0$, we get the equivalent control as

$$\begin{aligned} u_{eq}(t) = K_i x(t) - \left(G_i B_i\right)^{-1} G_i A_{di} x(t-d) \\ -(G_i B_i)^{-1} G_i B_{wi} \omega(t) - f(x(t), t). \end{aligned} \tag{3.14}$$

By substituting (3.14) into (3.2a)–(3.2c), the sliding mode dynamics can be obtained as

$$\begin{aligned} E\dot{x}(t) = (A_i + B_i K_i) x(t) + \left[I - B_i (G_i B_i)^{-1} G_i\right] A_{di} x(t-d) \\ + \left[I - B_i (G_i B_i)^{-1} G_i\right] B_{wi} \omega(t). \end{aligned} \tag{3.15}$$

For notational simplicity, we define

$$\begin{aligned} \tilde{A}_{di} &\triangleq \left[I - B_i (G_i B_i)^{-1} G_i\right] A_{di}, \quad \tilde{A}_i \triangleq A_i + B_i K_i, \\ \tilde{B}_{wi} &\triangleq \left[I - B_i (G_i B_i)^{-1} G_i\right] B_{wi}. \end{aligned}$$

Thus, the sliding mode dynamics in (3.15) and the controlled output equation in (3.2b) can be formulated as

$$E\dot{x}(t) = \tilde{A}_i x(t) + \tilde{A}_{di} x(t-d) + \tilde{B}_{wi} \omega(t), \tag{3.16a}$$

$$z(t) = C_i x(t) + C_{di} x(t-d) + D_{wi} \omega(t). \tag{3.16b}$$

The above analysis gives the first step of the SMC for the Markovian jump singular time-delay system in (3.1a)–(3.1c). Specifically, we design an integral-type switching function as (3.11) so that the dynamics restricted to the sliding surface (i.e. the sliding mode dynamics) has the form of (3.16a)–(3.16b). Thus, the remaining problems to be addressed in this chapter are as follows:

- $\mathcal{H}_\infty$ performance analysis of the sliding mode dynamics. Given all the system matrices in (3.1a)–(3.1c) and the matrices G_i and K_i in the switching function of (3.11), determine under what condition the sliding mode dynamics in (3.16a)–(3.16b) is stochastically admissible with a bounded $\mathcal{L}_2$ gain performance defined in the sense of Definition 3.2.2.
- SMC law synthesis. Synthesize an SMC law to drive the system state trajectories onto the predefined sliding surface $s(t) = 0$ in a finite time and maintain them there for all subsequent time.

By Theorem 3.3.1, we have the following result for dynamics (3.16a)–(3.16b).

Corollary 3.4.1 *Given a scalar $\gamma > 0$, the sliding mode dynamics in (3.16a)–(3.16b) is stochastically admissible with bounded $\mathcal{L}_2$ gain performance γ, if there exist matrices $\mathcal{Q} > 0$, $\mathcal{R} > 0$, $\mathcal{P}_i \triangleq \begin{bmatrix} \mathcal{P}_{11i} & \mathcal{P}_{12i} \\ 0 & \mathcal{P}_{22i} \end{bmatrix}$ (with $\mathcal{P}_{11i} > 0$), $\mathcal{W}_i \triangleq [\,\mathcal{W}_{1i} \quad 0_{n\times(n-r)}\,]$ such that for each $i \in \mathcal{S}$,*

$$\begin{bmatrix} \Omega_{11i} & \mathcal{P}_i\tilde{A}_{di} - \mathcal{W}_i & \mathcal{P}_i\tilde{B}_{\omega i} & d\mathcal{W}_i & d\tilde{A}_i^T\mathcal{R} & C_i^T \\ \star & -\mathcal{Q} & 0 & 0 & d\tilde{A}_{di}^T\mathcal{R} & C_{di}^T \\ \star & \star & -\gamma^2 I & 0 & d\tilde{B}_{wi}^T\mathcal{R} & D_{wi}^T \\ \star & \star & \star & -d\mathcal{R} & 0 & 0 \\ \star & \star & \star & \star & -d\mathcal{R} & 0 \\ \star & \star & \star & \star & \star & -I \end{bmatrix} < 0, \tag{3.17}$$

where

$$\Omega_{11i} \triangleq \mathcal{P}_i\tilde{A}_i + \tilde{A}_i^T\mathcal{P}_i^T + \mathcal{Q} + \sum_{j=1}^{N} \pi_{ij}\mathcal{P}_jE + \mathcal{W}_i + \mathcal{W}_i^T.$$

The following theorem is devoted to solving the parameter K_i in the switching function of (3.11).

Theorem 3.4.2 *Given a scalar $\gamma > 0$, the sliding mode dynamics in (3.16a)–(3.16b) is stochastically admissible with bounded $\mathcal{L}_2$ gain performance γ, if there exist matrices $\mathcal{Q} > 0$, $\mathcal{R} > 0$, $\mathcal{K}_i$, $\mathcal{P}_i \triangleq \begin{bmatrix} \mathcal{P}_{11i} & \mathcal{P}_{12i} \\ 0 & \mathcal{P}_{22i} \end{bmatrix}$ (with $\mathcal{P}_{11i} > 0$), $\mathcal{W}_i \triangleq [\,\mathcal{W}_{1i} \quad 0_{n\times(n-r)}\,]$ and a scalar $\sigma > 0$ such that for each $i \in \mathcal{S}$,*

$$\begin{bmatrix} \tilde{\Omega}_{11i} & \tilde{\Omega}_{12i} & \tilde{B}_{\omega i} & d\mathcal{W}_i & \tilde{\Omega}_{15i} & \mathcal{P}_iC_i^T & \mathcal{P}_i & \tilde{\Omega}_{18i} \\ \star & \tilde{\Omega}_{22i} & 0 & 0 & d\mathcal{P}_i\tilde{A}_{di}^T & \mathcal{P}_iC_{di}^T & 0 & 0 \\ \star & \star & -\gamma^2 I & 0 & d\tilde{B}_{wi}^T & D_{wi}^T & 0 & 0 \\ \star & \star & \star & \tilde{\Omega}_{44i} & 0 & 0 & 0 & 0 \\ \star & \star & \star & \star & -d\mathcal{R} & 0 & 0 & 0 \\ \star & \star & \star & \star & \star & -I & 0 & 0 \\ \star & \star & \star & \star & \star & \star & -\mathcal{Q} & 0 \\ \star & \star & \star & \star & \star & \star & \star & \tilde{\Omega}_{88i} \end{bmatrix} < 0, \tag{3.18}$$

where

$$\begin{cases} \tilde{\Omega}_{11i} \triangleq A_i\mathcal{P}_i^T + B_i\mathcal{K}_i + \mathcal{P}_iA_i^T + \mathcal{K}_i^TB_i^T + \pi_{ii}E\mathcal{P}_i^T + \mathcal{W}_i + \mathcal{W}_i^T, \\ \tilde{\Omega}_{12i} \triangleq \tilde{A}_{di}\mathcal{P}_i^T - \mathcal{W}_i, \\ \tilde{\Omega}_{22i} \triangleq -\mathcal{P}_i - \mathcal{P}_i^T + \mathcal{Q}, \\ \tilde{\Omega}_{44i} \triangleq -d\mathcal{P}_i - d\mathcal{P}_i^T + d\mathcal{R}, \\ \tilde{\Omega}_{15i} \triangleq d\mathcal{P}_iA_i^T + d\mathcal{K}_i^TB_i^T, \\ \tilde{\Omega}_{18i} \triangleq \big[\sqrt{\pi_{i1}}\mathcal{P}_i \quad \sqrt{\pi_{i2}}\mathcal{P}_i \quad \cdots \quad \sqrt{\pi_{i(i-1)}}\mathcal{P}_i \\ \qquad \sqrt{\pi_{i(i+1)}}\mathcal{P}_i \quad \cdots \quad \sqrt{\pi_{i(N-1)}}\mathcal{P}_i \quad \sqrt{\pi_{iN}}\mathcal{P}_i \big], \\ \tilde{\Omega}_{88i} \triangleq -\mathrm{diag}\left\{ \left(\mathcal{P}_1E + \sigma^{-1}\mathcal{I}\right), \left(\mathcal{P}_2E + \sigma^{-1}\mathcal{I}\right), \ldots, \left(\mathcal{P}_{i-1}E + \sigma^{-1}\mathcal{I}\right) \right. \\ \qquad \left. \left(\mathcal{P}_{i+1}E + \sigma^{-1}\mathcal{I}\right), \ldots, \left(\mathcal{P}_{N-1}E + \sigma^{-1}\mathcal{I}\right), \left(\mathcal{P}_NE + \sigma^{-1}\mathcal{I}\right) \right\}. \end{cases}$$

Moreover, if the above LMI conditions have a set of feasible solutions then the parametric matrices K_i in (3.11) can be computed by

$$K_i = \mathcal{K}_i\mathcal{P}_i^{-T}. \tag{3.19}$$

Proof. Letting $\mathcal{P}_i = P_i^{-1}$, $\mathcal{Q} \triangleq Q^{-1}$, $\mathcal{R} = R^{-1}$ and $\mathcal{W}_i \triangleq \mathcal{P}_iW_i\mathcal{P}_i^T$, and performing a congruence transformation on (3.17) by $\mathrm{diag}\left\{\mathcal{P}_i, \mathcal{P}_i, I, \mathcal{P}_i, \mathcal{R}, I\right\}$, we have

$$\begin{bmatrix} \bar{\Omega}_{11i} & \tilde{\Omega}_{12i} & \tilde{B}_{\omega i} & d\mathcal{W}_i & d\mathcal{P}_i\tilde{A}_i^T & \mathcal{P}_iC_i^T \\ \star & \bar{\Omega}_{22i} & 0 & 0 & d\mathcal{P}_i\tilde{A}_{di}^T & \mathcal{P}_iC_{di}^T \\ \star & \star & -\gamma^2 I & 0 & d\tilde{B}_{wi}^T & D_{wi}^T \\ \star & \star & \star & \bar{\Omega}_{44i} & 0 & 0 \\ \star & \star & \star & \star & -d\mathcal{R} & 0 \\ \star & \star & \star & \star & \star & -I \end{bmatrix} < 0, \tag{3.20}$$

where

$$\begin{cases} \bar{\Omega}_{11i} \triangleq \tilde{A}_i\mathcal{P}_i^T + \mathcal{P}_i\tilde{A}_i^T + \mathcal{P}_i\mathcal{Q}\mathcal{P}_i^T + \mathcal{W}_i + \mathcal{W}_i^T + \sum_{j=1}^{N} \pi_{ij}\mathcal{P}_i\mathcal{P}_jE\mathcal{P}_i^T, \\ \bar{\Omega}_{22i} \triangleq -\mathcal{P}_i\mathcal{Q}\mathcal{P}_i^T, \\ \bar{\Omega}_{44i} \triangleq -d\mathcal{P}_i\mathcal{R}\mathcal{P}_i^T, \end{cases}$$

and $\tilde{\Omega}_{12i}$ is defined in (3.18).

Also notice that

$$0 \le \left(\mathcal{P}_i - \mathcal{Q}\right)\mathcal{Q}\left(\mathcal{P}_i - \mathcal{Q}\right)^T = \mathcal{P}_i\mathcal{Q}\mathcal{P}_i^T - \mathcal{P}_i - \mathcal{P}_i^T + \mathcal{Q},$$

which implies

$$-\mathcal{P}_i\mathcal{Q}\mathcal{P}_i^T \leq -\mathcal{P}_i - \mathcal{P}_i^T + \mathcal{Q}.$$

Similarly, we have

$$-\mathcal{P}_i\mathcal{R}\mathcal{P}_i^T \leq -\mathcal{P}_i - \mathcal{P}_i^T + \mathcal{R}.$$

Moreover, noting that $\mathcal{P}_jE \geq 0$, it follows that there exists a sufficient small scalar $\sigma > 0$ such that $\mathcal{P}_jE + \sigma\mathcal{I} > 0$, where

$$\mathcal{I} = \begin{bmatrix} 0_{r\times r} & 0_{r\times(n-r)} \\ 0_{(n-r)\times r} & I_{(n-r)\times(n-r)} \end{bmatrix},$$

thus,

$$\begin{aligned}\left(\mathcal{P}_jE + \sigma\mathcal{I}\right)^{-1} &= \begin{bmatrix} \mathcal{P}_{11i} & 0 \\ 0 & \sigma I_{(n-r)\times(n-r)} \end{bmatrix}^{-1} \\ &= \begin{bmatrix} \mathcal{P}_{11i} & 0 \\ 0 & \sigma^{-1} I_{(n-r)\times(n-r)} \end{bmatrix} = \mathcal{P}_jE + \sigma^{-1}\mathcal{I}.\end{aligned}$$

Therefore, inequality (3.20) holds if

$$\begin{bmatrix} \hat{\Omega}_{11i} & \tilde{\Omega}_{12i} & \tilde{B}_{\omega i} & d\mathcal{W}_i & d\mathcal{P}_i\tilde{A}_i^T & \mathcal{P}_iC_i^T \\ \star & \tilde{\Omega}_{22i} & 0 & 0 & d\mathcal{P}_i\tilde{A}_{di}^T & \mathcal{P}_iC_{di}^T \\ \star & \star & -\gamma^2 I & 0 & d\tilde{B}_{wi}^T & D_{wi}^T \\ \star & \star & \star & \tilde{\Omega}_{44i} & 0 & 0 \\ \star & \star & \star & \star & -d\mathcal{R} & 0 \\ \star & \star & \star & \star & \star & -I \end{bmatrix} < 0,$$

where $\tilde{\Omega}_{12i}$, $\tilde{\Omega}_{22i}$, and $\tilde{\Omega}_{44i}$ are defined in (3.18), and

$$\hat{\Omega}_{11i} \triangleq \tilde{A}_i\mathcal{P}_i^T + \mathcal{P}_i\tilde{A}_i^T + \mathcal{P}_i\mathcal{Q}\mathcal{P}_i^T + \mathcal{W}_i + \mathcal{W}_i^T + \pi_{ii}E\mathcal{P}_i^T + \sum_{j=1, j\neq i}^{N} \pi_{ij}\mathcal{P}_i\left(\mathcal{P}_jE + \sigma\mathcal{I}\right)\mathcal{P}_i^T.$$

Moreover, letting $\mathcal{K}_i = K_i\mathcal{P}_i^T$ and by Schur complement, the above inequality is equivalent to (3.18). This completes the proof. ■

Remark 3.2 *Note that we introduced the term $\sigma\mathcal{I}(\sigma > 0)$ in the proof of Theorem 3.4.2. The reason is that $\mathcal{P}_jE$ is singular (and there is no inversion), while $\mathcal{P}_jE + \sigma\mathcal{I}$ is nonsingular and thus it has inversion.* ♦

3.4.2 SMC Law Design

In this section, we shall synthesize a discontinuous SMC law, by which the state trajectories of the Markovian jump singular time-delay system in (3.2a)–(3.2c) can be driven onto the predefined sliding surface $s(t) = 0$ in a finite time and then maintained there for all subsequent time.

Theorem 3.4.3 *Consider the Markovian jump singular time-delay system in (3.2a)–(3.2c). Suppose that the switching function is designed as (3.11) with K_i being solvable by (3.19), and matrices G_i in (3.11) are chosen such that G_iB_i are nonsingular. Then, the state trajectories of system (3.2a)–(3.2c) can be driven onto the sliding surface $s(t) = 0$ in a finite time by the following SMC law:*

$$u(t) = K_ix(t) - (G_iB_i)^{-1}G_iA_{di}x(t-d) - (\varrho + \eta\,\|x(t)\| + \mu\,\|\omega(t)\|)\,\text{sign}\left(B_i^TG_i^Ts(t)\right), \quad (3.21)$$

where ϱ is a positive constant which is adjustable, and μ is a positive constant which satisfies

$$\mu \geq \frac{\max_{i\in S}\left\{\sqrt{\lambda_{\max}\left(G_iB_{wi}B_{wi}^TG_i^T\right)}\right\}}{\min_{i\in S}\left\{\sqrt{\lambda_{\min}\left(G_iB_iB_i^TG_i^T\right)}\right\}}.$$

Proof. Suppose matrices G_i are chosen such that G_iB_i are nonsingular. Choose the following Lyapunov function:

$$W(t) = \frac{1}{2}s^T(t)s(t).$$

According to (3.13), we have

$$\dot{s}(t) = -G_iB_iK_ix(t) + G_iA_{di}x(t-d) + G_iB_{wi}\omega(t) + G_iB_iu(t) + G_iB_if(x(t),t). \quad (3.22)$$

Thus, taking the derivative of $W(t)$, and considering (3.22) and the SMC law designed in (3.21), we have

$$\begin{aligned}\dot{W}(t) = s^T(t)\dot{s}(t) = s^T(t)G_iB_i\big[&-K_ix(t) + (G_iB_i)^{-1}G_iA_{di}x(t-d)\\ &+(G_iB_i)^{-1}G_iB_{wi}\omega(t) + u(t) + f(x(t),t)\big]\end{aligned}$$

$$
\begin{aligned}
&= s^T(t)G_iB_i\left[(G_iB_i)^{-1}G_iB_{wi}\omega(t) + f(x(t),t)\right.\\
&\quad \left.-(\varrho + \eta\,\|x(t)\| + \mu\,\|\omega(t)\|)\,\mathrm{sign}\left(B_i^TG_i^Ts(t)\right)\right]\\
&\le \left\|s^T(t)G_iB_{wi}\right\|\,\|\omega(t)\| + \left\|s^T(t)G_iB_i\right\|\,\|f(x(t),t)\|\\
&\quad -(\varrho + \eta\,\|x(t)\| + \mu\,\|\omega(t)\|)\left|B_i^TG_i^Ts(t)\right|. \qquad (3.23)
\end{aligned}
$$

Substituting (3.21) into (3.23) and noting $\|B_i^TG_i^Ts(t)\| \le |B_i^TG_i^Ts(t)|$, we have

$$
\dot{W}(t) \le -\varrho\left\|B_i^TG_i^Ts(t)\right\| \le -\tilde{\varrho}W^{\frac{1}{2}}(t), \qquad (3.24)
$$

where

$$
\tilde{\varrho} \triangleq \sqrt{2}\varrho\min_{i\in S}\left\{\sqrt{\lambda_{\min}\left(G_iB_iB_i^TG_i^T\right)}\right\} > 0.
$$

It can be seen from (3.24) that there exists a time $t^* \le W^{1/2}(0)/\tilde{\varrho}$ such that $W(t) = 0$, and consequently $s(t) = 0$, for $t \ge t^*$. This means that the system state trajectories can reach onto the predefined sliding surface in a finite time, thereby completing the proof. ■

3.5 Illustrative Example

Example 3.5.1 Consider Markovian jump singular time-delay system (3.1a)–(3.1c) with two operating modes, that is, $N = 2$ and the following parameters:

$$
A_1 = \begin{bmatrix} 0.5023 & 2.0125 & 0.0150 \\ 0.3025 & 0.4004 & -4.0020 \\ -0.1002 & 0.3002 & -3.5001 \end{bmatrix}, \quad A_2 = \begin{bmatrix} 0.5005 & 0.5052 & -0.1002 \\ 0.1256 & -0.0552 & 0.3003 \\ 0.1033 & 1.0015 & -2.0045 \end{bmatrix},
$$

$$
A_{d1} = \begin{bmatrix} -0.1669 & 0.0802 & 1.6820 \\ -0.8162 & -0.9373 & 0.5936 \\ 2.0941 & 0.6357 & 0.7902 \end{bmatrix}, \quad A_{d2} = \begin{bmatrix} 0.1053 & -0.1948 & -0.6855 \\ 0.1586 & 0.0755 & -0.2684 \\ 0.7709 & -0.5266 & -1.1883 \end{bmatrix},
$$

$$
B_1 = \begin{bmatrix} 0.9 \\ 1.8 \\ 1.4 \end{bmatrix}, \quad B_2 = \begin{bmatrix} 1.5 \\ 0.9 \\ 1.1 \end{bmatrix}, \quad B_{w1} = \begin{bmatrix} 0.1 \\ 0.2 \\ 0.4 \end{bmatrix}, \quad B_{w2} = \begin{bmatrix} -0.6 \\ 0.5 \\ 0.8 \end{bmatrix},
$$

$$
C_1 = \begin{bmatrix} 0.8 & 0.3 & 0.9 \end{bmatrix}, \quad C_{d1} = \begin{bmatrix} 0.2486 & 0.1025 & -0.0410 \end{bmatrix},
$$

$$
C_2 = \begin{bmatrix} -0.5 & 0.2 & 0.3 \end{bmatrix}, \quad C_{d2} = \begin{bmatrix} -2.2476 & -0.5108 & 0.2492 \end{bmatrix},
$$

$$
E = \begin{bmatrix} 1 & 0 & 0 \\ 0 & 1 & 0 \\ 0 & 0 & 0 \end{bmatrix}, \quad \Pi = \begin{bmatrix} -0.3 & 0.3 \\ 0.8 & -0.8 \end{bmatrix}, \quad D_{w1} = 0.2, \quad D_{w2} = 0.5.
$$

In addition, $f(x(t), t) = 0.78\exp(-t)\sin(t)x_1(t)$ (so η can be chosen as $\eta = 0.78$), the time delay $d = 0.5$, and the disturbance input $\omega(t) = 1/(1+t^2)$.

Our aim here is to verify the effectiveness of the proposed theoretical results in the previous sections. By solving the LMI condition (3.18) in Theorem 3.4.2 by using LMI-Toolbox in the Matlab environment and then considering (3.19), we have

$$K_1 = \begin{bmatrix} -2.1356 & -1.7843 & 1.0213 \end{bmatrix},$$
$$K_2 = \begin{bmatrix} -1.2769 & -0.5120 & -0.0922 \end{bmatrix}.$$

Here, parameters G_1 and G_2 in (3.11) are chosen as

$$G_1 = \begin{bmatrix} 0.1107 & 0.2214 & 0.1722 \end{bmatrix},$$
$$G_2 = \begin{bmatrix} 0.5370 & 0.3222 & 0.3938 \end{bmatrix}.$$

Thus, the switching function in (3.11) is

$$s(t) = \begin{cases} s_1(t) = \begin{bmatrix} 0.1107 & 0.2214 & 0.0000 \end{bmatrix} x(t) \\ \quad - \int_0^t \begin{bmatrix} -1.4734 & -0.9559 & -0.7321 \end{bmatrix} x(\theta)d\theta, & i = 1, \\ s_2(t) = \begin{bmatrix} 0.5370 & 0.3222 & 0.0000 \end{bmatrix} x(t) \\ \quad - \int_0^t \begin{bmatrix} -1.6020 & -0.1347 & -0.8874 \end{bmatrix} x(\theta)d\theta, & i = 2. \end{cases}$$

Let $\varrho = 0.7748$, then the SMC law designed in (3.21) can be computed as

$$u(t) = \begin{cases} u_1(t) & = & \begin{bmatrix} -2.1356 & -1.7843 & 1.0213 \end{bmatrix} x(t) \\ & + & \begin{bmatrix} 0.2184 & -0.1206 & 0.6137 \end{bmatrix} x(t-d) \\ & - & \rho(t)\mathrm{sign}(s_1(t)), \quad i = 1, \\ u_2(t) & = & \begin{bmatrix} -1.2769 & -0.5120 & -0.0922 \end{bmatrix} x(t) \\ & + & \begin{bmatrix} 0.2690 & -0.1882 & -0.6035 \end{bmatrix} x(t-d) \\ & - & \rho(t)\mathrm{sign}(s_2(t)), \quad i = 2, \end{cases}$$

where

$$\rho(t) = 0.7748 + 0.78 \|x(t)\| + 0.2483 \|\omega(t)\|.$$

For a given initial condition of $\phi(t) = \begin{bmatrix} -0.8 & -1.2 & 1.1 \end{bmatrix}^T$, $t \in [-0.5, 0]$, the simulation results are given in Figures 3.1–3.2. Specifically, Figure 3.1 shows the states of the closed-loop system, and Figure 3.2 depicts the switching function $s(t)$.

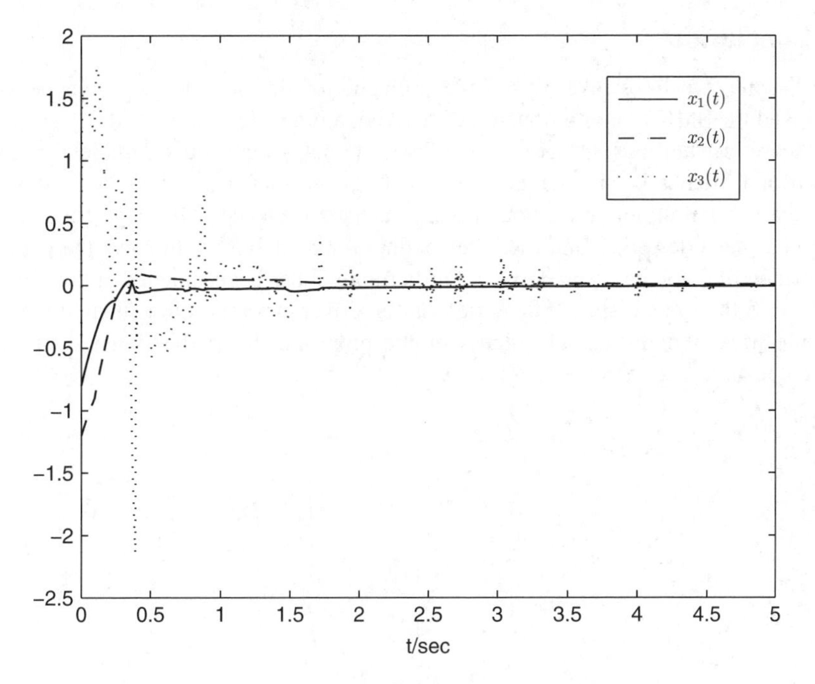

Figure 3.1 States of the closed-loop system

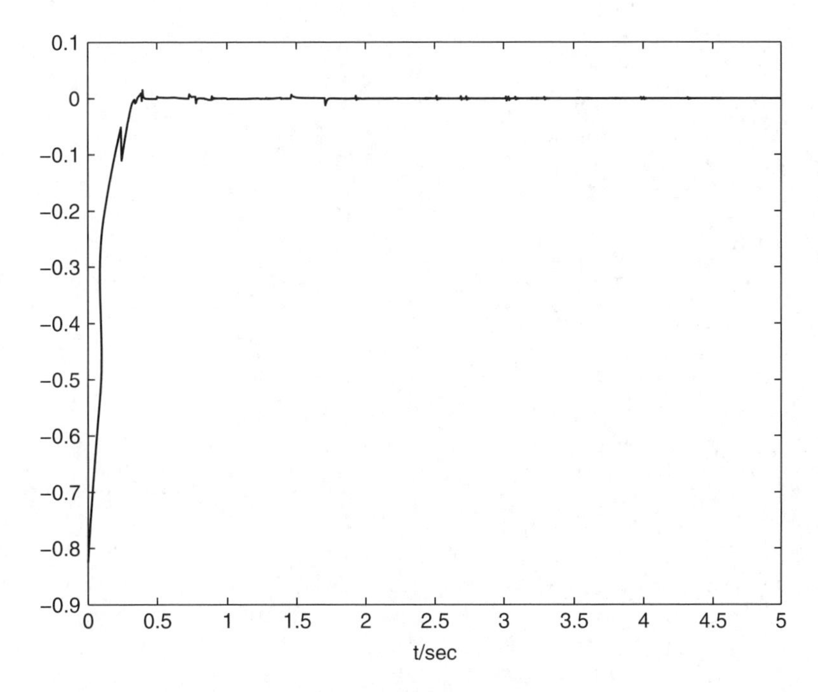

Figure 3.2 Switching function

3.6 Conclusion

In this chapter, we have investigated the problems of the bounded $\mathcal{L}_2$ gain performance analysis and the SMC of continuous-time Markovian jump singular time-delay systems. The major theoretical findings are as follows. First, the delay-dependent sufficient condition in the form of LMI has been established so as to ensure that the sliding mode dynamics is stochastically admissible with a bounded $\mathcal{L}_2$ gain performance. An integral-type switching function has been designed, and then the condition that enables us to solve the parameter in the switching function has been derived. Furthermore, it has been shown that, by synthesizing an SMC law, the system state trajectories can be driven onto the predefined sliding surface in a finite time. Finally, the usefulness of the proposed theory has been verified by the numerical results.

4

SMC of Markovian Jump Singular Systems with Stochastic Perturbation

4.1 Introduction

In this chapter, we will investigate the SMC design problem for Markovian jump singular systems with stochastic perturbation. The stochastic perturbation considered here is described as a Brownian motion, thus the overall dynamics is actually governed by an Itô stochastic differential equation with Markovian switching parameters and singularity, namely a Markovian jump singular stochastic system. There have been some results reported on SMC of stochastic systems [155, 156, 158] and Markovian jump stochastic systems [157], but the SMC problem for a Markovian jump singular stochastic system has not been fully investigated and still remains challenging. Due to the stochastic perturbation, the stability analysis methods proposed in Chapters 2 and 3 are not fully applicable in this chapter. The commonly used method of analyzing the stability of stochastic systems is based on the Itô formula. In [23], Boukas proposed a sufficient stability condition for a Markovian jump singular stochastic system, but the results are not all of strict LMI form since there exist some matrix equality constraints, which may cause problems in checking the conditions numerically.

We shall design an appropriate integral sliding surface, taking the singular matrix E into account. As a result, the sliding mode dynamics, described by a Markovian jump singular stochastic system, can be easily derived. The order of the resulting sliding mode dynamics is equivalent to that of the original system, which is convenient for analyzing its stochastic stability and the disturbance attenuation performance. A sufficient condition is proposed for the stochastic stability of the sliding mode dynamics in terms of strict LMIs, and by which the sliding surface can be designed. Following this, a discontinuous SMC law is synthesized to force the system state trajectories onto the sliding surface in a finite time. In addition, we also consider the disturbance attenuation problem when there exists an external disturbance in the sliding mode dynamics. A sufficient condition is established, which guarantees the sliding mode dynamics to be stochastically stable with an optimal $\mathcal{H}_\infty$ performance.

Sliding Mode Control of Uncertain Parameter-Switching Hybrid Systems, First Edition. Ligang Wu, Peng Shi and Xiaojie Su.

4.2 System Description and Preliminaries

Consider Markovian jump singular stochastic systems described by

$$Edx(t) = [A(\mathrm{r}_t)x(t) + B(\mathrm{r}_t)(u(t) + f(x, \mathrm{r}_t))]dt + D(\mathrm{r}_t)x(t)d\varpi(t), \tag{4.1}$$

where $\{\mathrm{r}_t, t \geq 0\}$ is a continuous-time Markov process on the probability space which has been defined in (2.2) of Chapter 2; $x(t) \in \mathbf{R}^n$ is the system state vector; $u(t) \in \mathbf{R}^m$ is the control input; $\varpi(t)$ is a one-dimensional Brownian motion satisfying $\mathbf{E}\{d\varpi(t)\} = 0$ and $\mathbf{E}\{d\varpi^2(t)\} = dt$. Matrix $E \in \mathbf{R}^{n\times n}$ may be singular, and we assume that $\mathrm{rank}(E) = r \leq n$. $A(\cdot)$, $B(\cdot)$; and $D(\cdot)$ are known real matrices with appropriate dimensions. $f(x, \mathrm{r}_t) \in \mathbf{R}^m$ are unknown nonlinear functions satisfying

$$\|f(x, \mathrm{r}_t)\| \leq \varepsilon(\mathrm{r}_t)\|x(t)\| \leq \varepsilon\|x(t)\|, \quad \mathrm{r}_t = i \in \mathcal{S}, \tag{4.2}$$

where $\varepsilon(\mathrm{r}_t) > 0$ are constant scalars and we define $\varepsilon \triangleq \max_{i\in\mathcal{S}}(\varepsilon_i)$.

For each possible value $\mathrm{r}_t = i \in \mathcal{S}$, $A(\mathrm{r}_t) = A_i$, $B(\mathrm{r}_t) = B_i$, $D(\mathrm{r}_t) = D_i$, and $f(x, \mathrm{r}_t) = f_i(x)$. Then, system (4.1) can be described by

$$Edx(t) = [A_i x(t) + B_i(u(t) + f_i(x))]dt + D_i x(t)d\varpi(t). \tag{4.3}$$

Assumption 4.1 *For each $i \in \mathcal{S}$, the pair (A_i, B_i) in (4.3) is controllable, and the matrix B_i has full column rank.*

The nominal system of (4.3) can be formulated as

$$Edx(t) = A_i x(t)dt + D_i x(t)d\varpi(t). \tag{4.4}$$

Definition 4.2.1 *The Markovian jump singular stochastic system in (4.4) is said to be stochastically stable if for any $x_0 \in \mathbf{R}^n$ and $\mathrm{r}_0 \in \mathcal{S}$, there exists a positive scalar $T(x_0, \mathrm{r}_0)$ such that*

$$\min_{t\to\infty} \mathbf{E}\left\{\int_0^t x^T(s, x_0, \mathrm{r}_0)x(s, x_0, \mathrm{r}_0)ds|(x_0, \mathrm{r}_0)\right\} \leq T(x_0, \mathrm{r}_0).$$

We first recall the following lemma [23].

Lemma 4.2.2 *The Markovian jump singular stochastic system in (4.4) is stochastically stable if there exist nonsingular matrices X_i, such that for $i \in \mathcal{S}$,*

$$E^T X_i = X_i^T E \geq 0,$$

$$A_i^T X_i + X_i^T A_i + D_i^T E^T X_i D_i + \sum_{j=1}^{N} \pi_{ij} E^T X_j < 0.$$

4.3 Integral SMC

4.3.1 Sliding Mode Dynamics Analysis

Design the following integral switching function:

$$s(t) = G_iEx(t) - \int_0^t G_i(A_i + B_iK_i)x(\tau)d\tau, \tag{4.5}$$

where for each $i \in S$, $G_i \in \mathbf{R}^{m\times n}$ and $K_i \in \mathbf{R}^{m\times n}$ are real matrices to be designed later. The matrix G_i is designed to satisfy that G_iB_i is nonsingular and $G_iD_i = 0$.

The solution of $Ex(t)$ is given as

$$Ex(t) = Ex(0) + \int_0^t [A_ix(\tau) + B_i(u(\tau) + f_i(x))]d\tau + \int_0^t D_ix(\tau)d\varpi(\tau). \tag{4.6}$$

It follows from (4.5) and (4.6) that

$$s(t) = G_iEx(0) + G_i \int_0^t [-B_iK_ix(\tau) + B_i(u(\tau) + f_i(x))]d\tau.$$

According to SMC theory, when the system state trajectories reach onto the sliding surface, it follows that $s(t) = 0$ and $\dot{s}(t) = 0$. Then, by $\dot{s}(t) = 0$, we obtain the equivalent control law as

$$u_{eq}(t) = K_ix(t) - f_i(x). \tag{4.7}$$

Thus, by substituting (4.7) into (4.3), the sliding mode dynamics can be obtained as

$$Edx(t) = (A_i + B_iK_i)x(t)dt + D_ix(t)d\varpi(t). \tag{4.8}$$

Now, we will analyze the stability of the sliding mode dynamics in (4.8) based on Lemma 4.2.2.

Proposition 4.3.1 *The sliding mode dynamics in (4.8) is stochastically stable if there exist nonsingular matrices X_i such that the following conditions hold for $i \in S$,*

$$E^TX_i = X_i^TE \geq 0, \tag{4.9a}$$

$$\left(A_i + B_iK_i\right)^T X_i + X_i^T \left(A_i + B_iK_i\right) + D_i^TE^TX_iD_i + \sum_{j=1}^{N} \pi_{ij}E^TX_j < 0. \tag{4.9b}$$

Remark 4.1 *Notice that the conditions in Proposition 4.3.1 are not all of strict LMI form owing to the matrix equality constraint of (4.9a). This may cause problems in checking the conditions numerically, since a matrix equality constraint is fragile and usually not satisfied perfectly. Therefore, the strict LMI conditions are more desirable than non-strict ones from the numerical point of view.* ♦

In the following, we present a new condition in terms of strict LMIs for the stochastic stability of the sliding mode dynamics in (4.8).

Proposition 4.3.2 *The sliding mode dynamics in (4.8) is stochastically stable if there exist matrices $P_i > 0$ and nonsingular matrices Q_i such that for $i \in \mathcal{S}$,*

$$\left(A_i + B_i K_i\right)^T \left(P_i E + R^T Q_i S^T\right) + \left(P_i E + R^T Q_i S^T\right)^T \left(A_i + B_i K_i\right) + D_i^T E^T P_i E D_i + \sum_{j=1}^{N} \pi_{ij} E^T P_j E < 0, \tag{4.10}$$

where $R \in \mathbf{R}^{(n-r)\times n}$ and $S \in \mathbf{R}^{n\times(n-r)}$ are matrices satisfying $RE = 0$ and $ES = 0$.

Proof. Letting $X_i \triangleq P_i E + R^T Q_i S^T$ in (4.10), we can obtain (4.9a)–(4.9b). Thus, by Proposition 4.3.1 we know that the sliding mode dynamics in (4.8) is stochastically stable. This completes the proof. ■

In the following, we present a strict LMI condition for solving parameter K_i in the switching function of (4.5). Before proceeding, we use the following lemma (i.e. Lemma 2.4.3 in Chapter 2) which will play a key role in the sequel.

Lemma 4.3.3 *Let P_i be symmetric matrices such that $E_L^T P_i E_L > 0$ and suppose Q_i is nonsingular. Then, $P_i E + R^T Q_i S^T$ is nonsingular and*

$$\left(P_i E + R^T Q_i S^T\right)^{-1} = \mathcal{P}_i E^T + S \mathcal{Q}_i R,$$

where $\mathcal{P}_i$ are symmetric matrices and $\mathcal{Q}_i$ are nonsingular matrices with

$$\mathcal{Q}_i = \left(S^T S\right)^{-1} Q_i^{-1} \left(RR^T\right)^{-1}, \quad E_R^T \mathcal{P}_i E_R = \left(E_L^T P_i E_L\right)^{-1}.$$

Theorem 4.3.4 *The sliding mode dynamics in (4.8) is stochastically stable if there exist symmetric matrices $\mathcal{P}_i \in \mathbf{R}^{n\times n}$, nonsingular matrices $\mathcal{Q}_i \in \mathbf{R}^{(n-r)\times(n-r)}$, and matrices $\mathcal{L}_i \in \mathbf{R}^{m\times n}$, $\mathcal{H}_i \in \mathbf{R}^{m\times(n-r)}$ such that for $i \in \mathcal{S}$,*

$$\begin{bmatrix} \Psi_{11i} & \mathcal{Z}_i^T D_i^T E_R & \Psi_{13i} \\ \star & -E_R^T \mathcal{P}_i E_R & 0 \\ \star & \star & -\Psi_{33i} \end{bmatrix} < 0, \tag{4.11}$$

where $\mathcal{Z}_i \triangleq \mathcal{P}_i E^T + S\mathcal{Q}_i R$ *and*

$$\begin{cases} \Psi_{11i} \triangleq A_i\mathcal{Z}_i + \mathcal{Z}_i^T A_i^T + B_i\left(\mathcal{L}_i E^T + \mathcal{H}_i R\right) + \left(\mathcal{L}_i E^T + \mathcal{H}_i R\right)^T B_i^T \\ \qquad + \pi_{ii}\left(E\mathcal{Z}_i + \mathcal{Z}_i^T E^T - E\mathcal{P}_i E^T\right), \\ \Psi_{33i} \triangleq \mathrm{diag}\left\{E_R^T\mathcal{P}_1 E_R, E_R^T\mathcal{P}_2 E_R, \ldots, E_R^T\mathcal{P}_{i-1}E_R, \right. \\ \qquad \left. E_R^T\mathcal{P}_{i+1}E_R, \ldots, E_R^T\mathcal{P}_N E_R\right\}, \\ \Psi_{13i} \triangleq \left[\sqrt{\pi_{i1}}\mathcal{Z}_i^T E_R \ \sqrt{\pi_{i2}}\mathcal{Z}_i^T E_R \ \cdots \ \sqrt{\pi_{i(i-1)}}\mathcal{Z}_i^T E_R \right. \\ \qquad \left. \sqrt{\pi_{i(i+1)}}\mathcal{Z}_i^T E_R \ \cdots \ \sqrt{\pi_{iN}}\mathcal{Z}_i^T E_R\right]. \end{cases}$$

Moreover, the parametric matrices K_i *in the switching function of (4.5) can be computed by*

$$\begin{aligned} K_i &\triangleq \left(\mathcal{L}_i E^T + \mathcal{H}_i R\right)\mathcal{Z}_i^{-1} \\ &= \left(\mathcal{L}_i E^T + \mathcal{H}_i R\right)\left(\mathcal{P}_i E^T + S\mathcal{Q}_i R\right)^{-1}. \end{aligned} \tag{4.12}$$

Proof. By Proposition 4.3.2 we know that the sliding mode dynamics in (4.8) is stochastically stable if there exist matrices $P_i > 0$ and nonsingular matrices Q_i, such that the conditions of (4.10) hold for $i \in \mathcal{S}$. Moreover, according to Lemma 4.3.3, we know that $P_i E + R^T Q_i S^T$ are nonsingular and $\mathcal{Z}_i \triangleq (P_i E + R^T Q_i S^T)^{-1} = \mathcal{P}_i E^T + S\mathcal{Q}_i R$. Now, performing a congruence transformation on (4.10) by matrices $\mathcal{Z}_i$, we have

$$\left(A_i + B_i K_i\right)\mathcal{Z}_i + \mathcal{Z}_i^T\left(A_i + B_i K_i\right)^T + \mathcal{Z}_i^T D_i^T E^T P_i E D_i \mathcal{Z}_i + \sum_{j=1}^{N}\pi_{ij}\mathcal{Z}_i^T E^T P_j E\mathcal{Z}_i < 0.$$

Letting $\mathcal{L}_i \triangleq K_i\mathcal{P}_i$ and $\mathcal{H}_i \triangleq K_i S\mathcal{Q}_i$, we have

$$\begin{aligned} &A_i\mathcal{Z}_i + \mathcal{Z}_i^T A_i^T + B_i\left(\mathcal{L}_i E^T + \mathcal{H}_i R\right) + \left(\mathcal{L}_i E^T + \mathcal{H}_i R\right)^T B_i^T + \mathcal{Z}_i^T D_i^T E_R\left(E_R^T\mathcal{P}_i E_R\right)^{-1}E_R^T D_i\mathcal{Z}_i \\ &\quad + \pi_{ii}\mathcal{Z}_i^T E_R\left(E_R^T\mathcal{P}_i E_R\right)^{-1}E_R^T\mathcal{Z}_i + \sum_{j=1, j\neq i}^{N}\pi_{ij}\mathcal{Z}_i^T E_R\left(E_R^T\mathcal{P}_j E_R\right)^{-1}E_R^T\mathcal{Z}_i < 0. \end{aligned} \tag{4.13}$$

But the following fact is true:

$$\begin{aligned} 0 &\leq \left[E_R^T\mathcal{Z}_i - \left(E_R^T\mathcal{P}_i E_R\right)E_L^T\right]^T\left(E_R^T\mathcal{P}_i E_R\right)^{-1}\left[E_R^T\mathcal{Z}_i - \left(E_R^T\mathcal{P}_i E_R\right)E_L^T\right] \\ &= \mathcal{Z}_i^T E_R\left(E_R^T\mathcal{P}_i E_R\right)^{-1}E_R^T\mathcal{Z}_i - E\mathcal{Z}_i - \mathcal{Z}_i^T E^T + E\mathcal{P}_i E^T. \end{aligned}$$

Considering $\pi_{ii} < 0$, thus we have

$$\pi_{ii}\mathcal{Z}_i^T E_R\left(E_R^T\mathcal{P}_i E_R\right)^{-1}E_R^T\mathcal{Z}_i \leq \pi_{ii}\left(E\mathcal{Z}_i + \mathcal{Z}_i^T E^T - E\mathcal{P}_i E^T\right).$$

Therefore, (4.13) holds if the following inequalities hold for $i \in \mathcal{S}$:

$$A_i\mathcal{Z}_i + \mathcal{Z}_i^T A_i^T + B_i\left(\mathcal{L}_iE^T + \mathcal{H}_iR\right) + \left(\mathcal{L}_iE^T + \mathcal{H}_iR\right)^T B_i^T + \mathcal{Z}_i^T D_i^T E_R\left(E_R^T\mathcal{P}_iE_R\right)^{-1}E_R^T D_i\mathcal{Z}_i$$

$$+\pi_{ii}\left(E\mathcal{Z}_i + \mathcal{Z}_i^T E^T - E\mathcal{P}_iE^T\right) + \sum_{j=1,j\neq i}^{N}\pi_{ij}\mathcal{Z}_i^T E_R\left(E_R^T\mathcal{P}_jE_R\right)^{-1}E_R^T\mathcal{Z}_i < 0. \tag{4.14}$$

By Schur complement, (4.14) is equivalent to (4.11). This completes the proof. ■

4.3.2 SMC Law Design

In the following, we shall design an SMC law, by which the state trajectories of the Markovian jump singular stochastic system in (4.1) can be driven onto the designed sliding surface $s(t) = 0$ in a finite time and maintained there for all subsequent time.

Theorem 4.3.5 *Consider the Markovian jump singular stochastic system in (4.1). Suppose that the switching function is given as (4.5) with K_i being solved by (4.12), and G_i is chosen to satisfy that G_iB_i is nonsingular and $G_iD_i = 0$. Then, the state trajectories of system (4.1) can be driven onto the sliding surface $s(t) = 0$ by the following SMC law:*

$$u(t) = K_ix(t) - (\lambda + \varepsilon\,\|x(t)\|)\,\mathrm{sign}\,(s(t)), \tag{4.15}$$

where $\lambda > 0$ is an adjustable scalar.

Proof. We choose G_i as $G_i = B_i^T Y_i$, where $Y_i > 0$ are matrices to be designed such that $G_iB_i = B_i^T Y_iB_i > 0$ for $i \in \mathcal{S}$. Choose the following Lyapunov function:

$$V(t) = \frac{1}{2}s^T(t)\left(B_i^T Y_iB_i\right)^{-1}s(t).$$

According to (4.5), we have

$$\begin{aligned}\dot{s}(t) &= G_iE\dot{x}(t) - G_i\left(A_i + B_iK_i\right)x(t)\\ &= G_i\left[A_ix(t) + B_i\left(u(t) + f_i(x)\right)\right] - G_i\left(A_i + B_iK_i\right)x(t).\end{aligned} \tag{4.16}$$

Substituting (4.15) into (4.16) yields

$$\dot{s}(t) = B_i^T Y_iB_i\left[-(\lambda + \varepsilon\,\|x(t)\|)\,\mathrm{sign}\,(s(t)) + f_i(t)\right]. \tag{4.17}$$

Thus, taking the derivation of $V(t)$ and considering (4.17), we have

$$\begin{aligned}\dot{V}(t) &= s^T(t)\left(B_i^T Y_iB_i\right)^{-1}\dot{s}(t)\\ &= s^T(t)\left[-(\lambda + \varepsilon\,\|x(t)\|)\,\mathrm{sign}\,(s(t)) + f_i(t)\right]\\ &\leq -\lambda\,\|s(t)\| < 0, \quad \text{for } \|s(t)\| \neq 0,\end{aligned}$$

which implies that the state trajectories of the system in (4.1) will be driven onto the sliding surface $s(t) = 0$ in a finite time. This completes the proof. ∎

In the implementation of the SMC law in (4.15), the upper bound scalar ε of $f_i(t)$ in (4.2) is required to be known *a priori*. If the value of ε is not available, we have to estimate it. In the following theorem, we shall consider this case, and first design an adaptive law to estimate ε, thus an adaptive SMC law will be presented for system (4.1).

Theorem 4.3.6 *Consider the Markovian jump singular stochastic system in (4.1), and assume that the exact value of the upper bound scalar ε is not available. Suppose that the switching function is given as (4.5) with K_i being solved by (4.12), and G_i is chosen to satisfy that G_iB_i are nonsingular and $G_iD_i = 0$. Then, the state trajectories of system (4.1) can be driven onto the sliding surface $s(t) = 0$ by the following adaptive SMC law:*

$$u(t) = K_ix(t) - (\lambda + \hat{\varepsilon}(t) \|x(t)\|) \operatorname{sign}(s(t)),$$

where $\hat{\varepsilon}(t)$ represents the estimate of ε, and the adaptive law is given as

$$\dot{\hat{\varepsilon}}(t) = \frac{1}{\delta} \|s(t)\| \, \|x(t)\|,$$

with $\varepsilon(0) = 0$, where $\delta > 0$ is an adjustable scalar.

Proof. Select the following Lyapunov function:

$$\hat{V}(t) = \frac{1}{2} \left[s^T(t) \left(B_i^T Y_i B_i \right)^{-1} s(t) + \delta \tilde{\varepsilon}^2(t) \right].$$

The rest of the proof can be followed along the same lines as the proof of Theorem 4.3.5. ∎

4.4 Optimal $\mathcal{H}_\infty$ Integral SMC

In this section, within the framework of the SMC problem, we shall further analyze the $\mathcal{H}_\infty$ performance for Markovian jump singular stochastic systems with an $\mathcal{L}_2$ external disturbance. Specifically, we will propose a sufficient condition by which the sliding mode dynamics of the controlled system is guaranteed to be stochastically stable with an $\mathcal{H}_\infty$ performance.

4.4.1 Performance Analysis and SMC Law Design

Consider the following singular stochastic systems with Markovian jump parameters and an external disturbance:

$$Edx(t) = \left[A_ix(t) + B_i\left(u(t) + f_i(x)\right) + F_i\omega(t)\right] dt + D_ix(t)d\varpi(t), \quad (4.18a)$$

$$z(t) = C_ix(t) + H_i\omega(t), \quad (4.18b)$$

where $\omega(t) \in \mathbf{R}^p$ is the disturbance input which belongs to $\mathcal{L}_2[0, \infty)$; $z(t) \in \mathbf{R}^q$ is the controlled output; C_i, F_i and H_i are real constant matrices. Unless other specified, the notations in (4.18a)–(4.18b) have the same meanings as those in (4.1).

Designing the same switching function as in (4.5) and employing the methods used in Section 4.3, we can obtain the following sliding mode dynamics:

$$Edx(t) = \left\{ \left(A_i + B_i K_i\right) x(t) + \left[I - B_i \left(G_i B_i\right)^{-1} G_i\right] F_i \omega(t) \right\} dt + D_i x(t) d\varpi(t). \tag{4.19}$$

Remark 4.2 *Notice from (4.19) that if matrix F_i in (4.18a)–(4.18b) satisfies the so-called matching condition – that is, there exist matrices $\tilde{F}_i$ satisfying $F_i = B_i \tilde{F}_i$ – it follows that the sliding mode dynamics in (4.19) becomes (4.8). This implies that the sliding mode dynamics is adaptive to the disturbance $\omega(t)$. In this case, the methods proposed in the previous section of this chapter can be applied directly. In the following, we assume that matrix F_i does not satisfy the matching condition, thus there will exist a disturbance $\omega(t)$ in the sliding mode dynamics, and the results will be sharply different from the matching case.* ♦

Definition 4.4.1 *Given a scalar $\gamma > 0$, the sliding mode dynamics in (4.19) is said to be stochastically stable with an $\mathcal{H}_\infty$ performance level γ, if it is stochastically stable with $\omega(t) = 0$, and under zero condition, for nonzero $\omega(t) \in \mathcal{L}_2[0, \infty)$, it holds that*

$$\mathbf{E}\left\{ \int_0^\infty z^T(t) z(t) dt \right\} < \gamma^2 \int_0^\infty \omega^T(t)\omega(t) dt. \tag{4.20}$$

Now, we will analyze the stability and the $\mathcal{H}_\infty$ performance of the sliding mode dynamics in (4.19).

Theorem 4.4.2 *Given a scalar $\gamma > 0$, the sliding mode dynamics in (4.19) is stochastically stable with an $\mathcal{H}_\infty$ performance level γ, if there exist nonsingular matrices $\mathcal{X}_i$ such that for $i \in \mathcal{S}$,*

$$E^T \mathcal{X}_i = \mathcal{X}_i^T E \geq 0, \tag{4.21a}$$

$$\begin{bmatrix} \Xi_{11i} & \Xi_{12i} & \mathcal{X}_i^T B_i \\ \star & \Xi_{22i} & 0 \\ \star & \star & -B_i^T E^T \mathcal{X}_i B_i \end{bmatrix} < 0, \tag{4.21b}$$

$$B_i^T E^T \mathcal{X}_i D_i = 0, \tag{4.21c}$$

where

$$\begin{cases} \Xi_{11i} \triangleq \mathcal{X}_i^T \left(A_i + B_i K_i\right) + \left(A_i + B_i K_i\right)^T \mathcal{X}_i + C_i^T C_i + D_i^T E^T \mathcal{X}_i D_i + \sum_{j=1}^N \pi_{ij} E^T \mathcal{X}_j, \\ \Xi_{12i} \triangleq C_i^T H_i + \mathcal{X}_i^T F_i, \\ \Xi_{22i} \triangleq -\gamma^2 I + H_i^T H_i + F_i^T E^T \mathcal{X}_i F_i. \end{cases}$$

Proof. Choose the following Lyapunov function:

$$W(x, \mathrm{r}_t) = x^T(t)E^T\mathcal{X}(\mathrm{r}_t)x(t),$$

where $\mathcal{X}^T(\mathrm{r}_t)E = E^T\mathcal{X}(\mathrm{r}_t) \geq 0$ and $\mathcal{X}(\mathrm{r}_t)$ (denoted by $\mathcal{X}_i$ when $\mathrm{r}_t = i$) are nonsingular matrices to be specified such that $B_i^T E^T \mathcal{X}_i B_i$ are positive definite for $i \in \mathcal{S}$.

Let $\mathcal{A}$ be the infinitesimal generator of the Markov process $\{(x(t), \mathrm{r}_t), t \geq 0\}$. Then, the average derivative emanating from point (x, i) at time t is given by the following expression:

$$\begin{aligned}
\mathcal{A}W(x,i) &= 2x^T(t)\mathcal{X}_i^T \left\{ (A_i + B_iK_i)\, x(t) + \left[I - B_i \left(G_iB_i \right)^{-1} G_i \right] F_i\omega(t) \right\} \\
&\quad + x^T(t)D_i^T E^T \mathcal{X}_i D_i x(t) + \sum_{j=1}^{N} \pi_{ij} x^T(t)E^T\mathcal{X}_j x(t) \\
&\leq x^T(t)\Upsilon_i x(t) + 2x^T(t)\mathcal{X}_i^T F_i\omega(t) + \omega^T(t)F_i^T E^T \mathcal{X}_i F_i \omega(t),
\end{aligned} \tag{4.22}$$

where

$$\begin{aligned}
\Upsilon_i &\triangleq \mathcal{X}_i^T (A_i + B_iK_i) + (A_i + B_iK_i)^T \mathcal{X}_i + D_i^T E^T \mathcal{X}_i D_i \\
&\quad + \mathcal{X}_i^T B_i \left(B_i^T E^T \mathcal{X}_i B_i \right)^{-1} B_i^T \mathcal{X}_i + \sum_{j=1}^{N} \pi_{ij} E^T \mathcal{X}_j.
\end{aligned}$$

Here, we choose $G_i = B_i^T E^T \mathcal{X}_i$ in the above derivation, which guarantees that $G_iB_i = B_i^T E^T \mathcal{X}_i B_i$ is nonsingular since $E^T\mathcal{X}_i > 0$. In addition, $B_i^T E^T \mathcal{X}_i D_i = 0$ are introduced due to $G_iD_i = 0$. Therefore, when $\omega(t) = 0$ in (4.22), it follows that $\mathcal{A}W(x, i) = x^T(t)\Upsilon_i x(t)$. By Schur complement, (4.21b) implies $\Upsilon_i < 0$ for $i \in \mathcal{S}$, thus,

$$\mathcal{A}W(x, i) < -\min_{i\in\mathcal{S}} \left\{ \lambda_{\min}(\Upsilon_i) \right\} \|x(t)\| .$$

Therefore, we know that the the sliding mode dynamics in (4.19) with $\omega(t) = 0$ is stochastically stable.

Now, we will establish the $\mathcal{H}_\infty$ performance. To this end, assume zero initial condition (that is, $x(0) = 0$, thus $W(0, \mathrm{r}_0) = 0$) and consider index:

$$\mathcal{J} = \mathbf{E}\left\{ \int_0^\infty \left[z^T(t)z(t) - \gamma^2\omega^T(t)\omega(t) \right] dt \right\}.$$

Dynkin's formula gives

$$\mathbf{E}\left\{ W(x, \mathrm{r}_t) \right\} - \mathbf{E}\left\{ W(0, \mathrm{r}_0) \right\} = \mathbf{E}\left\{ \int_0^\infty \mathcal{A}W(x, i)dt \right\},$$

and together with (4.22), we have

$$\begin{aligned}
\mathcal{J} &\leq \mathbf{E}\left\{\int_0^{\infty}\left[z^T(t)z(t)-\gamma^2\omega^T(t)\omega(t)\right]dt\right\}+\mathbf{E}\left\{W(x,\mathrm{r}_t)\right\}-\mathbf{E}\left\{W(0,\mathrm{r}_0)\right\} \\
&= \mathbf{E}\left\{\int_0^{\infty}\left[z^T(t)z(t)-\gamma^2\omega^T(t)\omega(t)+\mathcal{A}W(x,i)\right]dt\right\} \\
&\triangleq \mathbf{E}\left\{\int_0^{\infty}\begin{bmatrix}x(t)\\ \omega(t)\end{bmatrix}^T\begin{bmatrix}\Upsilon_i+C_i^TC_i & \Xi_{12i}\\ \star & \Xi_{22i}\end{bmatrix}\begin{bmatrix}x(t)\\ \omega(t)\end{bmatrix}dt\right\},
\end{aligned} \tag{4.23}$$

where Ξ_{12i} and Ξ_{22i} are defined in (4.21b). By Schur complement, (4.21b) implies

$$\begin{bmatrix}\Upsilon_i+C_i^TC_i & \Xi_{12i}\\ \star & \Xi_{22i}\end{bmatrix}<0.$$

Then $\mathcal{J}<0$ from (4.23), thus (4.20) holds. This completes the proof. ■

The following theorem will give a sufficient condition by which the sliding mode dynamics in (4.19) is guaranteed to be stochastically stable with an $\mathcal{H}_\infty$ performance level γ, and the switching function in (4.5) can be solved.

Theorem 4.4.3 *Given a scalar $\gamma>0$, the sliding mode dynamics in (4.19) is stochastically stable with an $\mathcal{H}_\infty$ performance level γ, if there exist symmetric matrices $\mathcal{P}_i\in\mathcal{R}^{n\times n}$, positive definite matrices $\mathcal{G}_i\in\mathcal{R}^{r\times r}$, $\mathcal{S}_i\in\mathcal{R}^{r\times r}$, nonsingular matrices $\mathcal{Q}_i\in\mathcal{R}^{(n-r)\times(n-r)}$, and matrices $\mathcal{L}_i\in\mathcal{R}^{m\times n}$, $\mathcal{H}_i\in\mathcal{R}^{m\times(n-r)}$ such that for $i\in\mathcal{S}$,*

$$\begin{bmatrix}
\tilde{\Xi}_{11i} & \tilde{\Xi}_{12i} & B_i & 0 & \mathcal{Z}_i^TC_i^T & \mathcal{Z}_i^TD_i^TE_R & \tilde{\Xi}_{17i}\\
\star & \tilde{\Xi}_{22i} & 0 & F_i^TE_R & 0 & 0 & 0\\
\star & \star & \tilde{\Xi}_{33i} & 0 & 0 & 0 & 0\\
\star & \star & \star & -\mathcal{S}_i & 0 & 0 & 0\\
\star & \star & \star & \star & -I & 0 & 0\\
\star & \star & \star & \star & \star & -\mathcal{S}_i & 0\\
\star & \star & \star & \star & \star & \star & -\tilde{\Xi}_{77i}
\end{bmatrix}<0, \tag{4.24a}$$

$$\mathcal{S}_i-E_R^T\mathcal{P}_iE_R=0, \tag{4.24b}$$

$$B_i^TE_R\mathcal{S}_iE_R^TD_i=0, \tag{4.24c}$$

$$\mathcal{G}_i\mathcal{S}_i=I, \tag{4.24d}$$

where

$$\begin{cases} \tilde{\Xi}_{11i} \triangleq A_i\mathcal{Z}_i + \mathcal{Z}_i^T A_i^T + B_i\left(\mathcal{L}_i E^T + \mathcal{H}_i R\right) + \left(\mathcal{L}_i E^T + \mathcal{H}_i R\right)^T B_i^T \\ \qquad + \pi_{ii}\left(E\mathcal{Z}_i + \mathcal{Z}_i^T E^T - E\mathcal{P}_i E^T\right), \\ \tilde{\Xi}_{12i} \triangleq \mathcal{Z}_i^T C_i^T H_i + F_i, \\ \tilde{\Xi}_{22i} \triangleq -\gamma^2 I + H_i^T H_i, \\ \tilde{\Xi}_{33i} \triangleq -B_i^T E_R \mathcal{G}_i E_R^T B_i, \\ \tilde{\Xi}_{77i} \triangleq \text{diag}\left\{\mathcal{S}_1, \mathcal{S}_2, \ldots, \mathcal{S}_{i-1}, \mathcal{S}_{i+1}, \ldots, \mathcal{S}_N\right\}, \\ \mathcal{Z}_i \triangleq \mathcal{P}_i E^T + S\mathcal{Q}_i R, \\ \tilde{\Xi}_{17i} \triangleq \left[\sqrt{\pi_{i1}}\mathcal{Z}_i^T E_R \ \sqrt{\pi_{i2}}\mathcal{Z}_i^T E_R \ \cdots \ \sqrt{\pi_{i(i-1)}}\mathcal{Z}_i^T E_R \ \sqrt{\pi_{i(i+1)}}\mathcal{Z}_i^T E_R \ \cdots \ \sqrt{\pi_{iN}}\mathcal{Z}_i^T E_R\right]. \end{cases}$$

Moreover, the parametric matrices K_i in the switching function of (4.5) can be computed by

$$\begin{aligned} K_i &\triangleq \left(\mathcal{L}_i E^T + \mathcal{H}_i R\right)\mathcal{Z}_i^{-1} \\ &= \left(\mathcal{L}_i E^T + \mathcal{H}_i R\right)\left(\mathcal{P}_i E^T + S\mathcal{Q}_i R\right)^{-1}. \end{aligned} \tag{4.25}$$

Proof. From Proposition 4.3.2, it can be seen that the sliding mode dynamics in (4.19) is stochastically stable with an $\mathcal{H}_\infty$ performance level γ, if there exist matrices $P_i > 0$ and nonsingular matrices Q_i such that for $i \in \mathcal{S}$,

$$\begin{bmatrix} \hat{\Xi}_{11i} & \hat{\Xi}_{12i} & Z_i^T B_i \\ \star & \hat{\Xi}_{22i} & 0 \\ \star & \star & -B_i^T E^T P_i E B_i \end{bmatrix} < 0, \tag{4.26}$$

$$B_i^T E^T P_i E D_i = 0, \tag{4.27}$$

where

$$\begin{cases} \hat{\Xi}_{11i} \triangleq Z_i^T(A_i + B_i K_i) + (A_i + B_i K_i)^T Z_i + C_i^T C_i + D_i^T E^T P_i E D_i + \sum_{j=1}^{N} \pi_{ij} E^T P_j E, \\ \hat{\Xi}_{12i} \triangleq C_i^T H_i + Z_i^T F_i, \\ \hat{\Xi}_{22i} \triangleq -\gamma^2 I + H_i^T H_i + F_i^T E^T P_i E F_i, \\ Z_i \triangleq P_i E + R^T Q_i S^T. \end{cases}$$

By Lemma 4.3.3, we know that $Z_i \triangleq P_iE + R^TQ_iS^T$ are nonsingular and $Z_i^{-1} = \mathcal{P}_iE^T + S\mathcal{Q}_iR \triangleq \mathcal{Z}_i$, where $\mathcal{P}_i$ are symmetric matrices and $\mathcal{Q}_i$ are nonsingular matrices with

$$E_R^T\mathcal{P}_iE_R = \left(E_L^TP_iE_L\right)^{-1} > 0.$$

Now, performing a congruence transformation on (4.26) by $\text{diag}\{\mathcal{Z}_i, I, I\}$, we have

$$\begin{bmatrix} \breve{\Xi}_{11i} & \tilde{\Xi}_{12i} & B_i \\ \star & \hat{\Xi}_{22i} & 0 \\ \star & \star & -B_i^TE^TP_iEB_i \end{bmatrix} < 0, \tag{4.28}$$

where $\tilde{\Xi}_{12i}$ are defined in (4.24a) and

$$\begin{aligned}\breve{\Xi}_{11i} \triangleq{}& (A_i + B_iK_i)\mathcal{Z}_i + \mathcal{Z}_i^T(A_i + B_iK_i)^T + \mathcal{Z}_i^TC_i^TC_i\mathcal{Z}_i + \mathcal{Z}_i^TD_i^TE^TP_iED_i\mathcal{Z}_i \\ &+ \sum_{j=1}^{N}\pi_{ij}\mathcal{Z}_i^TE^TP_jE\mathcal{Z}_i.\end{aligned}$$

Notice that

$$\pi_{ii}\mathcal{Z}_i^TE_R\left(E_R^T\mathcal{P}_iE_R\right)^{-1}E_R^T\mathcal{Z}_i \leq \pi_{ii}\left(E\mathcal{Z}_i + \mathcal{Z}_i^TE^T - E\mathcal{P}_iE^T\right).$$

Letting $\mathcal{L}_i \triangleq K_i\mathcal{P}_i$ and $\mathcal{H}_i \triangleq K_iS\mathcal{Q}_i$, we know that (4.28) holds if for $i \in \mathcal{S}$,

$$\begin{bmatrix} \tilde{\Xi}_{11i} & \tilde{\Xi}_{12i} & B_i & 0 & \mathcal{Z}_i^TC_i^T & \mathcal{Z}_i^TD_i^TE_R & \tilde{\Xi}_{17i} \\ \star & \tilde{\Xi}_{22i} & 0 & F_i^TE_R & 0 & 0 & 0 \\ \star & \star & \breve{\Xi}_{33i} & 0 & 0 & 0 & 0 \\ \star & \star & \star & -E_R^T\mathcal{P}_iE_R & 0 & 0 & 0 \\ \star & \star & \star & \star & -I & 0 & 0 \\ \star & \star & \star & \star & \star & -E_R^T\mathcal{P}_iE_R & 0 \\ \star & \star & \star & \star & \star & \star & -\tilde{\Xi}_{77i} \end{bmatrix} < 0, \tag{4.29}$$

where

$$\begin{aligned}\breve{\Xi}_{33i} &\triangleq -B_i^TE_R\left(E_R^T\mathcal{P}_iE_R\right)^{-1}E_R^TB_i \\ &= -B_i^TE_R\mathcal{S}_i^{-1}E_R^TB_i = -B_i^TE_R\mathcal{G}_iE_R^TB_i,\end{aligned}$$

where $\mathcal{G}_i = \mathcal{S}_i^{-1}$, and the other notations are defined in (4.24a). Furthermore, considering (4.24b) and (4.24d), inequality (4.24a) yields (4.29), and (4.24c) yields (4.27). This completes the proof. ∎

Now, by applying the same procedures as in Section 4.3, we design a discontinuous SMC law to drive the system state trajectories onto the predefined sliding surface in a finite time and maintain it there for all the subsequent time.

Theorem 4.4.4 *Consider the Markovian jump singular stochastic system in (4.1). Suppose that the switching function is given as (4.5) with K_i being solved by (4.25), and G_i are chosen as $G_i = B_i^T E_R^T \mathcal{G}_i E_R$, where $\mathcal{G}_i$ is the solution of (4.24a)–(4.24d). Then, the state trajectories of system (4.1) can be driven onto the sliding surface $s(t) = 0$ by the SMC law $u(t)$ designed in (4.15).*

4.4.2 Computational Algorithm

Notice that there exist three matrix equalities of (4.24b)–(4.24d) in Theorem 4.4.3, which can not be solved directly by applying the LMI procedures. In the following, we will propose some algorithms to solve them. Firstly, to solve (4.24b)–(4.24c), we consider the following matrix inequalities for scalars $\alpha > 0$ and $\beta > 0$,

$$\left(\mathcal{S}_i - E_R^T \mathcal{P}_i E_R\right)^T \left(\mathcal{S}_i - E_R^T \mathcal{P}_i E_R\right) \leq \alpha I, \quad \text{for } i \in \mathcal{S}, \tag{4.30}$$

$$\left(B_i^T E_R \mathcal{S}_i E_R^T D_i\right)^T \left(B_i^T E_R \mathcal{S}_i E_R^T D_i\right) \leq \beta I, \quad \text{for } i \in \mathcal{S}. \tag{4.31}$$

By Schur complement, (4.30) and (4.31) are respectively equivalent to

$$\begin{bmatrix} -\alpha I & \left(\mathcal{S}_i - E_R^T \mathcal{P}_i E_R\right)^T \\ \star & -I \end{bmatrix} \leq 0, \quad \text{for } i \in \mathcal{S}, \tag{4.32}$$

$$\begin{bmatrix} -\beta I & \left(B_i^T E_R \mathcal{S}_i E_R^T D_i\right)^T \\ \star & -I \end{bmatrix} \leq 0, \quad \text{for } i \in \mathcal{S}. \tag{4.33}$$

Therefore, when $\alpha > 0$ and $\beta > 0$ are chosen as two sufficiently small scalars, matrix equalities (4.24b) and (4.24c) can be solved through LMIs (4.32) and (4.33), respectively.

We use the CCL method [66] to solve (4.24d) by formulating it into a sequential optimization problem subject to LMI constraints. We suggest the following minimization problem involving LMI conditions instead of the original nonconvex feasibility problem in Theorem 4.4.3.

Problem SMDA (Sliding mode dynamics analysis):

$$\min \ \text{trace}\left(\sum_{i \in \mathcal{S}} \mathcal{G}_i \mathcal{S}_i\right)$$

$$\text{subject to (4.24a), (4.32)–(4.33) and for } i \in \mathcal{S},$$

$$\begin{bmatrix} \mathcal{G}_i & I \\ I & \mathcal{S}_i \end{bmatrix} \geq 0. \tag{4.34}$$

If the solution of the aforesaid minimization problem is Nr, then the conditions in Theorem 4.4.3 are solvable. Although it is still not possible to always find the global optimal solution, the proposed minimization problem is easier to solve than the original nonconvex feasibility problem. We suggest the following algorithm to solve Problem SMDA.

Algorithm SMDA

Step 1. Choose $\alpha > 0$ and $\beta > 0$ as sufficiently small scalars.
Step 2. Find a feasible set $(\mathcal{P}_i^{(0)}, \mathcal{G}_i^{(0)}, \mathcal{S}_i^{(0)}, \mathcal{Q}_i^{(0)}, \mathcal{L}_i^{(0)}, \mathcal{H}_i^{(0)})$ satisfying (4.24a), (4.32)–(4.33), and (4.34). Set $\kappa = 0$.
Step 3. Solve the following optimization problem

$$\min \ \text{trace}\left(\sum_{i\in\mathcal{S}} \left(\mathcal{G}_i^{(\kappa)}\mathcal{S}_i + \mathcal{G}_i\mathcal{S}_i^{(\kappa)}\right)\right)$$
$$\text{subject to} \ \ (4.24\text{a}), (4.32)\text{–}(4.33) \text{ and } (4.34),$$

and denote f^* as the optimized value.
Step 4. Substitute the obtained matrices $(\mathcal{P}_i, \mathcal{G}_i, \mathcal{S}_i, \mathcal{Q}_i, \mathcal{L}_i, \mathcal{H}_i)$ into (4.29). If (4.29) is satisfied, with

$$\left|f^* - 2Nr\right| < \varepsilon,$$

for a sufficiently small scalar $\varepsilon > 0$, then output the feasible solutions $(\mathcal{P}_i, \mathcal{G}_i, \mathcal{S}_i, \mathcal{Q}_i, \mathcal{L}_i, \mathcal{H}_i)$, so EXIT.
Step 5. If $\kappa > \mathbb{N}$ where $\mathbb{N}$ is the maximum number of iterations allowed, so EXIT.
Step 6. Set $\kappa = \kappa + 1$, $(\mathcal{P}_i^{(\kappa)}, \mathcal{G}_i^{(\kappa)}, \mathcal{S}_i^{(\kappa)}, \mathcal{Q}_i^{(\kappa)}, \mathcal{L}_i^{(\kappa)}, \mathcal{H}_i^{(\kappa)})$, and go to Step 3.

4.5 Illustrative Example

Example 4.5.1 Consider the Markovian jump singular stochastic system in (4.1) with two operating modes, that is, $N = 2$ and the following parameters:

$$A_1 = \begin{bmatrix} 1.5 & -1.0 & -1.2 \\ 1.3 & 1.6 & 1.1 \\ 0.6 & 0.8 & -0.8 \end{bmatrix}, \quad D_1 = \begin{bmatrix} 0.1 & 0.2 & 0.0 \\ 0.1 & 0.2 & 0.0 \\ 0.0 & 0.0 & 0.0 \end{bmatrix}, \quad B_1 = \begin{bmatrix} 1.0 \\ 0.5 \\ 0.4 \end{bmatrix},$$

$$A_2 = \begin{bmatrix} 0.5 & -0.6 & 0.7 \\ 1.2 & 2.4 & -0.4 \\ 0.6 & 0.2 & 1.5 \end{bmatrix}, \quad D_2 = \begin{bmatrix} 0.2 & 0.1 & 0.0 \\ 0.2 & 0.1 & 0.0 \\ 0.0 & 0.0 & 0.0 \end{bmatrix}, \quad B_2 = \begin{bmatrix} 0.8 \\ 1.0 \\ 0.4 \end{bmatrix}.$$

In addition, $f_1(x) = f_2(x) = 1.5\exp(-t)\sin(t)x(t)$ (thus, ε in (4.2) can be chosen as $\varepsilon = 1.5$) and

$$E = \begin{bmatrix} 1.0 & 0.0 & 0.0 \\ 0.0 & 0.4 & 0.0 \\ 0.0 & 0.0 & 0.0 \end{bmatrix}, \quad E_R = \begin{bmatrix} 1.0 & 0.0 \\ 0.0 & 1.0 \\ 0.0 & 0.0 \end{bmatrix}, \quad S = \begin{bmatrix} 0.0 \\ 0.0 \\ 1.0 \end{bmatrix},$$

$$E_L = \begin{bmatrix} 1.0 & 0.0 & 0.0 \\ 0.0 & 0.4 & 0.0 \end{bmatrix}, \quad \Pi = \begin{bmatrix} -0.6 & 0.6 \\ 0.8 & -0.8 \end{bmatrix}, \quad R = [\,0.0 \quad 0.0 \quad 1.0\,].$$

Here, we only simulate the results in Section 4.3. Our aim is to design an SMC law $u(t)$ in (4.15) such that the closed-loop system is stochastically stable. Solving the LMI conditions in Theorem 4.3.4, we obtain

$$\mathcal{P}_1 = \begin{bmatrix} 42.5930 & 11.8279 & 16.0462 \\ 11.8279 & 32.5262 & -8.0232 \\ 16.0462 & -8.0232 & 52.5856 \end{bmatrix}, \quad \mathcal{P}_2 = \begin{bmatrix} 39.6506 & 9.4146 & -9.2143 \\ 9.4146 & 36.7383 & 0.7701 \\ -9.2143 & 0.7701 & 44.6905 \end{bmatrix},$$

$$\mathcal{L}_1 = [\,-50.9447 \quad -186.4032 \quad 0.000\,], \quad \mathcal{L}_2 = [\,-29.8068 \quad -144.8451 \quad 0.000\,],$$

$$\mathcal{Q}_1 = 12.8865, \quad \mathcal{Q}_2 = -16.7176, \quad \mathcal{H}_1 = 2.7937, \quad \mathcal{H}_2 = 11.9350.$$

Thus, by (4.12) we have

$$K_1 = [0.3324 \quad -5.7983 \quad 0.2168],$$
$$K_2 = [0.0159 \quad -3.9317 \quad -0.7139].$$

Here, parameter G_i, $i \in \{1, 2\}$ in (4.5) can be chosen as $G_1 = G_2 = [\,1 \ -1 \ 1\,]$. Thus, G_iB_i are nonsingular and $G_iD_i = 0$ are guaranteed for $i \in \{1, 2\}$. By (4.5), the switching functions can be computed as

$$s(t) = \begin{cases} s_1(t) = [\,1.0 \quad -1.0 \quad 0.0\,]\,x(t) \\ \qquad - \displaystyle\int_0^t [\,1.0992 \quad -7.0185 \quad -2.9049\,]\,x(s)ds, & i = 1, \\ s_2(t) = [\,1.0 \quad -1.0 \quad 0.0\,]\,x(t) \\ \qquad - \displaystyle\int_0^t [\,-0.0968 \quad -3.5863 \quad 2.4572\,]\,x(s)ds, & i = 2. \end{cases}$$

Let the adjustable scalar λ be $\lambda = 0.5$, then the SMC law designed in (4.15) can be computed as

$$u(t) = \begin{cases} u_1(t) = [\,0.3324 \quad -5.7983 \quad 0.2168\,]\,x(t) \\ \qquad - (0.5 + 1.5\|x(t)\|)\text{sign}(s_1(t)), \quad i = 1, \\ \\ u_2(t) = [\,0.0159 \quad -3.9317 \quad -0.7139\,]\,x(t) \\ \qquad - (0.5 + 1.5\|x(t)\|)\text{sign}(s_2(t)), \quad i = 2. \end{cases}$$

To prevent the control signals from chattering, we replace sign($s(t)$) with $\frac{s(t)}{0.01+\|s(t)\|}$. By using the discretization approach [96], we simulate standard Brownian motion. Some initial parameters are given as follows: the simulation time $t \in [0, T^*]$ with $T^* = 8$, the normally distributed variance $\delta t = \frac{T^*}{N^*}$ with $N^* = 2^{11}$, step size $\Delta t = \rho\delta t$ with $\rho = 2$, and the number of discretized Brownian paths $p = 10$. The simulation results are given in Figures 4.1–4.7. Among them, Figures 4.1–4.3 are the simulation results along an individual discretized Brownian path. Figure 4.1 shows the states of the closed-loop system; Figure 4.2 depicts the switching function $s(t)$; and Figure 4.3 gives the SMC input $u(t)$. Figures 4.4–4.7 show the corresponding simulation results along 10 individual paths (dotted lines) and the average over 10 paths (solid line). Figures 4.4–4.6 show the states of the closed-loop system. Figure 4.7 depicts the switching function $s(t)$.

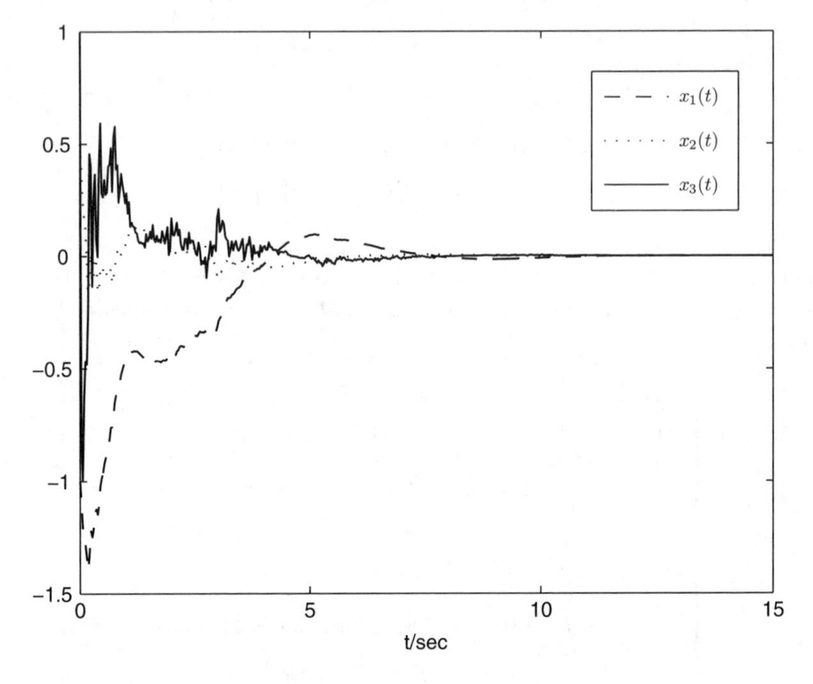

Figure 4.1 States of the closed-loop system

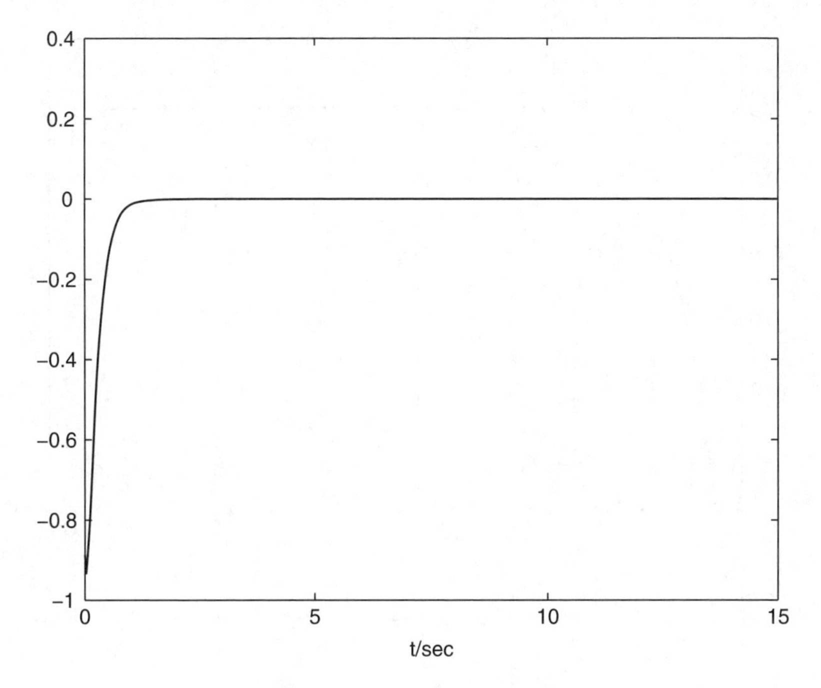

Figure 4.2 Switching function

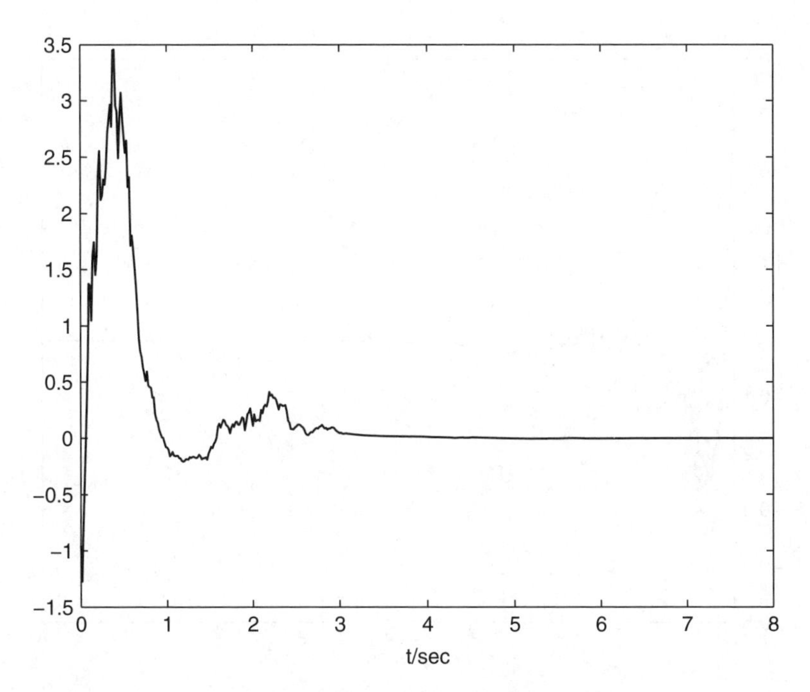

Figure 4.3 Control input

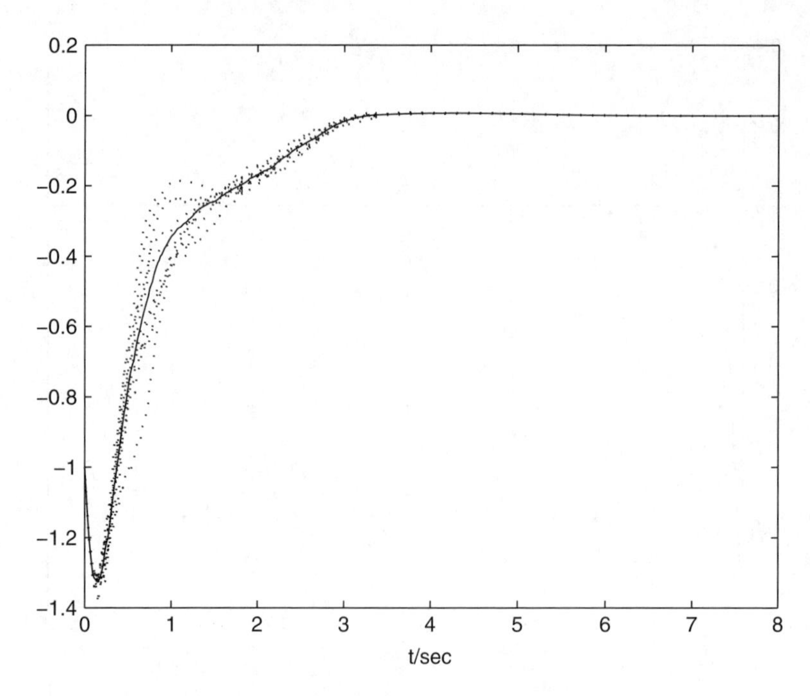

Figure 4.4 Individual paths and the average of the state of the closed-loop system: first component

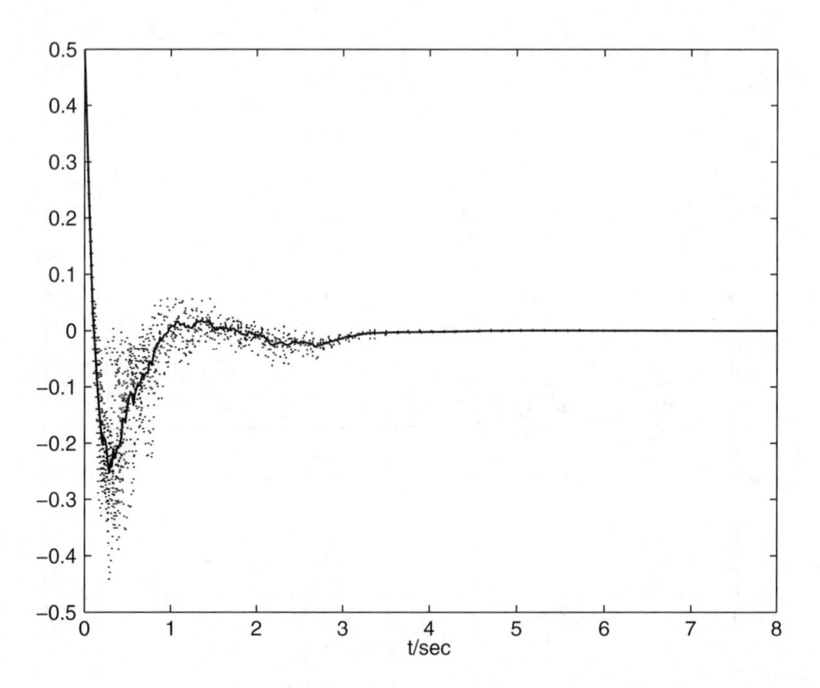

Figure 4.5 Individual paths and the average of the state of the closed-loop system: second component

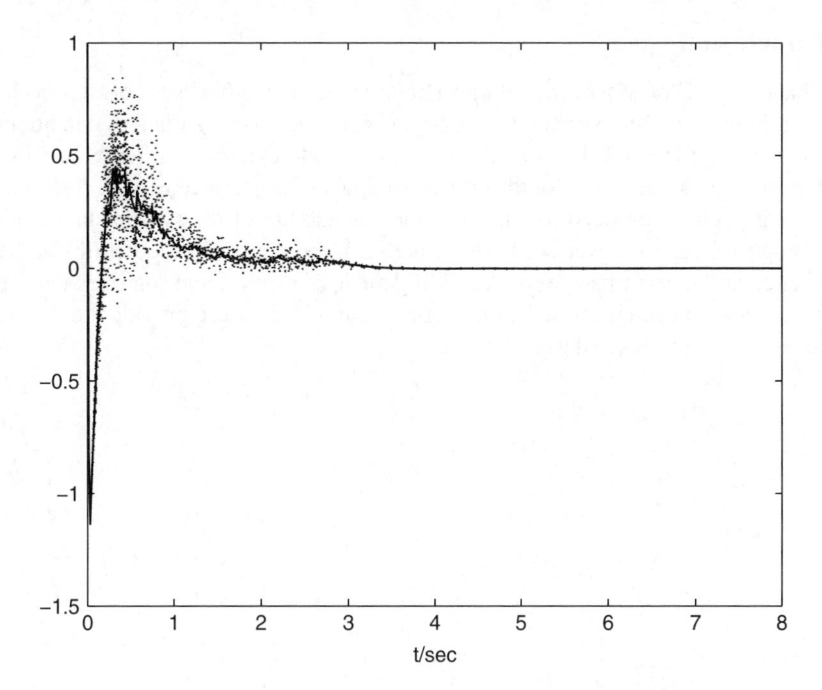

Figure 4.6 Individual paths and the average of the state of the closed-loop system: third component

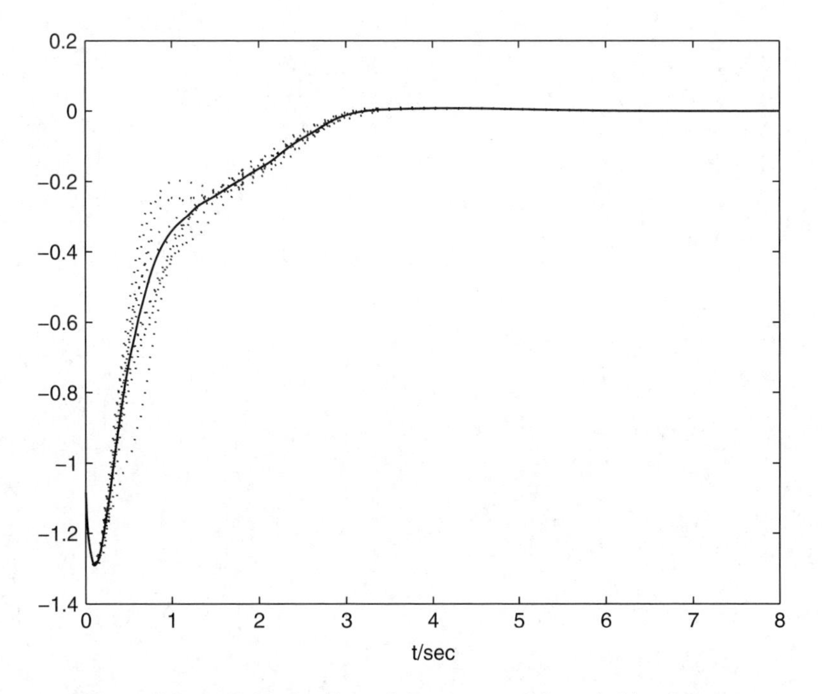

Figure 4.7 Individual paths and the average of the switching function

4.6 Conclusion

In this chapter, SMC of Markovian jump singular stochastic hybrid systems has been investigated. An integral sliding surface has been designed and some sufficient conditions have been proposed for the stochastic stability of sliding mode dynamics in terms of strict LMI. Also, an explicit parametrization of the desired sliding surface has been given. A sliding mode controller has been synthesized to guarantee the reachability of the system state trajectories to the sliding surface. Moreover, we have further analyzed the stochastic stability and $\mathcal{L}_2$ disturbance attenuation performance for the sliding mode dynamics, and some related sufficient conditions have also been proposed. A numerical example has been provided to illustrate the effectiveness of the proposed design scheme.

Part Two

SMC of Switched State-Delayed Hybrid Systems

5

Stability and Stabilization of Switched State-Delayed Hybrid Systems

5.1 Introduction

In the previous three chapters, we have solved SMC design problems for systems with Markovian switching parameters. From this chapter onward, we shall study the analysis and synthesis problems for another kind of parameter-switching systems, namely switched hybrid systems. In Chapter 1, we presented an overview of the recent developments in switched hybrid systems. The stability analysis problem for switched hybrid systems can be classified into two major categories: stability analysis under arbitrary switching and under restricted switchings. It was shown that switched systems may fail to preserve stability under arbitrary switching, but may be stable under restricted switching signals [131]. That is to say, the stability results under restricted switchings may have less conservativeness than those under arbitrary switching. Among the restricted switchings, the most famous concept is the average dwell time. In this book, we shall assume that the switching signal in the considered switched hybrid systems is not arbitrary, but is a restricted one having an average dwell time.

In this chapter, we shall investigate the stability analysis and stabilization problems for continuous- and discrete time switched hybrid systems with time-varying delays. For continuous-time system, the time-varying delay $d(t)$ is assumed to satisfy either (A1) $0 \leq d(t) \leq d$ and $\dot{d}(t) \leq \tau$ or (A2) $0 \leq d(t) \leq d$. By using the average dwell time approach and the piecewise Lyapunov function technique, two delay-dependent sufficient conditions are established for the exponential stability of the considered hybrid system with (A1) and (A2), respectively. Here, the slack matrix approach is applied to further reduce the conservativeness of the stability conditions caused by the time delay. For discrete-time system, the stability conditions are also derived by the average dwell time approach, and the results are all delay-dependent and thus less conservative. The stabilization problem is then solved by designing a memoryless state feedback controller, and an explicit expression for the desired

Sliding Mode Control of Uncertain Parameter-Switching Hybrid Systems, First Edition. Ligang Wu, Peng Shi and Xiaojie Su.

controller is given. The work in this chapter is an important foundation for the development of the SMC methodologies for switched hybrid systems in subsequent chapters.

5.2 Continuous-Time Systems

5.2.1 System Description

Consider the continuous-time switched state-delayed hybrid systems described by

$$\dot{x}(t) = A(\alpha(t))x(t) + A_d(\alpha(t))x(t - d(t)) + B(\alpha(t))u(t), \tag{5.1a}$$

$$x(t) = \phi(t), \quad t \in [-d, 0], \tag{5.1b}$$

where $x(t) \in \mathbf{R}^n$ is the state vector; $u(t) \in \mathbf{R}^m$ is the control input; $\phi(t) \in \mathbf{C}_{n,d}$ is a differentiable vector-valued initial function on $[-d, 0]$ for a known constant $d > 0$; $d(t)$ denotes the time-varying delay satisfying either (A1) $0 \leq d(t) \leq d$; and $\dot{d}(t) \leq \tau$ or (A2) $0 \leq d(t) \leq d$.

In system (5.1a), $\{(A(\alpha(t)), A_d(\alpha(t)), B(\alpha(t))) : \alpha(t) \in \mathcal{N}\}$ is a family of matrices parameterized by an index set $\mathcal{N} = \{1, 2, \dots, N\}$, and $\alpha(t) : \mathbf{R} \to \mathcal{N}$ is a piecewise constant function of time t called a switching signal. At a given time t, the value of $\alpha(t)$, denoted by α for simplicity, might depend on t or $x(t)$, or both, or may be generated by any other hybrid scheme. Therefore, the switched delayed hybrid system effectively switches among N subsystems with the switching sequence controlled by α. We assume that the value of α is unknown, but its instantaneous value is available in real time.

For each $\alpha = i \, (i \in \mathcal{N})$, we will denote the system matrices associated with mode i by $A(i) = A(\alpha)$, $A_d(i) = A_d(\alpha)$, and $B(i) = B(\alpha)$, where $A(i)$, $A_d(i)$, and $B(i)$ are constant matrices. Corresponding to the switching signal α, we have the switching sequence $\{(i_0, t_0), (i_1, t_1), \dots, (i_k, t_k), \dots, \mid i_k \in \mathcal{N}, \ k = 0, 1, \dots\}$ with $t_0 = 0$, which means that the i_kth subsystem is activated when $t \in [t_k, t_{k+1})$.

For the switching signal α, we revisit the average dwell time property from the following definition.

Definition 5.2.1 [129] *For any $T_2 > T_1 \geq 0$, let $N_\alpha(T_1, T_2)$ denote the number of switching of α over (T_1, T_2). If $N_\alpha(T_1, T_2) \leq N_0 + (T_2 - T_1)/T_a$ holds for $T_a > 0$, $N_0 \geq 0$, then T_a is called an average dwell time.*

Assumption 5.1 *The switching signal $\alpha(t)$ has an average dwell time.*

Definition 5.2.2 *The continuous-time switched state-delayed hybrid system in (5.1a)–(5.1b) with $u(t) = 0$ is said to be exponentially stable under $\alpha(t)$ if the solution $x(t)$ of the system satisfies*

$$\|x(t)\| \leq \eta \left\|x(t_0)\right\|_{\mathrm{C}} e^{-\lambda(t - t_0)}, \quad \forall t \geq t_0,$$

for constants $\eta \geq 1$ and $\lambda > 0$, and

$$\left\|x(t_0)\right\|_{\mathrm{C}} \triangleq \sup_{-d \leq \theta \leq 0} \left\{\left\|x(t_0 + \theta)\right\|, \left\|\dot{x}(t_0 + \theta)\right\|\right\}.$$

Remark 5.1 *By the average dwell time switching, we mean a class of switching signals such that the average time interval between consecutive switchings is at least T_a. Then, a basic problem for such systems is how to specify the minimal T_a and thereby get the admissible switching signals such that the system is stable and satisfies a prescribed performance if the system dynamics meets some conditions. As commonly used in the literature, we choose $N_0 = 0$ in Definition 5.2.1.* ♦

5.2.2 *Main Results*

In this section, we will establish an exponential stability condition for system (5.1a)–(5.1b) with $u(t) = 0$ by applying the average dwell time approach and the piecewise Lyapunov function technique, and give the following result.

Theorem 5.2.3 *For a given constant $\beta > 0$, suppose (A1) holds and there exist matrices $P(i) > 0$, $Q(i) > 0$, $R(i) > 0$, and $X(i)$, $Y(i)$ such that for $i \in \mathcal{N}$,*

$$\begin{bmatrix} \bar{\Pi}_{11}(i) & \bar{\Pi}_{12}(i) & dA^T(i)R(i) & dX(i) \\ \star & \bar{\Pi}_{22}(i) & dA_d^T(i)R(i) & dY(i) \\ \star & \star & -dR(i) & 0 \\ \star & \star & \star & -de^{-\beta d}R(i) \end{bmatrix} < 0, \tag{5.2}$$

where

$$\begin{cases} \bar{\Pi}_{11}(i) \triangleq P(i)A(i) + A^T(i)P(i) + Q(i) + \beta P(i) + X(i) + X^T(i), \\ \bar{\Pi}_{12}(i) \triangleq P(i)A_d(i) + Y^T(i) - X(i), \\ \bar{\Pi}_{22}(i) \triangleq -(1-\tau)e^{-\beta d}Q(i) - Y(i) - Y^T(i). \end{cases}$$

Then the switched system in (5.1a)–(5.1b) with $u(t) = 0$ is exponentially stable for any switching signal with average dwell time satisfying $T_a > T_a^* = \frac{\ln \mu}{\beta}$, where $\mu \geq 1$ and satisfies

$$P(i) \leq \mu P(j),\ Q(i) \leq \mu Q(j),\ R(i) \leq \mu R(j), \quad \forall i,j \in \mathcal{N}. \tag{5.3}$$

Moreover, an estimate of the state decay is given by

$$\|x(t)\| \leq \eta \, \|x(0)\|_C \, e^{-\lambda t}, \tag{5.4}$$

where

$$\begin{cases} \lambda \triangleq \dfrac{1}{2}\left(\beta - \dfrac{\ln \mu}{T_a}\right) > 0, \quad \eta \triangleq \sqrt{\dfrac{b}{a}} \geq 1, \\ a \triangleq \min\limits_{\forall i \in \mathcal{N}} \lambda_{\min}(P(i)), \\ b \triangleq \max\limits_{\forall i \in \mathcal{N}} \lambda_{\max}(P(i)) + d \max\limits_{\forall i \in \mathcal{N}} \lambda_{\max}(Q(i)) + \dfrac{d^2}{2} \max\limits_{\forall i \in \mathcal{N}} \lambda_{\max}(R(i)). \end{cases} \tag{5.5}$$

Proof. Choose a Lyapunov function of the following form:

$$V(x_t, \alpha) \triangleq V_1(x_t, \alpha) + V_2(x_t, \alpha) + V_3(x_t, \alpha),$$

with

$$\begin{cases} V_1(x_t, \alpha) \triangleq x^T(t)P(\alpha)x(t), \\ V_2(x_t, \alpha) \triangleq \int_{t-d(t)}^{t} e^{\beta(s-t)}x^T(s)Q(\alpha)x(s)ds, \\ V_3(x_t, \alpha) \triangleq \int_{-d}^{0}\int_{t+\theta}^{t} e^{\beta(s-t)}\dot{x}^T(s)R(\alpha)\dot{x}(s)dsd\theta, \end{cases} \tag{5.6}$$

where $P(\alpha) > 0$, $Q(\alpha) > 0$, and $R(\alpha) > 0$ are to be determined. Then, as with the solution of (5.1a)–(5.1b) for a fixed α, we have

$$\begin{aligned} \dot{V}_1(x_t, \alpha) &= 2x^T(t)P(\alpha)\dot{x}(t) \\ &= 2x^T(t)P(\alpha)\left(A(\alpha)x(t) + A_d(\alpha)x(t-d(t))\right), \end{aligned} \tag{5.7}$$

$$\begin{aligned} \dot{V}_2(x_t, \alpha) \leq& -\beta\int_{t-d(t)}^{t} e^{\beta(s-t)}x^T(s)Q(\alpha)x(s)ds + x^T(t)Q(\alpha)x(t) \\ &-(1-\tau)e^{-\beta d}x^T(t-d(t))Q(\alpha)x(t-d(t)), \end{aligned} \tag{5.8}$$

$$\begin{aligned} \dot{V}_3(x_t, \alpha) \leq& -\beta\int_{-d}^{0}\int_{t+\theta}^{t} e^{\beta(s-t)}\dot{x}^T(s)R(\alpha)\dot{x}(s)dsd\theta + d\dot{x}^T(t)R(\alpha)\dot{x}(t) \\ &- \int_{t-d(t)}^{t} e^{-\beta d}\dot{x}^T(s)R(\alpha)\dot{x}(s)ds. \end{aligned} \tag{5.9}$$

However, the Newton–Leibniz formula gives

$$x(t) - x(t-d(t)) = \int_{t-d(t)}^{t} \dot{x}(s)ds.$$

Then, for any appropriately dimensioned matrices $Z(\alpha) \triangleq \begin{bmatrix} \bar{X}(\alpha) \\ \bar{Y}(\alpha) \end{bmatrix}$, we have

$$2e^{-\beta d}\varphi^T(t)Z(\alpha)\left[x(t) - x(t-d(t)) - \int_{t-d(t)}^{t} \dot{x}(s)ds\right] = 0, \tag{5.10}$$

where $\varphi(t) \triangleq \begin{bmatrix} x(t) \\ x(t-d(t)) \end{bmatrix}$.

Considering (5.7)–(5.10), it follows that

$$\dot{V}(x_t,\alpha)+\beta V(x_t,\alpha)\leq \varphi^T(t)\left[\Pi(\alpha)+de^{-\beta d}Z(\alpha)R^{-1}(\alpha)Z^T(\alpha)\right]\varphi(t)$$
$$-\int_{t-d(t)}^{t}e^{-\beta d}\left[Z^T(\alpha)\varphi(t)+R(\alpha)\dot{x}(s)\right]^T$$
$$\times R^{-1}(\alpha)\left[Z^T(\alpha)\varphi(t)+R(\alpha)\dot{x}(s)\right]ds, \tag{5.11}$$

where

$$\Pi(\alpha)\triangleq\begin{bmatrix}\Pi_{11}(\alpha) & \Pi_{12}(\alpha)\\ \star & \Pi_{22}(\alpha)\end{bmatrix},$$

with

$$\begin{cases}\Pi_{11}(\alpha)\triangleq P(\alpha)A(\alpha)+A^T(\alpha)P(\alpha)+Q(\alpha)+\beta P(\alpha)\\ \qquad\qquad +e^{-\beta d}\bar{X}(\alpha)+e^{-\beta d}\bar{X}^T(\alpha)+dA^T(\alpha)R(\alpha)A(\alpha),\\ \Pi_{12}(\alpha)\triangleq P(\alpha)A_d(\alpha)-e^{-\beta d}\bar{X}(\alpha)+e^{-\beta d}\bar{Y}^T(\alpha)+dA^T(\alpha)R(\alpha)A_d(\alpha),\\ \Pi_{22}(\alpha)\triangleq -(1-\tau)e^{-\beta d}Q(\alpha)-e^{-\beta d}\left(\bar{Y}(\alpha)+\bar{Y}^T(\alpha)\right)+dA_d^T(\alpha)R(\alpha)A_d(\alpha).\end{cases}$$

Notice that, in (5.11),

$$\left[Z^T(\alpha)\varphi(t)+R(\alpha)\dot{x}(s)\right]^T R^{-1}(\alpha)\left[Z^T(\alpha)\varphi(t)+R(\alpha)\dot{x}(s)\right]\geq 0. \tag{5.12}$$

Performing a congruence transformation on (5.2) by diag $\left\{I,I,I,e^{\beta d}I\right\}$ and considering $X(\alpha)\triangleq e^{-\beta d}\bar{X}(\alpha)$, $Y(\alpha)\triangleq e^{-\beta d}\bar{Y}(\alpha)$, by Schur complement, (5.2) implies

$$\Pi(\alpha)+de^{-\beta d}Z(\alpha)R^{-1}(\alpha)Z^T(\alpha)<0. \tag{5.13}$$

Thus, it follows from (5.11)–(5.13) that

$$\dot{V}(x_t,\alpha)+\beta V(x_t,\alpha)\leq 0. \tag{5.14}$$

Now, for an arbitrary piecewise constant switching signal α, and for any $t>0$, we let $0<t_1<\cdots<t_k<\cdots$, $k=0,1,\ldots$, denote the switching points of α over the interval $(0,t)$. As mentioned earlier, the i_kth subsystem is activated when $t\in\left[t_k,t_{k+1}\right)$. Integrating (5.14) from t_k to t gives

$$V(x_t,\alpha)\leq e^{-\beta(t-t_k)}V(x_{t_k},\alpha(t_k)). \tag{5.15}$$

Using (5.3) and (5.6), at switching instant t_k, we have

$$V(x_{t_k},\alpha(t_k))\leq \mu V(x_{t_k^-},\alpha(t_k^-)). \tag{5.16}$$

Therefore, it follows from (5.15)–(5.16) and the relationship $\vartheta = N_\alpha(0,t) \le (t-0)/T_a$ that

$$\begin{aligned} V(x_t,\alpha) &\le e^{-\beta(t-t_k)}\mu V(x_{t_k^-},\alpha(t_k^-)) \le \cdots \\ &\le e^{-\beta(t-0)}\mu^{\vartheta} V(x_0,\alpha(0)), \\ &\le e^{-(\beta-\ln\mu/T_a)t} V(x_0,\alpha(0)). \end{aligned} \tag{5.17}$$

Notice from (5.6) that

$$V(x_t,\alpha) \ge a\,\|x(t)\|^2\,,\ V(x_0,\alpha(0)) \le b\,\|x(0)\|_{\mathrm{C}}^2\,, \tag{5.18}$$

where a and b are defined in (5.5). Combining (5.17)–(5.18) yields

$$\|x(t)\|^2 \le \frac{1}{a}V(x_t,\alpha) \le \frac{b}{a}e^{-(\beta-\ln\mu/T_a)t}\,\|x(0)\|_{\mathrm{C}}^2\,,$$

which implies (5.4). By Definition 5.2.1 with $t_0 = 0$, system (5.1a)–(5.1b) is exponentially stable. This completes the proof. ■

Remark 5.2 *Notice that Theorem 5.2.3 gives a delay-dependent sufficient condition for the exponential stability of system (5.1a)–(5.1b) with $u(t) = 0$. In the derivation of the delay-dependent result in Theorem 5.2.3, no model transformation was performed to system (5.1a)–(5.1b). Moreover, we introduced slack variables $\bar{X}(\alpha)$ and $\bar{Y}(\alpha)$, which helps avoid using bounding techniques and hence the possible conservativenes.* ♦

Remark 5.3 *Notice that there exist constraints $P(i) \le \mu P(j)$, $Q(i) \le \mu Q(j)$, and $R(i) \le \mu R(j)$, $\forall i,j \in \mathcal{N}$ in (5.3) of Theorem 5.2.3. So $\mu(>1)$ is only dependent upon (5.3), and it is independent of (5.2). In fact, μ can be found to have very many solutions, for example,*

$$\mu = \mu^* = \max\left\{ \sup_{i,j\in\mathcal{N}}\left(\frac{\lambda_{\max}(P(i))}{\lambda_{\min}(P(j))}\right),\ \sup_{i,j\in\mathcal{N}}\left(\frac{\lambda_{\max}(Q(i))}{\lambda_{\min}(Q(j))}\right),\ \sup_{i,j\in\mathcal{N}}\left(\frac{\lambda_{\max}(R(i))}{\lambda_{\min}(R(j))}\right)\right\},$$

and any value larger than μ^* can also be considered as a solution of μ. ♦

Remark 5.4 *When $\mu = 1$ in $T_a > T_a^* = \frac{\ln\mu}{\beta}$, we have $T_a > T_a^* = 0$, which means that the switching signal α can be arbitrary. In this case, (5.3) turns out to be $P(i) \le P(j)$, $Q(i) \le Q(j)$, and $R(i) \le R(j)$, $\forall i,j \in \mathcal{N}$. Thus the only possibility is $P(i) = P(j) = P$, $Q(i) = Q(j) = Q$, and $R(i) = R(j) = R$, $\forall i,j \in \mathcal{N}$, which implies that a common (that is, mode-independent) Lyapunov function is required for all subsystems.* ♦

Remark 5.5 *When $\mu > 1$ and $\beta \to 0$ in $T_a > T_a^* = \frac{\ln\mu}{\beta}$, we have $T_a \to \infty$, that is, there is no switching. Switched system (5.1a)–(5.1b) is effectively operating at one of the subsystems all* *the time. In this case, according to the proof of Theorem 5.2.3, the asymptotic stability result of system (5.1a)–(5.1b) coincides with Theorem 1 in [243] when delay $d(t) = d$ is constant.* ♦

Remark 5.6 *It should be pointed out that the methods used in this chapter for deriving the stability condition in Theorem 5.2.3 are different from that in [188], thus the obtained results are different. Since it introduced more slack matrices in Theorem 1 of [188], the condition becomes hard to apply to stabilization and controller synthesis problems. Our result in Theorem 5.2.3 overcomes the above difficulty, and this can be verified by the SMC problem presented in Chapter 7.* ♦

The result in Theorem 5.2.3 is based on (A1), but when considering (A2), we have the following theorem. The result can be obtained by employing the same techniques used as in the proof of Theorem 5.2.3, thus we omit the proof.

Theorem 5.2.4 *For a given constant $\beta > 0$, suppose (A2) holds and there exist matrices $P(i) > 0$, $R(i) > 0$, and $X(i)$, $Y(i)$ such that for $i \in \mathcal{N}$,*

$$\begin{bmatrix} \tilde{\Pi}_{11}(i) & \tilde{\Pi}_{12}(i) & dA^T(i)R(i) & dX(i) \\ \star & \tilde{\Pi}_{22}(i) & dA_d^T(i)R(i) & dY(i) \\ \star & \star & -dR(i) & 0 \\ \star & \star & \star & -de^{-\beta d}R(i) \end{bmatrix} < 0,$$

where

$$\begin{cases} \tilde{\Pi}_{11}(i) \triangleq P(i)A(i) + A^T(i)P(i) + X(i) + X^T(i) + \beta P(i), \\ \tilde{\Pi}_{12}(i) \triangleq P(i)A_d(i) + Y^T(i) - X(i), \\ \tilde{\Pi}_{22}(i) \triangleq -Y(i) - Y^T(i). \end{cases}$$

Then the switched system in (5.1a)–(5.1b) with $u(t) = 0$ is exponentially stable for any switching signal with average dwell time satisfying $T_a > T_a^* = \frac{\ln \mu}{\beta}$, where $\mu \geq 1$ and satisfies

$$P(i) \leq \mu P(j), \; R(i) \leq \mu R(j), \quad \forall i, j \in \mathcal{N}.$$

Moreover, an estimate of state decay is given by

$$\|x(t)\| \leq \eta \, \|x(0)\|_C \, e^{-\lambda t},$$

where

$$\begin{cases} \lambda \triangleq \dfrac{1}{2}\left(\beta - \dfrac{\ln \mu}{T_a}\right) > 0, \quad \eta \triangleq \sqrt{\dfrac{c}{a}} \geq 1, \\ a \triangleq \min\limits_{\forall i \in \mathcal{N}} \lambda_{\min}(P(i)), \\ c \triangleq \max\limits_{\forall i \in \mathcal{N}} \lambda_{\max}(P(i)) + \dfrac{d^2}{2} \max\limits_{\forall i \in \mathcal{N}} \lambda_{\max}(R(i)). \end{cases}$$

Remark 5.7 *Comparing the results in Theorems 5.2.3 and 5.2.4, we found that the result in Theorem 5.2.4 requires a weaker condition on the time-varying delay when compared with Theorem 5.2.3. To obtain Theorem 5.2.4, a Lyapunov function is chosen as follows:*

$$W(x_t, \alpha) \triangleq x^T(t)P(\alpha)x(t) + \int_{-d}^{0}\int_{t+\theta}^{t} e^{\beta(s-t)}\dot{x}^T(s)R(\alpha)\dot{x}(s)dsd\theta.$$

Notice that the derivative of the functional does not require bounding of the rate of delay $d(t)$. In particular, the delay in Theorem 5.2.3 is required to be differentiable, but the one in Theorem 5.2.4 may be non-differentiable with arbitrarily fast time-varying behavior. ♦

5.2.3 Illustrative Example

Example 5.2.5 Consider the switched delay system in (5.1a)–(5.1b) with $N = 2$ and the following system parameters:

$$A(1) = \begin{bmatrix} -0.4 & 0.2 \\ 0.2 & -0.3 \end{bmatrix}, \quad A_d(1) = \begin{bmatrix} -0.2 & 0.0 \\ 0.1 & -0.4 \end{bmatrix},$$

$$A(2) = \begin{bmatrix} -0.2 & 0.3 \\ 0.2 & -0.7 \end{bmatrix}, \quad A_d(2) = \begin{bmatrix} -0.3 & 0.1 \\ 0.0 & -0.2 \end{bmatrix},$$

and $d = 1.2$, $\beta = 0.5$, $\tau = 0.3$. It can be checked by using Theorem 1 of [243] that the above two subsystems are both asymptotically stable. We consider the average dwell time scheme, and set $\mu = 1.2 > 1$, thus $T_a > T_a^* = \frac{\ln \mu}{\beta} = 0.3646$ by (5.3). Solving LMIs (5.2)–(5.3), it follows that

$$P(1) = \begin{bmatrix} 3.0746 & -0.1222 \\ -0.1222 & 3.0759 \end{bmatrix}, \quad P(2) = \begin{bmatrix} 2.6710 & 0.0254 \\ 0.0254 & 3.5407 \end{bmatrix},$$

$$Q(1) = \begin{bmatrix} 0.0682 & -0.1614 \\ -0.1614 & 0.4278 \end{bmatrix}, \quad Q(2) = \begin{bmatrix} 0.0688 & -0.1637 \\ -0.1637 & 0.3622 \end{bmatrix},$$

$$R(1) = \begin{bmatrix} 2.7061 & -0.3964 \\ -0.3964 & 3.0228 \end{bmatrix}, \quad R(2) = \begin{bmatrix} 3.0755 & -0.5469 \\ -0.5469 & 2.6960 \end{bmatrix},$$

which means that the above switched system is exponentially stable. Taking $T_a = 1 > T_a^*$, and considering (5.4)–(5.5) yield $a = 2.6702$, $b = 6.6340$, $\eta = 1.5762$ and $\lambda = 0.1588$, thus

$$\|x(t)\| \leq 1.5762 \, \|x(0)\|_C \, e^{-0.1588t}.$$

5.3 Discrete-Time Systems

5.3.1 System Description

Consider a discrete-time switched system with time delays, which can be described by the following dynamical equations:

$$x(k+1) = A(\alpha(k))x(k) + A_d(\alpha(k))x(k-d(k)) + B(\alpha(k))u(k), \tag{5.19a}$$

$$x(k) = \phi(k), \quad k = -d_2, -d_2+1, \dots, 0, \tag{5.19b}$$

where $x(k) \in \mathbf{R}^n$ is the system state vector; $u(k) \in \mathbf{R}^m$ represents the control input; $\phi(k)$ is the initial condition; $\{(A(\alpha(k)), A_d(\alpha(k)), B(\alpha(k))) : \alpha(k) \in \mathcal{N}\}$ is a family of matrices parameterized by an index set $\mathcal{N} = \{1, 2, \dots, N\}$; and $\alpha(k) : \mathbf{Z}^+ \to \mathcal{N}$ is a piecewise constant function of time, called a switching signal, which takes its values in the finite set $\mathcal{N}$. At an arbitrary discrete time k, the value of $\alpha(k)$, denoted by α for simplicity, might depend on k or $x(k)$, or both, or may be generated by any other hybrid scheme. We assume that the sequence of subsystems in switching signal α is unknown *a priori*, but its instantaneous value is available in real time. For the switching time sequence $k_0 < k_1 < k_2 < \cdots$ of switching signal α, the holding time between $[k_l, k_{l+1}]$ is called the dwell time of the currently engaged subsystem, where $l \in \mathcal{N}$. The delay $d(k)$ satisfies $1 \le d_1 \le d(k) \le d_2$, where d_1 and d_2 are constant positive scalars representing the minimum and maximum delays, respectively.

Remark 5.8 *For each possible value $\alpha = i$, $i \in \mathcal{N}$, we will denote the system matrices associated with mode i by $A(i) = A(\alpha)$, $A_d(i) = A_d(\alpha)$, and $B(i) = B(\alpha)$, where $A(i)$, $A_d(i)$, and $B(i)$ are constant matrices. Corresponding to the switching signal α, we have the switching sequence $\{(i_0, k_0), (i_1, k_1), \dots, (i_l, k_l), \dots, \mid i_l \in \mathcal{N}, l = 0, 1, \dots\}$ with $k_0 = 0$, which means that the i_lth subsystem is activated when $k \in [k_l, k_{l+1})$.* ♦

For the switching signal α, we introduce the following definition.

Definition 5.3.1 *For a switching signal and any $k_i > k_j > k_0$, let $N_\alpha(k_j, k_i)$ be the switching numbers of α_k over the interval $[k_j, k_i]$. If for any given $N_0 > 0$ and $T_a > 0$, we have $N_\alpha(k_j, k_i) \le N_0 + (k_i - k_j)/T_a$, then T_a and N_0 are called average dwell time and the chatter bound, respectively.*

Here, we assume $N_0 = 0$ for simplicity as commonly used in the literature.

Assumption 5.2 *The switching signal $\alpha(k)$ has an average dwell time.*

Design a stabilization controller with the following general structure:

$$u(k) = K(\alpha)x(k), \tag{5.20}$$

where $K(\alpha) \in \mathbf{R}^{m \times n}$ are parameter matrices to be designed.

Substituting the stabilization controller in (5.20) into system (5.19a)–(5.19b), we obtain the closed-loop system as

$$x(k+1) = \tilde{A}(\alpha)x(k) + A_d(\alpha)x(k-d(k)), \tag{5.21a}$$

$$x(k) = \phi(k), \quad k = -d_2, -d_2+1, \ldots, 0, \tag{5.21b}$$

where $\tilde{A}(\alpha) \triangleq A(\alpha) + B(\alpha)K(\alpha)$.

Definition 5.3.2 *The discrete-time switched time-delay hybrid system in (5.19a)–(5.19b) with $u(k) = 0$ is said to be exponentially stable under α if the solution $x(k)$ satisfies*

$$\|x(k)\| \leq \eta \left\|x(k_0)\right\|_C \rho^{(k-k_0)}, \quad \forall k \geq k_0,$$

for constants $\eta \geq 1$ and $0 < \rho < 1$, and

$$\left\|x(k_0)\right\|_C \triangleq \underbrace{\{\|x(k+\theta)\|, \|\xi(k+\theta)\|\}}_{\sup_{-d_2 \leq \theta \leq 0}},$$

where $\xi(\theta) \triangleq x(\theta+1) - x(\theta)$.

Remark 5.9 *Notice that the phrase 'under α' appears in Definition 5.3.2. This serves to emphasize that all results obtained subsequently in this chapter are dependent on the switching signal α, and α is not an arbitrary switching signal but a restricted one having an average dwell time.* ♦

5.3.2 Main Results

First, we will use the piecewise Lyapunov technique and the average dwell time approach to propose a sufficient condition for the exponential stability of the discrete-time switched time-delay system in (5.19a)–(5.19b) with $u(k) = 0$. We have the following theorem.

Theorem 5.3.3 *Given a constant $0 < \beta < 1$, suppose that there exist matrices $P(i) > 0$, $Q(i) > 0$, $R(i) > 0$, $S_1(i) > 0$, and $S_2(i) > 0$, and matrices $L(i)$, $M(i)$, and $N(i)$ such that for $i \in \mathcal{N}$,*

$$\begin{bmatrix} \beta^{-(d_2+1)}\Phi(i) & d_2L(i) & (d_2-d_1)M(i) & d_2N(i) \\ \star & -d_2S_1(i) & 0 & 0 \\ \star & \star & -(d_2-d_1)S_1(i) & 0 \\ \star & \star & \star & -d_2S_2(i) \end{bmatrix} < 0, \tag{5.22}$$

where

$$\begin{cases} \Phi(i) \triangleq \begin{bmatrix} \Phi_{11}(i) & 0 & 0 \\ \star & -\beta^{d_2+1}Q(i) & 0 \\ \star & \star & -\beta^{d_2+1}R(i) \end{bmatrix} + \begin{bmatrix} A^T(i) \\ A_d^T(i) \\ 0 \end{bmatrix} P(i) \begin{bmatrix} A^T(i) \\ A_d^T(i) \\ 0 \end{bmatrix}^T \\ \quad + d_2\beta \begin{bmatrix} A^T(i) - I \\ A_d^T(i) \\ 0 \end{bmatrix} (S_1(i) + S_2(i)) \begin{bmatrix} A^T(i) - I \\ A_d^T(i) \\ 0 \end{bmatrix}^T \\ \quad + 2\beta^{d_2+1} \left\{ L(i) \begin{bmatrix} I \\ -I \\ 0 \end{bmatrix}^T + M(i) \begin{bmatrix} 0 \\ I \\ -I \end{bmatrix}^T + N(i) \begin{bmatrix} I \\ 0 \\ -I \end{bmatrix}^T \right\}, \\ \Phi_{11}(i) \triangleq -\beta P(i) + \beta R(i) + \beta(d_2 - d_1 + 1)Q(i). \end{cases}$$

Then the discrete-time switched time-delay system in (5.19a)–(5.19b) with $u(k) = 0$ is exponentially stable for any switching signal with average dwell time satisfying $T_a > T_a^* = \text{ceil}\left(-\frac{\ln \mu}{\ln \beta}\right)$, where $\mu \geq 1$ satisfies

$$\begin{aligned} &P(i) \leq \mu P(j), \quad Q(i) \leq \mu Q(j), \quad R(i) \leq \mu R(j), \\ &S_1(i) \leq \mu S_1(j), \quad S_2(i) \leq \mu S_2(j), \quad \forall i, j \in \mathcal{N}. \end{aligned} \tag{5.23}$$

Proof. Choose a Lyapunov function of the following form:

$$\begin{cases} V(x,\alpha) \triangleq \sum_{i=1}^{5} V_i(x,\alpha), \\ V_1(x,\alpha) \triangleq x^T(k)P(\alpha)x(k), \\ V_2(x,\alpha) \triangleq \sum_{l=k-d(k)}^{k-1} \beta^{k-l} x^T(l)Q(\alpha)x(l), \\ V_3(x,\alpha) \triangleq \sum_{l=k-d_2}^{k-1} \beta^{k-l} x^T(l)R(\alpha)x(l), \\ V_4(x,\alpha) \triangleq \sum_{s=-d_2+1}^{-d_1} \sum_{l=k+s}^{k-1} \beta^{k-l} x^T(l)Q(\alpha)x(l), \\ V_5(x,\alpha) \triangleq \sum_{s=-d_2}^{-1} \sum_{l=k+s}^{k-1} \beta^{k-l} \xi^T(l)S(\alpha)\xi(l), \end{cases} \tag{5.24}$$

where $\xi(k) \triangleq x(k+1) - x(k)$, $S(\alpha) \triangleq S_1(\alpha) + S_2(\alpha)$, and $P(\alpha) > 0$, $Q(\alpha) > 0$, $R(\alpha) > 0$, $S_1(\alpha) > 0$, and $S_2(\alpha) > 0$ are real matrices. For $k \in [k_l, k_{l+1})$, we define

$$\Delta V_j(x(k), \alpha) \triangleq V_j(x(k+1), \alpha) - V_j(x(k), \alpha), \quad j = 1, 2, 3, 4, 5,$$

thus $\Delta V(x, \alpha) = \sum_{i=1}^{5} \Delta V_i(x, \alpha)$ with

$$\Delta V_1(x, \alpha) = x^T(k+1)P(\alpha)x(k+1) - x^T(k)P(\alpha)x(k), \tag{5.25}$$

$$\begin{aligned}\Delta V_2(x, \alpha) \leq &-(1-\beta) \sum_{l=k-d(k)}^{k-1} \beta^{k-l} x^T(l)Q(\alpha)x(l) \\ &+ \sum_{l=k+1-d_2}^{k-d_1} \beta^{k+1-l} x^T(l)Q(\alpha)x(l) + \beta x^T(k)Q(\alpha)x(k) \\ &-\beta^{d_2+1} x^T(k-d(k))Q(\alpha)x(k-d(k)),\end{aligned} \tag{5.26}$$

$$\begin{aligned}\Delta V_3(x, \alpha) = &-(1-\beta) \sum_{l=k-d_2}^{k-1} \beta^{k-l} x^T(l)R(\alpha)x(l) + \beta x^T(k)R(\alpha)x(k) \\ &-\beta^{d_2+1} x^T(k-d_2)R(\alpha)x(k-d_2),\end{aligned} \tag{5.27}$$

$$\begin{aligned}\Delta V_4(x, \alpha) = &-(1-\beta) \sum_{s=-d_2+1}^{-d_1} \sum_{l=k+s}^{k-1} \beta^{k-l} x^T(l)Q(\alpha)x(l) + \beta(d_2 - d_1)x^T(k)Q(\alpha)x(k) \\ &- \sum_{l=k+1-d_2}^{k-d_1} \beta^{k+1-l} x^T(l)Q(\alpha)x(l),\end{aligned} \tag{5.28}$$

$$\begin{aligned}\Delta V_5(x, \alpha) \leq &-(1-\beta) \sum_{s=-d_2}^{-1} \sum_{l=k+s}^{k-1} \beta^{k-l} \xi^T(l) \left(S_1(\alpha_k) + S_2(\alpha)\right) \xi(l) \\ &+ d_2 \beta \xi^T(k) \left(S_1(\alpha) + S_2(\alpha_k)\right) \xi(k) \\ &- \beta^{d_2+1} \left[\sum_{l=k-d_2}^{k-1} \xi^T(l)S_2(\alpha)\xi(l) + \sum_{l=k-d(k)}^{k-1} \xi^T(l)S_1(\alpha)\xi(l) \right. \\ &\left. + \sum_{l=k-d_2}^{k-d(k)-1} \xi^T(l)S_1(\alpha)\xi(l) \right].\end{aligned} \tag{5.29}$$

Moreover, for $\zeta(k) \triangleq \begin{bmatrix} x^T(k) & x^T(k-d(k)) & x^T(k-d_2) \end{bmatrix}^T$ and any appropriately dimensioned matrices $L(\alpha)$, $M(\alpha)$, and $N(\alpha)$, the following equations are true:

$$\left.\begin{aligned} 2\beta^{d_2+1}\zeta^T(k)L(\alpha)\left[x(k)-x(k-d(k))-\sum_{l=k-d(k)}^{k-1}\xi(l)\right] &= 0 \\ 2\beta^{d_2+1}\zeta^T(k)M(\alpha)\left[x(k-d(k))-x(k-d_2)-\sum_{l=k-d_2}^{k-d(k)-1}\xi(l)\right] &= 0 \\ 2\beta^{d_2+1}\zeta^T(k)N(\alpha)\left[x(k)-x(k-d_2)-\sum_{l=k-d_2}^{k-1}\xi(l)\right] &= 0 \end{aligned}\right\}. \tag{5.30}$$

Considering (5.25)–(5.29) and (5.30), we have

$$\begin{aligned} &\Delta V(x,\alpha)+(1-\beta)V(x,\alpha) \\ &\leq \zeta^T(k)\Big\{\Phi(\alpha)+\beta^{d_2+1}\Big[d_2L(\alpha)S_1^{-1}(\alpha)L^T(\alpha) \\ &+\left(d_2-d_1\right)M(\alpha)S_1^{-1}(\alpha)M^T(\alpha)+d_2N(\alpha)S_2^{-1}(\alpha)N^T(\alpha)\Big]\Big\}\zeta(k) \\ &-\beta^{d_2+1}\left[\sum_{l=k-d(k)}^{k-1}\Gamma_1^TS_1^{-1}(\alpha)\Gamma_1+\sum_{l=k-d_2}^{k-d(k)-1}\Gamma_2^TS_1^{-1}(\alpha)\Gamma_2+\sum_{l=k-d_2}^{k-1}\Gamma_3^TS_2^{-1}(\alpha)\Gamma_3\right], \end{aligned}$$

where $\Phi(\alpha)$ is defined in (5.22) and

$$\begin{cases} \Gamma_1 \triangleq S_1(\alpha)\xi(l)+L^T(\alpha)\zeta(k), \quad \Gamma_2 \triangleq S_1(\alpha)\xi(l)+M^T(\alpha)\zeta(k), \\ \Gamma_3 \triangleq S_2(\alpha)\xi(l)+N^T(\alpha)\zeta(k). \end{cases}$$

Moreover, it can be seen from (5.22) that

$$\begin{aligned} \Phi(\alpha)+\beta^{d_2+1}\Big[&d_2L(\alpha)S_1^{-1}(\alpha)L^T(\alpha)+\left(d_2-d_1\right)M(\alpha)S_1^{-1}(\alpha)M^T(\alpha) \\ &+d_2N(\alpha)S_2^{-1}(\alpha)N^T(\alpha)\Big]<0. \end{aligned}$$

Then we have

$$\Delta V(x(k),\alpha(k))+(1-\beta)V(x(k),\alpha(k))<0, \quad \forall k\in[k_l,k_{l+1}). \tag{5.31}$$

Now, for an arbitrary piecewise constant switching signal α_k, and for any $k>0$, we let $k_0<k_1<\cdots<k_l<\cdots$, $l=1,\ldots,$ denote the switching points of α_k over the interval $(0,k)$.

As mentioned earlier, the i_lth subsystem is activated when $k \in [k_l, k_{l+1})$. Therefore, for $k \in [k_l, k_{l+1})$, it holds from (5.31) that

$$V(x(k), \alpha(k)) < \beta^{k-k_l} V(x(k_l), \alpha(k_l)). \tag{5.32}$$

Using (5.23) and (5.24), at switching instant t_k, we have

$$V(x(k_l), \alpha(k_l)) \leq \mu V(x(k_l), \alpha(k_{l-1})). \tag{5.33}$$

Therefore, it follows from (5.32)–(5.33) and the relationship $\vartheta = N_\alpha(0, k) \leq (k - k_0)/T_a$ that

$$\begin{aligned} V(x(k), \alpha(k)) &\leq \beta^{k-k_l} \mu V(x(k_l), \alpha(k_{l-1})) \\ &\leq \cdots \\ &\leq \beta^{(k-k_0)} \mu^{\vartheta} V(x(k_0), \alpha(k_0)) \\ &\leq (\beta \mu^{1/T_a})^{(k-k_0)} V(x(k_0), \alpha(k_0)). \end{aligned} \tag{5.34}$$

Note from (5.24) that there exist two positive constants a and b $(a \leq b)$ such that

$$V(x(k), \alpha(k)) \geq a \|x(k)\|^2, \quad V(x(k_0), \alpha(k_0)) \leq b \|x(k_0)\|_{\mathrm{C}}^2. \tag{5.35}$$

Combining (5.34) and (5.35) yields

$$\begin{aligned} \|x(k)\|^2 &\leq \frac{1}{a} V(x(k), \alpha(k)) \\ &\leq \frac{b}{a} \left(\beta \mu^{1/T_a}\right)^{(k-k_0)} \|x(k_0)\|_{\mathrm{C}}^2. \end{aligned}$$

Furthermore, letting $\rho \triangleq \sqrt{\beta \mu^{1/T_a}}$, it follows that

$$\|x(k)\| \leq \sqrt{\frac{b}{a}} \rho^{(k-k_0)} \|x(k_0)\|_{\mathrm{C}}.$$

By Definition 5.3.2, we know that if $0 < \rho < 1$, that is, $T_a > T_a^* = \text{ceil}\left(-\frac{\ln \mu}{\ln \beta}\right)$, the discrete-time switched time-delay system in (5.19a)–(5.19b) with $u(k) = 0$ is exponentially stable, where function ceil(h) represents rounding real number h to the nearest integer greater than or equal to h. The proof is completed. ∎

Remark 5.10 *In Theorem 5.3.3, the parameter β plays a key role in controlling the lower bound of the average dwell time, which can be seen from $T_a > T_a^* = \text{ceil}\left(-\frac{\ln \mu}{\ln \beta}\right)$. Specifically, if β is given a smaller value, the lower bound of the average dwell time becomes smaller with a fixed μ, which may result in the instability of the system.* ♦

Remark 5.11 *Note that when* $\mu = 1$ *in* $T_a > T_a^* = \text{ceil}\left(-\frac{\ln\mu}{\ln\beta}\right)$ *we have* $T_a > T_a^* = 0$*, which means that the switching signal* $\alpha(k)$ *can be arbitrary. In this case, (5.23) turns out to be* $P(i) = P(j) = P$, $Q(i) = Q(j) = P$, $R(i) = R(j) = P$, $S_1(i) = S_1(j) = S_1$, $S_2(i) = S_2(j) = S_2$, $\forall i, j \in \mathcal{N}$, *and the proposed approach becomes a quadratic one thus conservative. In this case, the system in (5.19a)–(5.19b) with* $u(k) = 0$ *turns out to be*

$$x(k+1) = Ax(k) + A_d x(k - d(k)), \tag{5.36a}$$

$$x(k) = \phi(k), \quad k = -d_2, -d_2 + 1, \dots, 0, \tag{5.36b}$$

and we have the following result for the system in (5.36a)–(5.36b). ♦

Corollary 5.3.4 *The discrete-time time-delay system in (5.36a)–(5.36b) is asymptotically stable if there exist matrices* $P > 0$, $Q > 0$, $R > 0$, $S_1 > 0$, *and* $S_2 > 0$, *and matrices L, M, and N such that*

$$\begin{bmatrix} \Psi & d_2 L & (d_2 - d_1) M & d_2 N \\ \star & -d_2 S_1 & 0 & 0 \\ \star & \star & -(d_2 - d_1) S_1 & 0 \\ \star & \star & \star & -d_2 S_2 \end{bmatrix} < 0,$$

where

$$\begin{cases} \Psi \triangleq \begin{bmatrix} \Psi_{11} & 0 & 0 \\ \star & -Q & 0 \\ \star & \star & -R \end{bmatrix} + \begin{bmatrix} A^T \\ A_d^T \\ 0 \end{bmatrix} P \begin{bmatrix} A^T \\ A_d^T \\ 0 \end{bmatrix}^T \\ \quad + \begin{bmatrix} A^T - I \\ A_d^T \\ 0 \end{bmatrix} d_2 (S_1 + S_2) \begin{bmatrix} A^T - I \\ A_d^T \\ 0 \end{bmatrix}^T \\ \quad + 2\left\{ L \begin{bmatrix} I \\ -I \\ 0 \end{bmatrix}^T + M \begin{bmatrix} 0 \\ I \\ -I \end{bmatrix}^T + N \begin{bmatrix} I \\ 0 \\ -I \end{bmatrix}^T \right\}, \\ \Psi_{11} \triangleq -P + R + (d_2 - d_1 + 1) Q. \end{cases}$$

Proof. To prove the above result, we choose the following Lyapunov function:

$$\begin{aligned} W(x) \triangleq\; & x^T(k) P x(k) + \sum_{l=k-d(k)}^{k-1} x^T(l) Q x(l) + \sum_{l=k-d_2}^{k-1} x^T(l) R x(l) \\ & + \sum_{s=-d_2+1}^{-d_1} \sum_{l=k+s}^{k-1} x^T(l) Q x(l) + \sum_{s=-d_2}^{-1} \sum_{l=k+s}^{k-1} \xi^T(l) (S_1 + S_2) \xi(l), \end{aligned}$$

where $\xi(k) \triangleq x(k+1) - x(k)$, and $P > 0$, $Q > 0$, $R > 0$, $S_1 > 0$, and $S_2 > 0$ are real matrices to be determined. The remaining processes can be followed along the same lines as for the proof of Theorem 5.3.3, and we omit the details. ■

Notice that there exist two product terms between the Lyapunov matrices (i.e. $P(i)$ and $S_1(i) + S_2(i)$) and the system matrices $A(i)$ in the condition of Theorem 5.3.3, which will cause some problems with the solution of the stabilization control synthesis problem. In the following, a subsequent result is given in order to facilitate the control design procedure.

Corollary 5.3.5 *Given a constant* $0 < \beta < 1$, *suppose that there exist matrices* $P(i) > 0$ *and* $Q(i) > 0$ *such that for* $i \in \mathcal{N}$,

$$\begin{bmatrix} \Phi_{11}(i) + A^T(i)P(i)A(i) & A^T(i)P(i)A_d(i) \\ \star & -\beta^{d_2+1}Q(i) + A_d^T(i)P(i)A_d(i) \end{bmatrix} < 0,$$

where

$$\Phi_{11}(i) \triangleq -\beta P(i) + \beta(d_2 - d_1 + 1)Q(i).$$

Then the discrete-time switched time-delay system in (5.19a)–(5.19b) with $u(k) = 0$ *is exponentially stable for any switching signal with average dwell time satisfying* $T_a > T_a^* = \text{ceil}\left(-\frac{\ln \mu}{\ln \beta}\right)$, *where* $\mu \geq 1$ *satisfies*

$$P(i) \leq \mu P(j), \quad Q(i) \leq \mu Q(j), \quad \forall i, j \in \mathcal{N}.$$

Now, based on the above corollary, we consider the stabilization problem for system (5.19a)–(5.19b).

Theorem 5.3.6 *Given a constant* $0 < \beta < 1$, *the system in (5.19a)–(5.19b) is stabilizable, that is, the closed-loop system in (5.21a)–(5.21b) is exponentially stable under the control input* $u(k)$ *in (5.20), if there exist matrices* $\mathcal{P}(i) > 0$, $\mathcal{Q}(i) > 0$, *and* $\mathcal{Y}(i)$ *such that for* $i \in \mathcal{N}$,

$$\begin{bmatrix} \tilde{\Phi}_{11}(i) & 0 & \mathcal{P}(i)A^T(i) + \mathcal{Y}^T(i)B^T(i) \\ \star & -\beta^{d_2+1}\mathcal{Q}(i) & \mathcal{P}(i)A_d^T(i) \\ \star & \star & -\mathcal{P}(i) \end{bmatrix} < 0, \tag{5.37}$$

where

$$\tilde{\Phi}_{11}(i) \triangleq -\beta\mathcal{P}(i) + \beta(d_2 - d_1 + 1)\mathcal{Q}(i).$$

Then the discrete-time switched time-delay system in (5.19a)–(5.19b) is exponentially stabilizable for any switching signal with average dwell time satisfying $T_a > T_a^* = \text{ceil}\left(-\frac{\ln \mu}{\ln \beta}\right)$, *where* $\mu \geq 1$ *satisfies*

$$\mathcal{P}(i) \leq \mu\mathcal{P}(j), \quad \mathcal{Q}(i) \leq \mu\mathcal{Q}(j), \quad \forall i, j \in \mathcal{N}. \tag{5.38}$$

In this case, a stabilizing state feedback controller can be chosen by

$$u(k) = K(i)x(k) = \mathcal{Y}(i)\mathcal{P}^{-1}(i)x(k). \tag{5.39}$$

Proof. By Schur complement, it can be seen from Corollary 5.3.5 that the closed-loop system in (5.21a)–(5.21b) is exponentially stable if there matrices $P(i) > 0$ and $Q(i) > 0$ such that for $i \in \mathcal{N}$,

$$\begin{bmatrix} \Phi_{11}(i) & 0 & \tilde{A}^T(i)P(i) \\ \star & -\beta^{d_2+1}Q(i) & A_d^T(i)P(i) \\ \star & \star & -P(i) \end{bmatrix} < 0, \tag{5.40}$$

where the switching signal has an average dwell time satisfying $T_a > T_a^* = \text{ceil}\left(-\frac{\ln\mu}{\ln\beta}\right)$, where $\mu \geq 1$ satisfies

$$P(i) \leq \mu P(j), \quad Q(i) \leq \mu Q(j), \quad \forall i, j \in \mathcal{N}.$$

Performing a congruence transformation on (5.40) by diag $\{\mathcal{P}(i), \mathcal{P}(i), \mathcal{P}(i)\}$ (where $\mathcal{P}(i) = P^{-1}(i)$) and letting $\mathcal{Q}(i) \triangleq \mathcal{P}(i)Q(i)\mathcal{P}(i)$, we have

$$\begin{bmatrix} \tilde{\Phi}_{11}(i) & 0 & \mathcal{P}(i)\tilde{A}^T(i) \\ \star & -\beta^{d_2+1}\mathcal{Q}(i) & \mathcal{P}(i)A_d^T(i) \\ \star & \star & -\mathcal{P}(i) \end{bmatrix} < 0,$$

where $\tilde{\Phi}_{11}(i)$ is defined in (5.37). Moreover, we define $\mathcal{Y}(i) = K(i)\mathcal{P}(i)$, we have (5.37), and we know that $K(i) = \mathcal{Y}(i)\mathcal{P}^{-1}(i)$. The proof is completed. ■

5.3.3 Illustrative Example

Example 5.3.7 (Stability analysis) Consider the system in (5.19a)–(5.19b) with $N = 2$, and its parameters are given as follows:

$$A(1) = \begin{bmatrix} 0.2 & 0.1 & -0.01 \\ 0.1 & 0.2 & -0.1 \\ 0.2 & -0.06 & -0.13 \end{bmatrix}, \quad A_d(1) = \begin{bmatrix} 0.06 & -0.2 & -0.15 \\ 0.04 & -0.01 & 0.36 \\ 0.2 & 0.1 & -0.07 \end{bmatrix},$$

$$A(2) = \begin{bmatrix} 0.3 & -0.1 & -0.3 \\ -0.04 & 0.2 & 0.2 \\ 0.1 & -0.05 & -0.2 \end{bmatrix}, \quad A_d(2) = \begin{bmatrix} -0.04 & 0.05 & -0.2 \\ -0.2 & 0.1 & -0.1 \\ 0.06 & -0.1 & -0.03 \end{bmatrix}.$$

and $d_1 = 1$, $d_2 = 2$, $\beta = 0.8$. We consider the average dwell time approach proposed in this chapter, and set $\mu = 1.5 > 1$, thus $T_a > T_a^* = \text{ceil}\left(-\frac{\ln\mu}{\ln\beta}\right) = 2$. Solving LMI (5.22) with

Table 5.3.7 Upper bound of d_2 (denoted by $\hat{d}_2$) for different β

β	0.5	0.6	0.7	0.8	0.9
$\hat{d}_2$	1.3344	1.7345	2.2247	2.9278	4.1299

(5.23), we can obtain a feasible solution of $\big(P(1), P(2), Q(1), Q(2), R(1), R(2), S_1(1), S_1(2), S_2(1), S_2(2), L(1), L(2), M(1), M(2), N(1), N(2)\big)$. Therefore, we can conclude that the above discrete-time switched system is exponentially stable.

In addition, for $d_1 = 1$, $\mu = 1.5$, and $\tau = 0.6$, considering different β, the upper bound of d_2 for different cases are listed in Table 5.3.7.

Example 5.3.8 (Stabilization problem) Consider the system in (5.19a)–(5.19b) with $N = 2$, and the system parameters are given as follows:

$$A(1) = \begin{bmatrix} -0.9 & 0.2 & -0.2 \\ 0.2 & -0.1 & 0.3 \\ -0.3 & 0.1 & 0.3 \end{bmatrix}, \quad A_d(1) = \begin{bmatrix} 0.2 & 0 & 0.1 \\ 0.1 & 0.3 & 0.1 \\ 0.3 & 0.1 & 0.2 \end{bmatrix}, \quad B(1) = \begin{bmatrix} 0 \\ 0 \\ 2 \end{bmatrix},$$

$$A(2) = \begin{bmatrix} -0.8 & -0.1 & -0.2 \\ 0.2 & -0.1 & 0.3 \\ 0.2 & -0.1 & 0.2 \end{bmatrix}, \quad A_d(2) = \begin{bmatrix} 0.2 & 0.1 & 0 \\ 0.1 & 0.2 & 0.1 \\ 0.1 & 0.1 & 0.3 \end{bmatrix}, \quad B(2) = \begin{bmatrix} 0 \\ 0 \\ 2 \end{bmatrix},$$

with $d(k) = 2.5 + (-1)^k/2$ (thus $d_1 = 1$, $d_2 = 3$), and suppose that $\mu = 1.5$, $\beta = 0.9$, and $x(\theta) = \begin{bmatrix} -0.3 & 1.0 & -0.8 \end{bmatrix}^T$, $\theta = -3, -2, -1, 0$.

The switching signal is given in Figure 5.1 (which is generated randomly; here, '1' and '2' represent the first and second subsystems, respectively). The state trajectories of the open-loop system are shown in Figure 5.2, from which we can see that the open-loop system is not stable. In this situation, we will design a state feedback stabilization controller such that the closed-loop system is stable. To this end, by solving the LMI conditions in Theorem 5.3.6, we obtain

$$K_1 = \begin{bmatrix} 0.1480 & -0.0307 & -0.1965 \end{bmatrix},$$
$$K_2 = \begin{bmatrix} -0.2096 & 0.0475 & -0.1501 \end{bmatrix}.$$

The state trajectories of the closed-loop system are shown in Figure 5.3.

5.4 Conclusion

The stability analysis and stabilization problems have been investigated for continuous- and discrete-time switched hybrid systems with time-varying delay. By using the average dwell

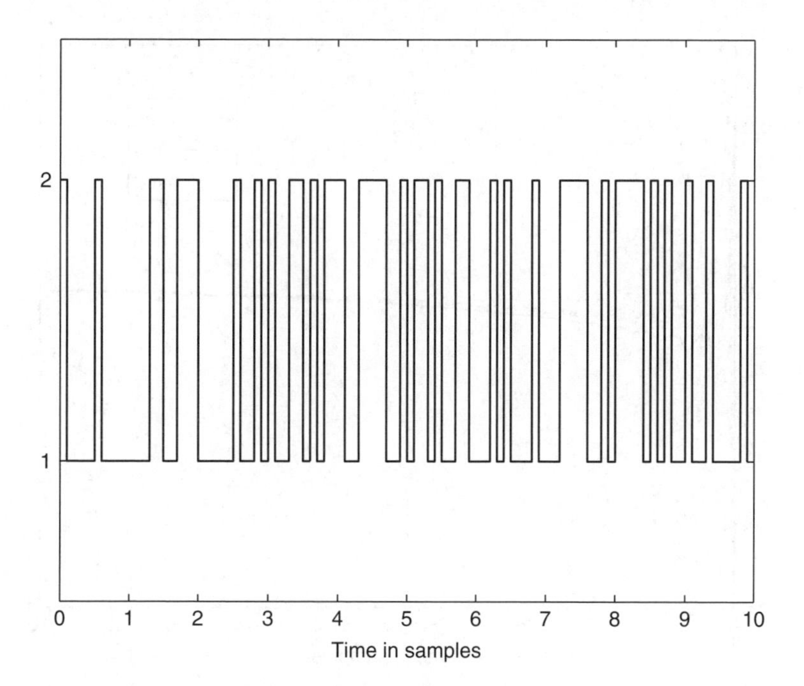

Figure 5.1 Switching signal

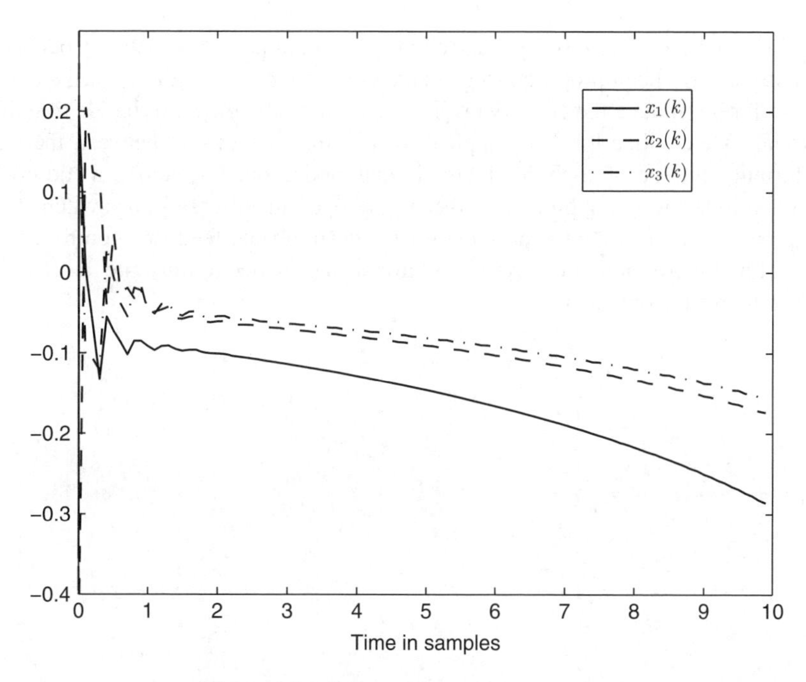

Figure 5.2 States of the open-loop system

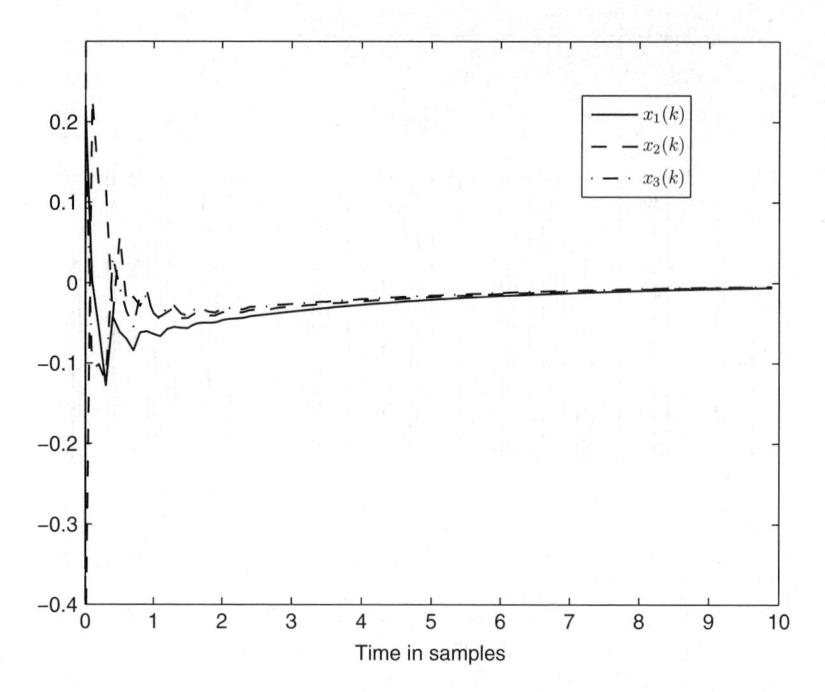

Figure 5.3 States of the closed-loop system

time approach and the piecewise Lyapunov function technique, some delay-dependent sufficient conditions have been proposed to guarantee the exponential stability of the considered systems. To further reduce the conservativeness caused by the time-varying delays, the slack matrix variables technique has been applied to seek the relationship between the Newton–Leibniz formula, instead of applying the traditional model transformation. In addition, a stabilization controller design approach has been developed for discrete-time switched delayed hybrid systems, and an explicit expression for the desired state feedback controller has also been given. Finally, two numerical examples have been provided to illustrate the effectiveness of the theoretic results obtained.

6

Optimal DOF Control of Switched State-Delayed Hybrid Systems

6.1 Introduction

In practice, there are always some system states that cannot be measured, so the unmeasurable state components can not be used for designing feedback control. Output feedback control is an effective control strategy to deal with systems with inaccessible state components. However, limited results have been reported on output feedback control of switched hybrid systems. In this chapter, the DOF control problem is studied for continuous-time switched hybrid systems with time-varying delays. This chapter is divided into two parts. First, we will investigate the optimal $\mathcal{L}_2$-$\mathcal{L}_\infty$ DOF control for switched hybrid systems with time-varying delays, and the weighted $\mathcal{L}_2$-$\mathcal{L}_\infty$ performance is first defined. By using the average dwell time approach and the piecewise Lyapunov function technique, a delay-dependent sufficient condition is proposed to assure the closed-loop system to be exponentially stable with a weighted $\mathcal{L}_2$-$\mathcal{L}_\infty$ performance (i.e. the existence condition for the desired $\mathcal{L}_2$-$\mathcal{L}_\infty$ DOF controller). Meanwhile, a decay estimate is explicitly given for quantifying the convergence rate of the dynamics of the closed-loop system. Since the proposed existence condition contains some product terms between the Lyapunov matrices and the system matrices – and more importantly, all these matrices are mode-dependent – it is difficult to use them directly to establish the DOF controller solvability condition by the linearizing variable transforms approach. A decoupling technique is used by introducing a slack mode-independent matrix variable, and a new existence condition is obtained, by which the corresponding solvability condition for the desired $\mathcal{L}_2$-$\mathcal{L}_\infty$ DOF controller design is then established. In the second part, we will study the guaranteed cost DOF controller design for the continuous-time switched hybrid system with both discrete and neutral delays. A sufficient delay-dependent condition is first proposed by means of LMIs, which guarantees the closed-loop system exponentially stable with a certain bound for the prescribed cost function. The corresponding solvability conditions for the desired guaranteed cost DOF controller is then established. In both parts, since the obtained DOF solvability conditions are not all expressed in terms of strict LMI, the CCL method is exploited to cast

Sliding Mode Control of Uncertain Parameter-Switching Hybrid Systems, First Edition. Ligang Wu, Peng Shi and Xiaojie Su.

them into sequential minimization problems subject to LMI constraints, which can be readily solved by using standard numerical software.

6.2 Optimal $\mathcal{L}_2$-$\mathcal{L}_\infty$ DOF Controller Design

6.2.1 System Description and Preliminaries

Consider a class of switched state-delayed hybrid systems of the form:

$$\dot{x}(t) = A(\alpha(t))x(t) + A_d(\alpha(t))x(t-d(t)) + B(\alpha(t))u(t) + B_1(\alpha(t))\omega(t), \tag{6.1a}$$

$$y(t) = C(\alpha(t))x(t) + C_d(\alpha(t))x(t-d(t)) + D_1(\alpha(t))\omega(t), \tag{6.1b}$$

$$z(t) = E(\alpha(t))x(t) + E_d(\alpha(t))x(t-d(t)), \tag{6.1c}$$

$$x(t) = \phi(t), \quad t \in [-d, 0], \tag{6.1d}$$

where $x(t) \in \mathbf{R}^n$ is the system state vector; $u(t) \in \mathbf{R}^m$ is the control input; $y(t) \in \mathbf{R}^p$ is the measured output; $z(t) \in \mathbf{R}^q$ is the controlled output; $\omega(t) \in \mathbf{R}^l$ is the disturbance input which belongs to $\mathcal{L}_2[0,\infty)$. $\{(A(\alpha), A_d(\alpha), B(\alpha), B_1(\alpha), C(\alpha), C_d(\alpha), D_1(\alpha), E(\alpha), E_d(\alpha)) : \alpha(t) \in \mathcal{N}\}$ is a family of matrices parameterized by an index set $\mathcal{N} = \{1, 2, \dots, N\}$, and $\alpha(t) : \mathbf{R} \to \mathcal{N}$ (denoted by α for simplicity) is the switching signal defined as the same in Chapter 5. $d(t)$ denotes the time-varying delays satisfying $0 \le d(t) \le d$ and $\dot{d}(t) \le \tau$ for known constants d and τ and $\phi(t)$ is a differentiable vector-valued initial function on $[-d, 0]$. For each possible value $\alpha \in \mathcal{N}$, we will denote the system matrices associated with mode i by $A(i) = A(\alpha)$, $A_d(i) = A_d(\alpha)$, $B(i) = B(\alpha)$, $B_1(i) = B_1(\alpha)$, $C(i) = C(\alpha)$, $C_d(i) = C_d(\alpha)$, $D_1(i) = D_1(\alpha)$, $E(i) = E(\alpha)$, and $E_d(i) = E_d(\alpha)$.

Assuming that some state components of system (6.1a) are not available, we now seek to design a DOF controller of general structure described by

$$\dot{x}_c(t) = A_c(\alpha)x_c(t) + B_c(\alpha)y(t), \tag{6.2a}$$

$$u(t) = C_c(\alpha)x_c(t) + D_c(\alpha)y(t), \tag{6.2b}$$

where $x_c(t) \in \mathbf{R}^n$ is the DOF controller state vector; $A_c(\alpha)$, $B_c(\alpha)$, $C_c(\alpha)$, and $D_c(\alpha)$ are appropriately dimensioned matrices to be determined.

Augmenting the model of (6.1a)–(6.1d) to include the states of the DOF controller dynamics in (6.2a)–(6.2b), we obtain the following closed-loop system:

$$\dot{\xi}(t) = \bar{A}(\alpha)\xi(t) + \bar{A}_d(\alpha)K\xi(t-d(t)) + \bar{B}(\alpha)\omega(t), \tag{6.3a}$$

$$z(t) = \bar{C}(\alpha)\xi(t) + \bar{C}_d(\alpha)K\xi(t-d(t)), \tag{6.3b}$$

$$\xi(t) = \varphi(t), \quad t \in [-d, 0], \tag{6.3c}$$

where $\xi(t) \triangleq \begin{bmatrix} x(t) \\ x_c(t) \end{bmatrix}$, $K \triangleq \begin{bmatrix} I & 0 \end{bmatrix}$ and

$$\begin{cases} \bar{A}(\alpha) \triangleq \begin{bmatrix} A(\alpha) + B(\alpha)D_c(\alpha)C(\alpha) & B(\alpha)C_c(\alpha) \\ B_c(\alpha)C(\alpha) & A_c(\alpha) \end{bmatrix}, \\ \bar{A}_d(\alpha) \triangleq \begin{bmatrix} A_d(\alpha) + B(\alpha)D_c(\alpha)C_d(\alpha) \\ B_c(\alpha)C_d(\alpha) \end{bmatrix}, \\ \bar{B}(\alpha) \triangleq \begin{bmatrix} B_1(\alpha) + B(\alpha)D_c(\alpha)D_1(\alpha) \\ B_c(\alpha)D_1(\alpha) \end{bmatrix}, \\ \bar{C}(\alpha) \triangleq \begin{bmatrix} E(\alpha) & 0 \end{bmatrix}, \quad \bar{C}_d(\alpha) \triangleq E_d(\alpha). \end{cases} \tag{6.4}$$

Before proceeding, we give the following definitions.

Definition 6.2.1 *The closed-loop system in (6.3a)–(6.3c) with $\omega(t) = 0$ is said to be exponentially stable under α if its solution $\xi(t)$ satisfies*

$$\|\xi(t)\| \leq \eta \, \|\xi(t_0)\|_C \, e^{-\lambda(t-t_0)}, \quad \forall t \geq t_0,$$

where $\eta \geq 1$ and $\lambda > 0$ are two real constants, and

$$\|\xi(t_0)\|_C \triangleq \sup_{-d \leq \theta \leq 0} \left\{ \|\xi(t_0 + \theta)\|, \|\dot{\xi}(t_0 + \theta)\| \right\}.$$

Definition 6.2.2 *For $\beta > 0$ and $\gamma > 0$, the closed-loop system in (6.3a)–(6.3c) is said to be exponentially stable with a weighted $\mathcal{L}_2$-$\mathcal{L}_\infty$ performance γ, if under α it is exponentially stable with $\omega(t) = 0$, and under zero initial condition, that is, $\varphi(t) = 0$, $t \in [-d, 0]$, for any nonzero $\omega(t) \in \mathcal{L}_2[0, \infty)$, it holds that*

$$\sup_{\forall t} \left\{ e^{-\beta t} z^T(t) z(t) \right\} < \gamma^2 \int_0^\infty \omega^T(t)\omega(t)dt. \tag{6.5}$$

Therefore, the $\mathcal{L}_2$-$\mathcal{L}_\infty$ DOF control problem can be formulated as follows: for switched state-delayed hybrid systems (6.1a)–(6.1d) and a prescribed performance level $\gamma > 0$, determine DOF controllers in the form of (6.2a)–(6.2b) such that the resulting closed-loop system in (6.3a)–(6.3c) is exponentially stable with a weighted $\mathcal{L}_2$-$\mathcal{L}_\infty$ performance level γ.

6.2.2 Main Results

First, we will investigate the exponential stability and the weighted $\mathcal{L}_2$-$\mathcal{L}_\infty$ performance for the closed-loop system (6.3a)–(6.3c), and give the following result.

Theorem 6.2.3 *Given constants $\beta > 0$ and $\gamma > 0$, suppose that there exist matrices $P(i) > 0$, $Q(i) > 0$, $R(i) > 0$, $X(i)$, $Y(i)$, and $Z(i)$ such that for $i \in \mathcal{N}$,*

$$\begin{bmatrix} \Pi_1(i) + \Pi_2(i) + \Pi_2^T(i) & d\Pi_3^T(i)K^T R(i) & d\Pi_4^T(i) \\ \star & -dR(i) & 0 \\ \star & \star & -de^{-\beta d}R(i) \end{bmatrix} < 0, \tag{6.6a}$$

$$\begin{bmatrix} -P(i) & 0 & \bar{C}^T(i) \\ \star & -P(i) & K^T\bar{C}_d^T(i) \\ \star & \star & -\frac{1}{2}\gamma^2 I \end{bmatrix} < 0, \tag{6.6b}$$

where $\tilde{d} \triangleq -(1-\tau)e^{-\beta d}$ and

$$\begin{cases} \Pi_1(i) \triangleq \begin{bmatrix} P(i)\bar{A}(i) + \bar{A}^T(i)P(i) + K^T Q(i)K + \beta P(i) & P(i)\bar{A}_d(i) & P(i)\bar{B}(i) \\ \star & -\tilde{d}Q(i) & 0 \\ \star & \star & -I \end{bmatrix}, \\ \Pi_2(i) \triangleq \begin{bmatrix} \Pi_4^T(i)K & -\Pi_4^T(i) & 0 \end{bmatrix}, \\ \Pi_3(i) \triangleq \begin{bmatrix} \bar{A}(i) & \bar{A}_d(i) & \bar{B}(i) \end{bmatrix}, \\ \Pi_4(i) \triangleq \begin{bmatrix} X^T(i) & Y^T(i) & Z^T(i) \end{bmatrix}. \end{cases}$$

Then, the closed-loop system in (6.3a)–(6.3c) is exponentially stable with a weighted $\mathcal{L}_2$-$\mathcal{L}_\infty$ performance level γ for any switching signal with average dwell time satisfying $T_a > T_a^* = \frac{\ln \mu}{\beta}$, where $\mu \geq 1$ satisfies

$$P(i) \leq \mu P(j), \quad Q(i) \leq \mu Q(j), \quad R(i) \leq \mu R(j), \quad \forall i, j \in \mathcal{N}. \tag{6.7}$$

Moreover, an estimate of the state decay is given by

$$\|\xi(t)\| \leq \eta \left\|\xi_0\right\|_C e^{-\lambda t}, \tag{6.8}$$

where

$$\begin{cases} \lambda \triangleq \frac{1}{2}\left(\beta - \frac{\ln \mu}{T_a}\right) > 0, \quad \eta \triangleq \sqrt{\frac{b}{a}} \geq 1, \\ a \triangleq \min_{\forall i \in \mathcal{N}} \lambda_{\min}(P(i)), \\ b \triangleq \max_{\forall i \in \mathcal{N}} \lambda_{\max}(P(i)) + d \max_{\forall i \in \mathcal{N}} \lambda_{\max}(Q(i)) + \frac{d^2}{2} \max_{\forall i \in \mathcal{N}} \lambda_{\max}(R(i)). \end{cases} \tag{6.9}$$

Proof. Choose a Lyapunov function of the following form:

$$
\begin{aligned}
V(\xi_t, \alpha) \triangleq \xi^T(t)P(\alpha)\xi(t) + \int_{t-d(t)}^{t} e^{\beta(s-t)}\xi^T(s)K^T Q(\alpha)K\xi(s)ds \\
+ \int_{-d}^{0}\int_{t+\theta}^{t} e^{\beta(s-t)}\dot{\xi}^T(s)K^T R(\alpha)K\dot{\xi}(s)dsd\theta,
\end{aligned}
\tag{6.10}
$$

where $P(\alpha) > 0$, $Q(\alpha) > 0$, and $R(\alpha) > 0$, $\alpha \in \mathcal{N}$ are to be determined. Then, as with the solution of system (6.3a)–(6.3c) for a fixed α, it follows that

$$
\begin{aligned}
\dot{V}(\xi_t, \alpha) \leq\ & 2\xi^T(t)P(\alpha)[\bar{A}(\alpha)\xi(t) + \bar{A}_d(\alpha)K\xi(t-d(t)) + \bar{B}(\alpha)\omega(t)] \\
& + \xi^T(t)K^T Q(\alpha)K\xi(t) + d\dot{\xi}^T(t)K^T R(\alpha)K\dot{\xi}(t) \\
& - (1-\tau)e^{-\beta d}\xi^T(t-d(t))K^T Q(\alpha)K\xi(t-d(t)) \\
& - \int_{t-d(t)}^{t} \beta e^{\beta(s-t)}\xi^T(s)K^T Q(\alpha)K\xi(s)ds \\
& - \int_{t-d(t)}^{t} e^{-\beta d}\dot{\xi}^T(s)K^T R(\alpha)K\dot{\xi}(s)ds \\
& - \int_{-d}^{0}\int_{t+\theta}^{t} \beta e^{\beta(s-t)}\dot{\xi}^T(s)K^T R(\alpha)K\dot{\xi}(s)dsd\theta.
\end{aligned}
\tag{6.11}
$$

On the other hand, Newton–Leibniz formula gives

$$
\xi(t) - \xi(t-d(t)) = \int_{t-d(t)}^{t} \dot{\xi}(s)ds.
$$

Then for any appropriately dimensioned matrices $W(\alpha) \triangleq \begin{bmatrix} X(\alpha) \\ Y(\alpha) \end{bmatrix}$, we have

$$
2\psi^T(t)W(\alpha)K\left[\xi(t) - \xi(t-d(t)) \quad \int_{t-d(t)}^{t} \dot{\xi}(s)ds\right] - 0,
\tag{6.12}
$$

where $\psi(t) \triangleq \begin{bmatrix} \xi(t) \\ K\xi(t-d(t)) \end{bmatrix}$.

First, we will show the stability of the closed-loop system (6.3a)–(6.3c) with $\omega(t) = 0$. By (6.11)–(6.12), we have

$$
\begin{aligned}
\dot{V}(\xi_t, \alpha) + \beta V(\xi_t, \alpha) \leq\ & \psi^T(t)\left[\Sigma(\alpha) + de^{\beta d}W(\alpha)R^{-1}(\alpha)W^T(\alpha)\right]\psi(t) \\
& - e^{\beta d}\int_{t-d(t)}^{t} \left[W^T(\alpha)\psi(t) + e^{-\beta d}R(\alpha)K\dot{\xi}(s)\right]^T R^{-1}(\alpha) \\
& \times \left[W^T(\alpha)\psi(t) + e^{-\beta d}R(\alpha)K\dot{\xi}(s)\right] ds,
\end{aligned}
\tag{6.13}
$$

where

$$\begin{cases} \Sigma(\alpha) \triangleq \Sigma_1(\alpha) + \Sigma_2(\alpha) + \Sigma_2^T(\alpha) + d\Sigma_3^T(\alpha)K^T R(\alpha) K \Sigma_3(\alpha), \\ \Sigma_1(\alpha) \triangleq \begin{bmatrix} P(\alpha)\bar{A}(\alpha) + \bar{A}^T(\alpha)P(\alpha) + K^T Q(\alpha) K + \beta P(\alpha) & P(\alpha)\bar{A}_d(\alpha) \\ \star & -\tilde{d}Q(\alpha) \end{bmatrix}, \\ \Sigma_2(\alpha) \triangleq \begin{bmatrix} W(\alpha)K & -W(\alpha) \end{bmatrix}, \quad \Sigma_3(\alpha) \triangleq \begin{bmatrix} \bar{A}(\alpha) & \bar{A}_d(\alpha) \end{bmatrix}. \end{cases}$$

By Schur complement, LMI (6.6) implies

$$\Sigma(\alpha) + de^{\beta d} W(\alpha) R^{-1}(\alpha) W^T(\alpha) < 0, \tag{6.14}$$

and noting

$$\begin{aligned} &\int_{t-d(t)}^{t} \left[W^T(\alpha)\psi(t) + e^{-\beta d} R(\alpha) K \dot{\xi}(s)\right]^T \\ &\times R^{-1}(\alpha) \left[W^T(\alpha)\psi(t) + e^{-\beta d} R(\alpha) K \dot{\xi}(s)\right] ds \geq 0. \end{aligned} \tag{6.15}$$

Thus considering (6.13)–(6.15), we have

$$\dot{V}(\xi_t, \alpha) + \beta V(\xi_t, \alpha) \leq 0. \tag{6.16}$$

For an arbitrary piecewise constant switching signal α, and for any $t > 0$, we let $0 = t_0 < t_1 < \cdots < t_k < \cdots$, $k = 1, 2, \ldots$, denote the switching points of α over the interval $(0, t)$. As mentioned earlier, the i_kth subsystem is activated when $t \in [t_k, t_{k+1})$. Integrating (6.16) from t_k to t gives

$$V(\xi_t, \alpha) \leq e^{-\beta(t-t_k)} V(\xi_{t_k}, \alpha(t_k)). \tag{6.17}$$

Using (6.7) and (6.10), at switching instant t_k, we have

$$V(\xi_{t_k}, \alpha(t_k)) \leq \mu V\left(\xi_{t_k^-}, \alpha\left(t_k^-\right)\right), \tag{6.18}$$

where t_k^- denotes the left limitation of t_k. Therefore, it follows from (6.17)–(6.18), and noting $\vartheta = N_\alpha(0, t) \leq (t - 0)/T_a$, that

$$\begin{aligned} V(\xi_t, \alpha) &\leq e^{-\beta(t-t_k)} \mu V\left(\xi_{t_k^-}, \alpha\left(t_k^-\right)\right) \\ &\leq \cdots \\ &\leq e^{-\beta(t-0)} \mu^{\vartheta} V(\xi_0, \alpha(0)) \\ &\leq e^{-(\beta - \ln\mu/T_a)t} V(\xi_0, \alpha(0)). \end{aligned} \tag{6.19}$$

Notice from (6.10) that $V(\xi_t, \alpha) \geq a \|\xi(t)\|^2$ and $V(\xi_0, \alpha(0)) \leq b \|\xi(0)\|_C^2$, where a and b are defined in (6.9). Considering (6.19) yields

$$\|\xi(t)\|^2 \leq \frac{1}{a} V(\xi_t, \alpha) \leq \frac{b}{a} e^{-(\beta - \ln \mu / T_a)t} \|\xi(0)\|_C^2,$$

which implies (6.8). By Definition 6.2.1 with $t_0 = 0$, we know that the closed-loop system (6.3a)–(6.3c) with $\omega(t) = 0$ is exponentially stable.

Now, we will establish the weight $\mathcal{L}_2$-$\mathcal{L}_\infty$ performance for the closed-loop system in (6.3a)–(6.3c). For any appropriately dimensioned matrix $\bar{W}(\alpha)$, we have

$$2\bar{\psi}^T(t)\bar{W}(\alpha)K\left[\xi(t) - \xi(t - d(t)) - \int_{t-d(t)}^{t} \dot{\xi}(s)ds\right] = 0, \tag{6.20}$$

where $\bar{\psi}(t) \triangleq \begin{bmatrix} \xi(t) \\ K\xi(t-d(t)) \\ \omega(t) \end{bmatrix}$ and $\bar{W}(\alpha) \triangleq \begin{bmatrix} X(\alpha) \\ Y(\alpha) \\ Z(\alpha) \end{bmatrix}$. Considering (6.11) and (6.20), we have

$$\begin{aligned} &\dot{V}(\xi_t, \alpha) + \beta V(\xi_t, \alpha) - \omega^T(t)\omega(t) \\ &\quad \leq \bar{\psi}^T(t)\left[\Pi(\alpha) + de^{\beta d}\bar{W}(\alpha)R^{-1}(\alpha)\bar{W}^T(\alpha)\right]\bar{\psi}(t) \\ &\quad - e^{\beta d}\int_{t-d(t)}^{t} \left[\bar{W}^T(\alpha)\bar{\psi}(t) + e^{-\beta d}R(\alpha)K\dot{\xi}(s)\right]^T \\ &\quad \times R^{-1}(\alpha)\left[\bar{W}^T(\alpha)\bar{\psi}(t) + e^{-\beta d}R(\alpha)K\dot{\xi}(s)\right] ds, \end{aligned} \tag{6.21}$$

where $\Pi(\alpha) \triangleq \Pi_1(\alpha) + \Pi_2(\alpha) + \Pi_2^T(\alpha) + d\Pi_3^T(\alpha)K^T R(\alpha)K\Pi_3(\alpha)$ with $\Pi_1(\alpha)$, $\Pi_2(\alpha)$ and $\Pi_3(\alpha)$ defined in (6.6). Note that

$$\begin{aligned} &\int_{t-d(t)}^{t} \left[\bar{W}^T(\alpha)\bar{\psi}(t) + e^{-\beta d}R(\alpha)K\dot{\xi}(s)\right]^T \\ &\times R^{-1}(\alpha)\left[\bar{W}^T(\alpha)\bar{\psi}(t) + e^{-\beta d}R(\alpha)K\dot{\xi}(s)\right] ds \geq 0. \end{aligned} \tag{6.22}$$

By Schur complement, LMI (6.6) implies

$$\Pi(\alpha) + de^{\beta d}\bar{W}(\alpha)R^{-1}(\alpha)\bar{W}^T(\alpha) < 0. \tag{6.23}$$

Thus, considering (6.21)–(6.23), we have

$$\dot{V}(\xi_t, \alpha) + \beta V(\xi_t, \alpha) - \omega^T(t)\omega(t) \leq 0. \tag{6.24}$$

Let $\Gamma(t) \triangleq -\omega^T(t)\omega(t)$, then (6.24) can be rewritten as

$$\dot{V}(\xi_t, \alpha) \leq -\beta V(\xi_t, \alpha) - \Gamma(t). \tag{6.25}$$

As in the proof of stability above, integrating (6.25) from t_k to t gives

$$V(\xi_t, \alpha) \leq e^{-\beta(t-t_k)} V(\xi_{t_k}, \alpha(t_k)) - \int_{t_k}^{t} e^{-\beta(t-s)}\Gamma(s)ds. \tag{6.26}$$

Therefore, it follows from (6.18) and (6.26) and the relationship $\vartheta = N_\alpha(0,t) \leq (t-0)/T_a$ that

$$\begin{aligned} V(\xi_t, \alpha) &\leq \mu e^{-\beta(t-t_k)} V\left(\xi_{t_k^-}, \alpha\left(t_k^-\right)\right) - \int_{t_k}^{t} e^{-\beta(t-s)}\Gamma(s)ds \\ &\leq \mu^{\vartheta} e^{-\beta t} V(\xi_0, \alpha(0)) - \mu^{\vartheta} \int_0^{t_1} e^{-\beta(t-s)}\Gamma(s)ds \\ &\quad -\mu^{\vartheta-1} \int_{t_1}^{t_2} e^{-\beta(t-s)}\Gamma(s)ds - \cdots - \mu^0 \int_{t_k}^{t} e^{-\beta(t-s)}\Gamma(s)ds \\ &= e^{-\beta t + N_\alpha(0,t)\ln\mu} V(\xi_0, \alpha(0)) - \int_0^t e^{-\beta(t-s)+N_\alpha(s,t)\ln\mu}\Gamma(s)ds. \end{aligned} \tag{6.27}$$

Under zero initial condition, (6.27) implies

$$V(\xi_t, \alpha) \leq \int_0^t e^{-\beta(t-s)+N_\alpha(s,t)\ln\mu}\omega^T(s)\omega(s)ds. \tag{6.28}$$

Multiplying both sides of (6.28) by $e^{-N_\alpha(0,t)\ln\mu}$ yields

$$\begin{aligned} e^{-N_\alpha(0,t)\ln\mu} V(\xi_t, \alpha) &\leq \int_0^t e^{-\beta(t-s)-N_\alpha(0,s)\ln\mu}\omega^T(s)\omega(s)ds \\ &\leq \int_0^t e^{-\beta(t-s)}\omega^T(s)\omega(s)ds \\ &\leq \int_0^t \omega^T(s)\omega(s)ds. \end{aligned} \tag{6.29}$$

Notice that $N_\alpha(0,t) \leq t/T_a$ and $T_a > T_a^* = \ln\mu/\beta$, we have $N_\alpha(0,t)\ln\mu \leq \beta t$. Thus, (6.29) implies

$$e^{-\beta t}V(\xi_t,\alpha) \leq \int_0^t \omega^T(s)\omega(s)ds. \tag{6.30}$$

Moreover, according to (6.10) and (6.30), we have

$$\begin{aligned} e^{-\beta t}\xi^T(t)P(\alpha)\xi(t) &\leq e^{-\beta t}V(\xi_t,\alpha) \\ &\leq \int_0^t \omega^T(s)\omega(s)ds \leq \int_0^\infty \omega^T(t)\omega(t)dt. \end{aligned}$$

Thus, for any time $t = t^\star \geq 0$, we have

$$e^{-\beta t^\star}\xi^T(t^\star)P(\alpha)\xi(t^\star) \leq \int_0^\infty \omega^T(t)\omega(t)dt. \tag{6.31}$$

Since $t^\star$ denotes any time, it is also true that

$$e^{-\beta t^\star}\xi^T(t^\star - d(t^\star))P(\alpha)\xi(t^\star - d(t^\star)) \leq \int_0^\infty \omega^T(t)\omega(t)dt. \tag{6.32}$$

From inequalities (6.31) and (6.32), we have

$$\begin{aligned} &e^{-\beta t^\star}\begin{bmatrix} \xi(t^\star) \\ \xi(t^\star - d(t^\star)) \end{bmatrix}^T \begin{bmatrix} P(\alpha) & 0 \\ 0 & P(\alpha) \end{bmatrix} \begin{bmatrix} \xi(t^\star) \\ \xi(t^\star - d(t^\star)) \end{bmatrix} \\ &\leq 2\int_0^\infty \omega^T(t)\omega(t)dt. \end{aligned} \tag{6.33}$$

Using Schur complement again, LMI (6.6) yields

$$\begin{bmatrix} P(\alpha) & 0 \\ 0 & P(\alpha) \end{bmatrix} > 2\gamma^{-2}\begin{bmatrix} \bar{C}^T(\alpha) \\ K^T\bar{C}_d^T(\alpha) \end{bmatrix} \begin{bmatrix} \bar{C}(\alpha) & \bar{C}_d(\alpha)K \end{bmatrix}. \tag{6.34}$$

Combining (6.33) with (6.34) gives

$$\begin{aligned} 2\int_0^\infty \omega^T(t)\omega(t)dt &> 2\gamma^{-2}e^{-\beta t^\star}\left[\bar{C}(\alpha)\xi(t^\star) + \bar{C}_d(\alpha)K\xi(t^\star - d(t^\star))\right]^T \\ &\quad \times \left[\bar{C}(\alpha)\xi(t^\star) + \bar{C}_d(\alpha)K\xi(t^\star - d(t^\star))\right] \\ &= 2\gamma^{-2}e^{-\beta t^\star}z^T(t^\star)z(t^\star), \end{aligned}$$

that is, for any $t^\star \geq 0$,

$$e^{-\beta t^\star} z^T(t^\star)z(t^\star) < \gamma^2 \int_0^\infty \omega^T(t)\omega(t)dt.$$

Taking the supremum over $t^\star \geq 0$ yields (6.5), thus the weight $\mathcal{L}_2$-$\mathcal{L}_\infty$ performance has been established. The proof is completed. ∎

Now, we are in a position to present a solution to the $\mathcal{L}_2$-$\mathcal{L}_\infty$ DOF control problem.

Theorem 6.2.4 *Consider the switched state-delayed hybrid systems in (6.1a)–(6.1d). For given constants $\beta > 0$ and $\gamma > 0$, suppose that there exist matrices $\mathcal{P}_1(i) > 0$, $\mathcal{P}_3(i) > 0$, $\mathcal{Q}_1(i) > 0$, $\mathcal{Q}_3(i) > 0$, $\mathcal{R}(i) > 0$, $R(i) > 0$, $\mathcal{P}_2(i)$, $\mathcal{Q}_2(i)$, $\mathcal{X}_1(i)$, $\mathcal{X}_2(i)$, $\mathcal{Y}(i)$, $\mathcal{Z}(i)$, $\mathcal{A}_c(i)$, $\mathcal{B}_c(i)$, $\mathcal{C}_c(i)$, $\mathcal{D}_c(i)$, $\mathcal{F}$, $\mathcal{G}$, and $\mathcal{K}$ such that for $i \in \mathcal{N}$,*

$$\begin{bmatrix} \Phi_{11}(i) & \Phi_{12}(i) & \Phi_{13}(i) & \Phi_{14}(i) & \Phi_{15}(i) & \Phi_{16}(i) & \Phi_{17}(i) & d\mathcal{X}_1(i) \\ \star & \Phi_{22}(i) & \Phi_{23}(i) & \Phi_{24}(i) & \Phi_{25}(i) & \Phi_{26}(i) & \Phi_{27}(i) & d\mathcal{X}_2(i) \\ \star & \star & \Phi_{33}(i) & \Phi_{34}(i) & \Phi_{15}(i) & \Phi_{16}(i) & 0 & 0 \\ \star & \star & \star & \Phi_{44}(i) & \Phi_{25}(i) & \Phi_{26}(i) & 0 & 0 \\ \star & \star & \star & \star & \Phi_{55}(i) & -\mathcal{Z}^T(i) & \Phi_{57}(i) & d\mathcal{Y}(i) \\ \star & \star & \star & \star & \star & -I & \Phi_{67}(i) & d\mathcal{Z}(i) \\ \star & \star & \star & \star & \star & \star & -d\mathcal{R}(i) & 0 \\ \star & \star & \star & \star & \star & \star & \star & -de^{-\beta d}R(i) \end{bmatrix} < 0, \quad (6.35a)$$

$$\begin{bmatrix} -\mathcal{P}_1(i) & -\mathcal{P}_2(i) & 0 & 0 & E^T(i) \\ \star & -\mathcal{P}_3(i) & 0 & 0 & \mathcal{G}^T E^T(i) \\ \star & \star & -\mathcal{P}_1(i) & -\mathcal{P}_2(i) & E_d^T(i) \\ \star & \star & \star & -\mathcal{P}_3(i) & \mathcal{G}^T E_d^T(i) \\ \star & \star & \star & \star & -\frac{1}{2}\gamma^2 I \end{bmatrix} < 0, \quad (6.35b)$$

$$\mathcal{P}(i) \triangleq \begin{bmatrix} \mathcal{P}_1(i) & \mathcal{P}_2(i) \\ \star & \mathcal{P}_3(i) \end{bmatrix} > 0, \quad (6.35c)$$

$$\mathcal{Q}(i) \triangleq \begin{bmatrix} \mathcal{Q}_1(i) & \mathcal{Q}_2(i) \\ \star & \mathcal{Q}_3(i) \end{bmatrix} > 0, \quad (6.35d)$$

$$R(i)\mathcal{R}(i) = I. \quad (6.35e)$$

Then there exists a DOF controller in the form of (6.2a)–(6.2b), such that the closed-loop system in (6.3a)–(6.3c) is exponentially stable with a weighted $\mathcal{L}_2$-$\mathcal{L}_\infty$ performance γ for any switching signal with average dwell time satisfying $T_a > T_a^* = \frac{\ln \mu}{\beta}$, where $\mu \geq 1$ satisfies

$$\mathcal{P}(i) \leq \mu\mathcal{P}(j), \quad \mathcal{Q}(i) \leq \mu\mathcal{Q}(j), \quad \mathcal{R}(i) \leq \mu\mathcal{R}(j), \quad \forall i,j \in \mathcal{N}. \quad (6.36)$$

Moreover, a desired $\mathcal{L}_2$-$\mathcal{L}_\infty$ DOF controller realization is given by

$$\begin{cases} \mathcal{A}_c(i) \triangleq \mathcal{F}^T A(i)\mathcal{G} + \mathcal{F}^T B(i)D_c(i)C(i)\mathcal{G} + F_4^T B_c(i)C(i)\mathcal{G} \\ \qquad + \mathcal{F}^T B(i)C_c(i)G_4 + F_4^T A_c(i)G_4, \\ \mathcal{B}_c(i) \triangleq \mathcal{F}^T B(i)D_c(i) + F_4^T B_c(i), \\ \mathcal{C}_c(i) \triangleq D_c(i)C(i)\mathcal{G} + C_c(i)G_4, \\ \mathcal{D}_c(i) \triangleq D_c(i), \end{cases} \tag{6.37}$$

where

$$\begin{cases} \Phi_{11}(i) \triangleq \mathcal{F}^T A(i) + \mathcal{B}_c(i)C(i) + A^T(i)\mathcal{F} + C^T(i)\mathcal{B}_c^T(i) \\ \qquad + \mathcal{Q}_1(i) + \beta\mathcal{P}_1(i) + \mathcal{X}_1(i) + \mathcal{X}_1^T(i), \\ \Phi_{12}(i) \triangleq \mathcal{A}_c(i) + A^T(i) + C^T(i)\mathcal{D}_c^T(i)B^T(i) + \mathcal{Q}_2(i) + \beta\mathcal{P}_2(i) + \mathcal{X}_2^T(i), \\ \Phi_{22}(i) \triangleq A(i)\mathcal{G} + B(i)\mathcal{C}_c(i) + \mathcal{G}^T A^T(i) + \mathcal{C}_c^T(i)B^T(i) + \mathcal{Q}_3(i) + \beta\mathcal{P}_3(i), \end{cases}$$

and

$$\begin{cases} \Phi_{13}(i) \triangleq \mathcal{P}_1(i) - \mathcal{F}^T + A^T(i)\mathcal{F} + C^T(i)\mathcal{B}_c^T(i), \\ \Phi_{23}(i) \triangleq \mathcal{P}_2^T(i) - I + \mathcal{A}_c^T(i), \\ \Phi_{33}(i) \triangleq -\mathcal{F} - \mathcal{F}^T, \\ \Phi_{14}(i) \triangleq \mathcal{P}_2(i) - \mathcal{K} + A^T(i) + C^T(i)\mathcal{D}_c^T(i)B^T(i), \\ \Phi_{24}(i) \triangleq \mathcal{P}_3(i) - \mathcal{G} + \mathcal{G}^T A^T(i) + \mathcal{C}_c^T(i)B^T(i), \\ \Phi_{34}(i) \triangleq -\mathcal{K} - I, \\ \Phi_{44}(i) \triangleq -\mathcal{G} - \mathcal{G}^T, \\ \Phi_{15}(i) \triangleq \mathcal{F}^T A_d(i) + \mathcal{B}_c(i)C_d(i) - \mathcal{X}_1(i) + \mathcal{Y}^T(i), \\ \Phi_{25}(i) \triangleq A_d(i) + B(i)\mathcal{D}_c(i)C_d(i) - \mathcal{X}_2(i), \\ \Phi_{55}(i) \triangleq -\tilde{d}\mathcal{Q}_1(i) - \mathcal{Y}(i) - \mathcal{Y}^T(i), \\ \Phi_{16}(i) \triangleq \mathcal{F}^T B_1(i) + \mathcal{B}_c(i)D_1(i) + \mathcal{Z}^T(i), \\ \Phi_{26}(i) \triangleq B_1(i) + B(i)\mathcal{D}_c(i)D_1(i), \\ \Phi_{17}(i) \triangleq dA^T(i) + dC^T(i)\mathcal{D}_c^T(i)B^T(i), \\ \Phi_{27}(i) \triangleq d\mathcal{G}^T A^T(i) + d\mathcal{C}_c^T(i)B^T(i), \\ \Phi_{57}(i) \triangleq dA_d^T(i) + dC_d^T(i)\mathcal{D}_c^T(i)B^T(i), \\ \Phi_{67}(i) \triangleq dB_1^T(i) + dD_1^T(i)\mathcal{D}_c^T(i)B^T(i). \end{cases}$$

Proof. Introducing a slack matrix F, it is not difficult to see that the conditions in Theorem 6.2.3 are satisfied if there exist matrices $P(i) > 0$, $Q(i) > 0$, $R(i) > 0$ $X(i)$, $Y(i)$, $Z(i)$, and F such that (6.6b) and the following conditions hold:

$$\begin{bmatrix} \breve{\Pi}_1(i) + \breve{\Pi}_2(i) + \breve{\Pi}_2^T(i) & d\breve{\Pi}_3^T(i)K^T & d\breve{\Pi}_4^T(i) \\ \star & -dR^{-1}(i) & 0 \\ \star & \star & -de^{-\beta d}R(i) \end{bmatrix} < 0, \tag{6.38}$$

where

$$\begin{cases} \breve{\Pi}_1(i) & \triangleq \begin{bmatrix} \Psi_1(i) & P(i) - F^T + \bar{A}^T(i)F & F^T\bar{A}_d(i) & F^T\bar{B}(i) \\ \star & -F - F^T & F^T\bar{A}_d(i) & F^T\bar{B}(i) \\ \star & \star & -(1-\tau)e^{-\beta d}Q(i) & 0 \\ \star & \star & \star & -I \end{bmatrix}, \\ \breve{\Pi}_2(i) & \triangleq \begin{bmatrix} \breve{\Pi}_4^T(i)K & 0 & -\breve{\Pi}_4^T(i) & 0 \end{bmatrix}, \\ \breve{\Pi}_3(i) & \triangleq \begin{bmatrix} \bar{A}(i) & 0 & \bar{A}_d(i) & \bar{B}(i) \end{bmatrix}, \\ \breve{\Pi}_4(i) & \triangleq \begin{bmatrix} X^T(i) & 0 & Y^T(i) & Z^T(i) \end{bmatrix}, \\ \Psi_1(i) & \triangleq F^T\bar{A}(i) + \bar{A}^T(i)F + K^TQ(i)K + \beta P(i). \end{cases}$$

The condition in (6.38) implies (6.6a) in Theorem 6.2.3. To show this, we perform a projection transformation to (6.38) by diag $\{\Lambda(i), R(i), I\}$ with

$$\Lambda(i) \triangleq \begin{bmatrix} I & 0 & 0 \\ \bar{A}(i) & \bar{A}_d(i) & \bar{B}(i) \\ 0 & I & 0 \\ 0 & 0 & I \end{bmatrix},$$

and thus imply (6.6a). Notice that if the condition in (6.38) holds, then matrix F is nonsingular, so we can let the matrix F be partitioned as

$$F \triangleq \begin{bmatrix} F_1 & F_2 \\ F_4 & F_3 \end{bmatrix}, \quad G = F^{-1} \triangleq \begin{bmatrix} G_1 & G_2 \\ G_4 & G_3 \end{bmatrix}. \tag{6.39}$$

As we are considering a full-order DOF controller, F_4 and G_4 are both square. Without loss of generality, we assume that F_4 and G_4 are nonsingular (if not, F_4 and G_4 may be perturbed respectively by matrices ΔF_4 and ΔG_4 with sufficiently small norms such that $F_4 + \Delta F_4$ and $G_4 + \Delta G_4$ are nonsingular and satisfy (6.38)). Then, we can define the following matrices, which are also nonsingular,

$$\mathcal{J}_F \triangleq \begin{bmatrix} F_1 & I \\ F_4 & 0 \end{bmatrix}, \quad \mathcal{J}_G \triangleq \begin{bmatrix} I & G_1 \\ 0 & G_4 \end{bmatrix}. \tag{6.40}$$

Notice that

$$F\mathcal{J}_G = \mathcal{J}_F, \quad G\mathcal{J}_F = \mathcal{J}_G, \quad F_1G_1 + F_2G_4 = I. \tag{6.41}$$

Performing a congruence transformation on (6.38) by matrix diag $\{\mathcal{J}_1, I, I\}$ with $\mathcal{J}_1 \triangleq$ diag $\{\mathcal{J}_G, \mathcal{J}_G, I, I\}$, we have

$$\begin{bmatrix} \tilde{\Pi}_1(i) + \tilde{\Pi}_2(i) + \tilde{\Pi}_2^T(i) & d\tilde{\Pi}_3^T(i)K^T & d\tilde{\Pi}_4^T(i) \\ \star & -dR^{-1}(i) & 0 \\ \star & \star & -de^{-\beta d}R(i) \end{bmatrix} < 0, \ i \in \mathcal{N}, \tag{6.42}$$

where

$$\begin{cases} \tilde{\Pi}_1(i) \triangleq \begin{bmatrix} \Upsilon_1(i) & \Upsilon_2(i) & \mathcal{J}_G^T F^T \bar{A}_d(i) & \mathcal{J}_G^T F^T \bar{B}(i) \\ \star & -\mathcal{J}_G^T(F + F^T)\mathcal{J}_G & \mathcal{J}_G^T F^T \bar{A}_d(i) & \mathcal{J}_G^T F^T \bar{B}(i) \\ \star & \star & -\tilde{d}Q(i) & 0 \\ \star & \star & \star & -I \end{bmatrix}, \\ \tilde{\Pi}_2(i) \triangleq \begin{bmatrix} \tilde{\Pi}_4^T(i)K & 0 & -\tilde{\Pi}_4^T(i) & 0 \end{bmatrix}, \\ \tilde{\Pi}_3(i) \triangleq \begin{bmatrix} \bar{A}(i)\mathcal{J}_G & 0 & \bar{A}_d(i) & \bar{B}(i) \end{bmatrix}, \\ \tilde{\Pi}_4(i) \triangleq \begin{bmatrix} \mathcal{X}_1^T(i) & \mathcal{X}_2^T(i) & 0 & \mathcal{Y}^T(i) & \mathcal{Z}^T(i) \end{bmatrix}, \\ \Upsilon_1(i) \triangleq \mathcal{J}_G^T(F^T\bar{A}(i) + \bar{A}^T(i)F + K^TQ(i)K + \beta P(i))\mathcal{J}_G, \\ \Upsilon_2(i) \triangleq \mathcal{J}_G^T(P(i) - F^T + \bar{A}^T(i)F)\mathcal{J}_G. \end{cases} \tag{6.43}$$

Define the following matrices:

$$\begin{cases} \mathcal{F} \triangleq F_1, \ \mathcal{G} \triangleq G_1, \ \mathcal{K} \triangleq F_1^T G_1 + F_4^T G_4, \ \mathcal{R}(i) \triangleq R^{-1}(i), \\ \mathcal{P}(i) \triangleq \mathcal{J}_G^T P(i) \mathcal{J}_G \triangleq \begin{bmatrix} \mathcal{P}_1(i) & \mathcal{P}_2(i) \\ \star & \mathcal{P}_3(i) \end{bmatrix} > 0, \\ \mathcal{Q}(i) \triangleq \begin{bmatrix} I \\ \mathcal{G}^T \end{bmatrix} Q(i) \begin{bmatrix} I \\ \mathcal{G}^T \end{bmatrix}^T = \begin{bmatrix} \mathcal{Q}_1(i) & \mathcal{Q}_2(i) \\ \star & \mathcal{Q}_3(i) \end{bmatrix} > 0, \end{cases} \tag{6.44}$$

and

$$\begin{cases} \mathcal{A}_c(i) \triangleq F_1^T A(i) G_1 + F_1^T B(i) D_c(i) C(i) G_1 + F_4^T B_c(i) C(i) G_1 \\ \qquad + F_1^T B(i) C_c(i) G_4 + F_4^T A_c(i) G_4, \\ \mathcal{B}_c(i) \triangleq F_1^T B(i) D_c(i) + F_4^T B_c(i), \\ \mathcal{C}_c(i) \triangleq D_c(i) C(i) G_1 + C_c(i) G_4, \\ \mathcal{D}_c(i) \triangleq D_c(i). \end{cases} \tag{6.45}$$

LMI (6.42) implies (6.35a) by considering (6.4), (6.39)–(6.41), and (6.43)–(6.45). Moreover, performing a congruence transformation on (6.6b) by matrix diag $\{\mathcal{J}_G, \mathcal{J}_G, I\}$, we have

$$\begin{bmatrix} -\mathcal{J}_G^T P(i)\mathcal{J}_G & 0 & \mathcal{J}_G^T \bar{C}^T(i) \\ \star & -\mathcal{J}_G^T P(i)\mathcal{J}_G & \mathcal{J}_G^T K^T \bar{C}_d^T(i) \\ \star & \star & -\frac{1}{2}\gamma^2 I \end{bmatrix} < 0, \; i \in \mathcal{N},$$

which is (6.35b) by noting (6.4) and (6.44). In addition, considering the conditions in (6.7) together (6.44) implies (6.36). Finally, considering (6.45) together with (6.44) yields (6.37). This completes the proof. ∎

Remark 6.1 *In the proof of Theorem 6.2.4, we used conditions (6.38) and (6.6b), and not (6.6a)–(6.6b), to solve the $\mathcal{L}_2$-$\mathcal{L}_\infty$ DOF control problem. The reason is that there is no product term between the parameter-dependent Lyapunov matrices and the system dynamic matrices in (6.38); this separation is crucial to solving the $\mathcal{L}_2$-$\mathcal{L}_\infty$ DOF control problem.* ♦

Remark 6.2 *To solve the parameters of the DOF controller in (6.37), matrices F_4 and G_4 should be available in advance, and they can be obtained by taking any full rank factorization of $F_4^T G_4 = \mathcal{K} - \mathcal{F}^T \mathcal{G}$ (derived from $\mathcal{K} \triangleq F_1^T G_1 + F_4^T G_4$).* ♦

Note that the obtained conditions in Theorem 6.2.4 are not all of LMI form because of (6.35e), which can not be solved directly using LMI procedures. Now, using the CCL method [66], we suggest the following minimization problem involving LMI conditions instead of the original nonconvex feasibility problem formulated in Theorem 6.2.4.

Problem DOFC-SDS (DOF control of switched delayed systems):

$$\min \operatorname{trace}\left(\sum\nolimits_{i\in\mathcal{N}} R(i)\mathcal{R}(i)\right)$$

subject to (6.35a)–(6.35d), (6.36) and

$$\begin{bmatrix} R(i) & I \\ I & \mathcal{R}(i) \end{bmatrix} \geq 0, \quad \forall i \in \mathcal{N}. \tag{6.46}$$

If the solution of the aforesaid minimization problem is Nn, then the conditions in Theorem 6.2.4 are solvable. We suggest the following algorithm to solve Problem DOFC-SDS.

Algorithm DOFC-SDS

Step 1. Find a feasible set $(\mathcal{P}_1^{(0)}(i), \mathcal{P}_3^{(0)}(i), \mathcal{Q}_1^{(0)}(i), \mathcal{Q}_3^{(0)}(i), \mathcal{R}^{(0)}(i), R^{(0)}(i), \mathcal{P}_2^{(0)}(i), \mathcal{Q}_2^{(0)}(i), \mathcal{X}_1^{(0)}(i), \mathcal{X}_2^{(0)}(i), \mathcal{Y}^{(0)}(i), \mathcal{Z}^{(0)}(i), \mathcal{A}_c^{(0)}(i), \mathcal{B}_c^{(0)}(i), \mathcal{C}_c^{(0)}(i), \mathcal{D}_c^{(0)}(i), \mathcal{F}^{(0)}, \mathcal{G}^{(0)}, \mathcal{K}^{(0)})$ satisfying (6.35a)–(6.35d), (6.36), and (6.46). Set $\kappa = 0$.

Step 2. Solve the following optimization problem:

$$\min \operatorname{trace}\left(\sum_{i\in\mathcal{N}}\left[R^{(\kappa)}(i)\mathcal{R}(i)+R(i)\mathcal{R}^{(\kappa)}(i)\right]\right)$$

$$\text{subject to (6.35a)–(6.35d), (6.36), and (6.46)}$$

and denote f^* as the optimized value.

Step 3. Substitute the obtained matrix variables $(\mathcal{P}_1(i), \mathcal{P}_3(i), \mathcal{Q}_1(i), \mathcal{Q}_3(i), \mathcal{R}(i), R(i), \mathcal{P}_2(i), \mathcal{Q}_2(i), \mathcal{X}_1(i), \mathcal{X}_2(i), \mathcal{Y}(i), \mathcal{Z}(i), \mathcal{A}_c(i), \mathcal{B}_c(i), \mathcal{C}_c(i), \mathcal{D}_c(i), \mathcal{F}, \mathcal{G}, \mathcal{K})$ into (6.42). If (6.42) is satisfied, with

$$|f^* - 2Nn| < \delta,$$

for a sufficiently small scalar $\delta > 0$, then output the feasible solutions $(\mathcal{P}_1(i), \mathcal{P}_3(i), \mathcal{Q}_1(i), \mathcal{Q}_3(i), \mathcal{R}(i), R(i), \mathcal{P}_2(i), \mathcal{Q}_2(i), \mathcal{X}_1(i), \mathcal{X}_2(i), \mathcal{Y}(i), \mathcal{Z}(i), \mathcal{A}_c(i), \mathcal{B}_c(i), \mathcal{C}_c(i), \mathcal{D}_c(i), \mathcal{F}, \mathcal{G}, \mathcal{K})$, and EXIT.

Step 4. If $\kappa > \mathbb{N}$ where $\mathbb{N}$ is the maximum number of iterations allowed, so EXIT.

Step 5. Set $\kappa = \kappa + 1$. Let $(\mathcal{P}_1^{(\kappa)}(i), \mathcal{P}_3^{(\kappa)}(i), \mathcal{Q}_1^{(\kappa)}(i), \mathcal{Q}_3^{(\kappa)}(i), \mathcal{R}^{(\kappa)}(i), R^{(\kappa)}(i), \mathcal{P}_2^{(\kappa)}(i), \mathcal{Q}_2^{(\kappa)}(i), \mathcal{X}_1^{(\kappa)}(i), \mathcal{X}_2^{(\kappa)}(i), \mathcal{Y}^{(\kappa)}(i), \mathcal{Z}^{(\kappa)}(i), \mathcal{A}_c^{(\kappa)}(i), \mathcal{B}_c^{(\kappa)}(i), \mathcal{C}_c^{(\kappa)}(i), \mathcal{D}_c^{(\kappa)}(i), \mathcal{F}^{(\kappa)}, \mathcal{G}^{(\kappa)}, \mathcal{K}^{(\kappa)}) = (\mathcal{P}_1(i), \mathcal{P}_3(i), \mathcal{Q}_1(i), \mathcal{Q}_3(i), \mathcal{R}(i), R(i), \mathcal{P}_2(i), \mathcal{Q}_2(i), \mathcal{X}_1(i), \mathcal{X}_2(i), \mathcal{Y}(i), \mathcal{Z}(i), \mathcal{A}_c(i), \mathcal{B}_c(i), \mathcal{C}_c(i), \mathcal{D}_c(i), \mathcal{F}, \mathcal{G}, \mathcal{K})$, and go to Step 2.

6.2.3 *Illustrative Example*

Example 6.2.5 Consider system (6.1a)–(6.1d) with $N = 2$ and the following parameters:

$$A(1) = \begin{bmatrix} -0.9 & 0.2 & -0.2 \\ 0.2 & -0.6 & 0.3 \\ -0.3 & 0.1 & -0.1 \end{bmatrix}, \quad A_d(1) = \begin{bmatrix} 0.2 & 0.0 & 0.1 \\ 0.1 & 0.3 & 0.1 \\ 0.3 & 0.1 & 0.2 \end{bmatrix}, \quad B_1(1) = \begin{bmatrix} 0.3 \\ 0.5 \\ 0.2 \end{bmatrix},$$

$$A(2) = \begin{bmatrix} -0.8 & -0.1 & -0.2 \\ 0.2 & -0.7 & 0.3 \\ 0.2 & -0.1 & 0.1 \end{bmatrix}, \quad A_d(2) = \begin{bmatrix} 0.2 & 0.1 & 0.0 \\ 0.1 & 0.2 & 0.1 \\ 0.1 & 0.1 & 0.3 \end{bmatrix}, \quad B_1(2) = \begin{bmatrix} 0.4 \\ 0.2 \\ 0.3 \end{bmatrix},$$

$$B(1) = \begin{bmatrix} 1.0 \\ 0.5 \\ 2.0 \end{bmatrix}, \quad \begin{array}{l} C(1) = \begin{bmatrix} -1.2 & 1.5 & 0.9 \end{bmatrix}, \quad C_d(1) = \begin{bmatrix} 0.3 & 0.1 & 0.2 \end{bmatrix}, \\ C(2) = \begin{bmatrix} -1.0 & 1.2 & 0.5 \end{bmatrix}, \quad C_d(2) = \begin{bmatrix} 0.1 & 0.3 & 0.4 \end{bmatrix}, \end{array}$$

$$B(2) = \begin{bmatrix} 0.5 \\ 0.7 \\ 1.5 \end{bmatrix}, \quad \begin{array}{l} E(1) = \begin{bmatrix} 0.8 & 1.0 & 0.5 \end{bmatrix}, \quad E_d(1) = \begin{bmatrix} 0.2 & 0.3 & 0.1 \end{bmatrix}, \quad D_1(1) = 0.2, \\ E(2) = \begin{bmatrix} 0.6 & 1.2 & 0.3 \end{bmatrix}, \quad E_d(2) = \begin{bmatrix} 0.3 & 0.4 & 0.2 \end{bmatrix}, \quad D_1(2) = 0.1, \end{array}$$

and $d(t) = 0.9 + 0.3\sin(t)$, $\beta = 0.5$. A straightforward calculation gives $d = 1.2$ and $\tau = 0.3$. It can be checked that the switched state-delayed hybrid system in (6.1a)–(6.1d) with $u(t) = 0$ and the above parameters is unstable for switching signal given in Figure 6.1 (which is

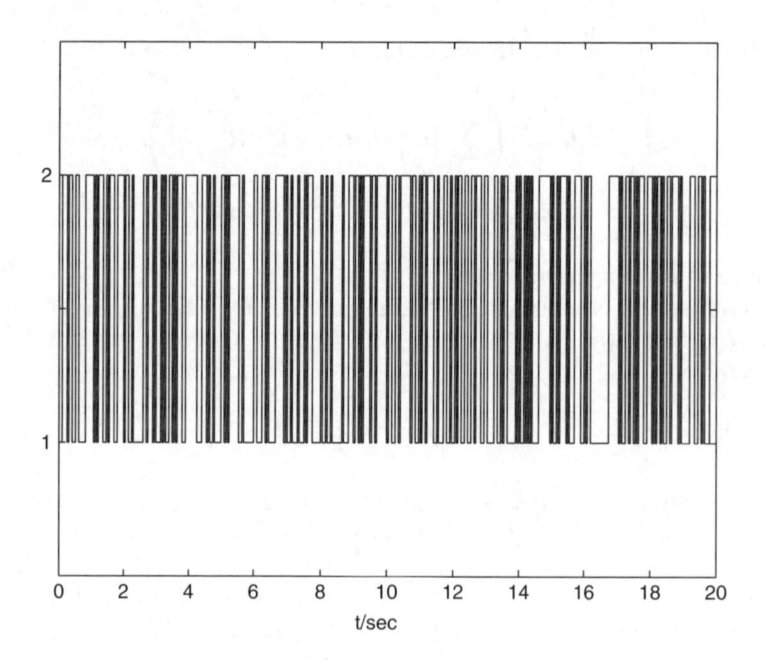

Figure 6.1 Switching signal

generated randomly; here, '1' and '2' represent the first and second subsystems, respectively) – the states of open-loop system are shown in Figure 6.2 with the initial condition given by $x(t) = \begin{bmatrix} -1.0 & 0.5 & 1.0 \end{bmatrix}^T, t \in [-1.2, 0]$.

Our aim is to design an $\mathcal{L}_2$-$\mathcal{L}_\infty$ DOF controller in the form of (6.2a)–(6.2b), such that the closed-loop system is exponentially stable with a weighted $\mathcal{L}_2$-$\mathcal{L}_\infty$ performance. Setting $\mu = 1.01$ (in this case, $T_a > T_a^* = \frac{\ln \mu}{\beta} = 0.0199$, thus we can choose $T_a \geq 0.02$) and solving Problem DOFC-SDS using Algorithm DOFC-SDS, it follows that the minimized feasible γ is $\gamma^* = 1.1726$, and

$$\mathcal{F} = \begin{bmatrix} 6.3445 & -0.8569 & 2.6942 \\ -6.9729 & 6.7830 & -2.0758 \\ -2.3134 & 1.3527 & 3.1193 \end{bmatrix}, \quad \mathcal{G} = \begin{bmatrix} 0.9194 & -0.0674 & -0.4138 \\ -0.0227 & 0.4423 & 0.1605 \\ -0.2669 & -0.2131 & 0.6869 \end{bmatrix},$$

$$\mathcal{K} = \begin{bmatrix} 0.6087 & -0.0500 & -0.5661 \\ 0.5629 & 0.9938 & 0.6335 \\ 0.5812 & -0.4143 & 0.0915 \end{bmatrix},$$

$$\mathcal{A}_c(1) = \begin{bmatrix} -0.3414 & 0.6311 & 1.3370 \\ -0.2710 & -0.9813 & -1.1610 \\ -0.3865 & 0.1436 & -0.1777 \end{bmatrix}, \quad \mathcal{B}_c(1) = \begin{bmatrix} 6.3128 \\ -9.1566 \\ -3.2113 \end{bmatrix},$$

$$\mathcal{A}_c(2) = \begin{bmatrix} -0.5660 & 0.5966 & 1.3975 \\ 0.1049 & -0.9344 & -1.0946 \\ 0.1107 & 0.1730 & 0.1076 \end{bmatrix}, \quad \mathcal{B}_c(2) = \begin{bmatrix} 8.7955 \\ -10.6525 \\ -4.1636 \end{bmatrix},$$

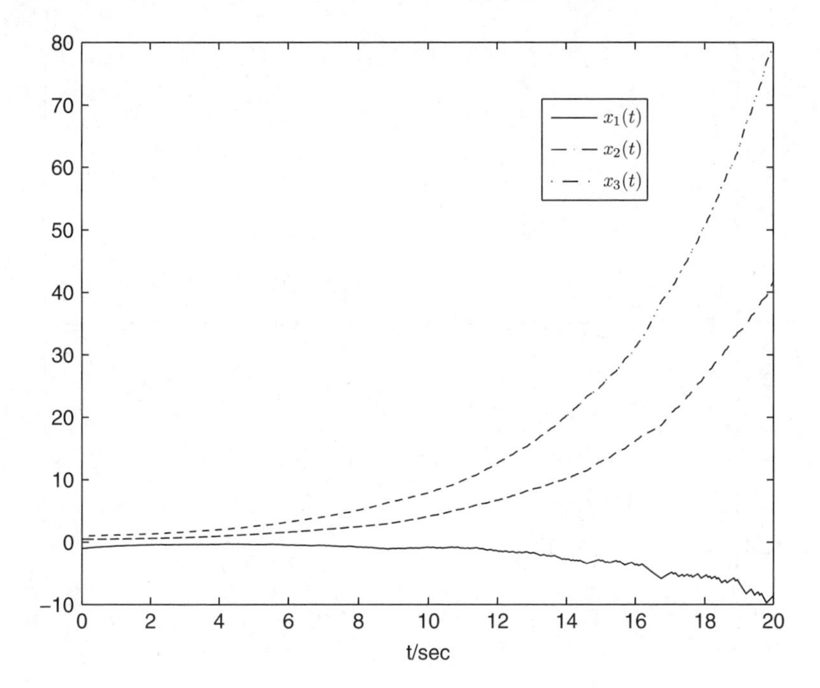

Figure 6.2 States of the open-loop system

$$\mathcal{C}_c(1) = \begin{bmatrix} -0.1523 & -0.1327 & -0.5952 \end{bmatrix}, \; \mathcal{D}_c(1) = -0.5461,$$
$$\mathcal{C}_c(2) = \begin{bmatrix} -0.2882 & -0.0546 & -0.6316 \end{bmatrix}, \; \mathcal{D}_c(2) = -0.9756.$$

Setting $F_4 = I$, then $G_4 = \mathcal{K} - \mathcal{F}^T\mathcal{G}$ from $F_4^T G_4 = \mathcal{K} - \mathcal{F}^T\mathcal{G}$ in Remark 6.2.3. Therefore, by (6.37) we have

$$A_c(1) = \begin{bmatrix} -2.0397 & -0.5763 & -0.6922 \\ 1.0149 & -2.6626 & -1.6111 \\ -1.9295 & -3.3293 & -3.4423 \end{bmatrix}, \quad B_c(1) = \begin{bmatrix} 5.3467 \\ -6.2948 \\ 1.1004 \end{bmatrix},$$
$$C_c(1) = \begin{bmatrix} 0.2827 & 0.5932 & 0.2838 \end{bmatrix}, \quad D_c(1) = -0.5461,$$
$$A_c(2) = \begin{bmatrix} -0.8577 & 1.3584 & 0.1620 \\ -0.2279 & -4.1920 & -1.7676 \\ -1.2900 & -3.9589 & -2.6826 \end{bmatrix}, \quad B_c(2) = \begin{bmatrix} 3.7427 \\ -4.4584 \\ 0.2980 \end{bmatrix},$$
$$C_c(2) = \begin{bmatrix} 0.3541 & 0.6607 & 0.4052 \end{bmatrix}, \quad D_c(2) = -0.9756.$$

To show the effectiveness of the designed $\mathcal{L}_2$-$\mathcal{L}_\infty$ DOF controller through simulation, let the exogenous disturbance input be $\omega(t) = \exp(-t)\sin(t)$. Figure 6.3 gives the states of the closed-loop system, and Figure 6.4 depicts the states of the DOF controller.

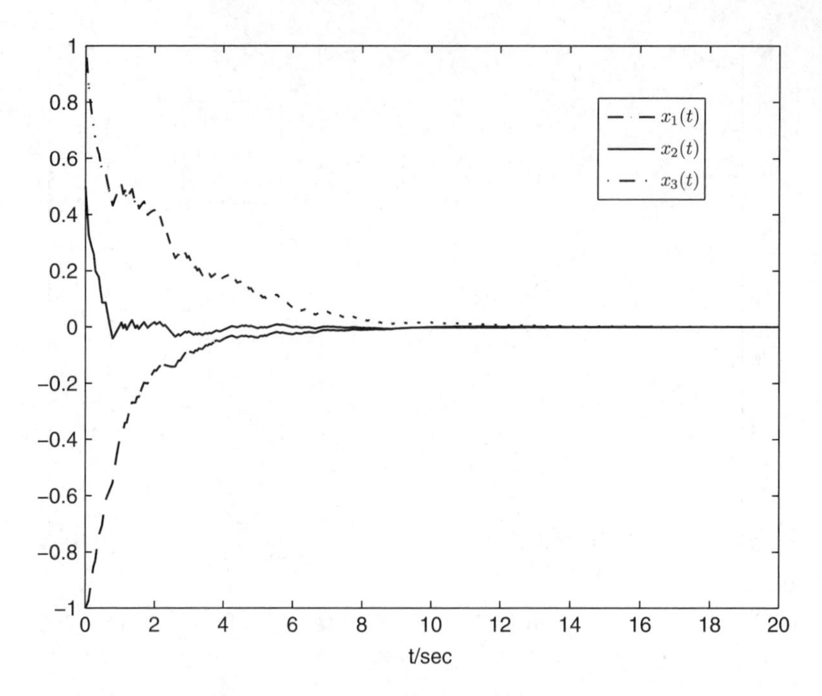

Figure 6.3 States of the closed-loop system

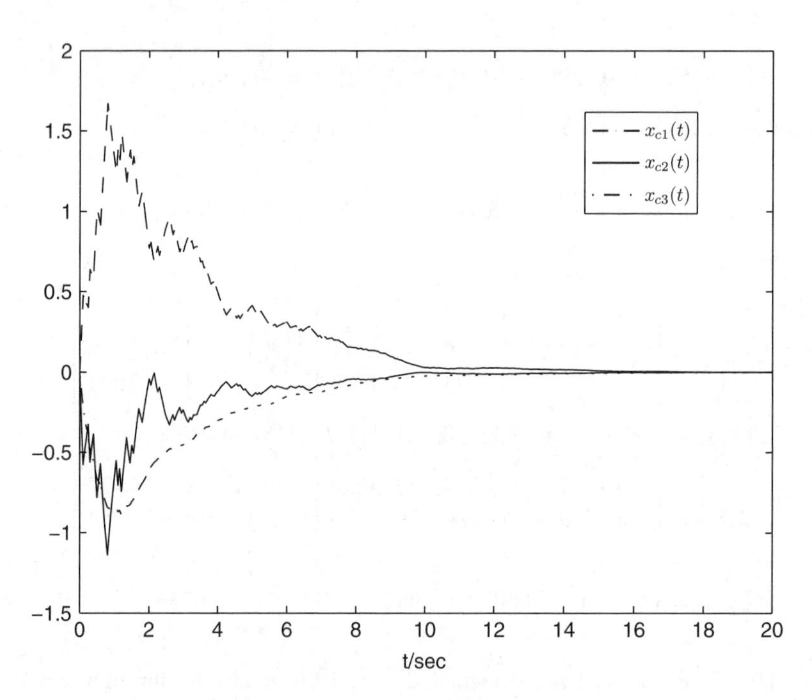

Figure 6.4 States of the DOF controller

6.3 Guaranteed Cost DOF Controller Design

6.3.1 System Description and Preliminaries

Consider a class of switched neutral delay systems of the form

$$\dot{x}(t) = A(\alpha(t))x(t) + A_d(\alpha(t))x(t-d(t)) + A_h(\alpha(t))\dot{x}(t-h) + B(\alpha(t))u(t), \tag{6.47a}$$

$$y(t) = C(\alpha(t))x(t) + C_d(\alpha(t))x(t-d(t)), \tag{6.47b}$$

$$x(t) = \phi(t), \quad t \in \left[-\bar{d}, 0\right], \tag{6.47c}$$

where $x(t) \in \mathbf{R}^n$ is the system state vector, $y(t) \in \mathbf{R}^p$ is the measured output, and $u(t) \in \mathbf{R}^m$ is the control input. $\alpha(t) : \mathbf{R} \to \mathcal{N} = \{1, 2, \dots, N\}$ (denoted by α for simplicity) is the switching signal defined as the same in the previous section. $h \geq 0$ is the constant neutral delay and $d(t)$ denotes the time-varying delays satisfying $0 \leq d(t) \leq d$ and $\dot{d}(t) \leq \tau < 1$, where d and τ are two known constants. $\phi(t)$ is a differentiable vector-valued initial function on $\left[-\bar{d}, 0\right]$ with $\bar{d} \triangleq \max\{d, h\}$.

For each possible value $\alpha = i$ $(i \in \mathcal{N})$, we denote the system matrices associated with mode i by $A(i) = A(\alpha)$, $A_d(i) = A_d(\alpha)$, $A_h(i) = A_h(\alpha)$, $B(i) = B(\alpha)$, $C(i) = C(\alpha)$, and $C_d(i) = C_d(\alpha)$, where $A(i)$, $A_d(i)$, $A_h(i)$, $B(i)$, $C(i)$, and $C_d(i)$ are constant matrices. Corresponding to the switching signal α, we have the switching sequence $\{(i_0, t_0), (i_1, t_1), \dots, (i_k, t_k), \dots, \mid i_k \in \mathcal{N},\ k = 0, 1, \dots\}$ with $t_0 = 0$, which means that the i_kth subsystem is activated when $t \in [t_k, t_{k+1})$.

Here, we are interested in designing a DOF controller of a general structure described by

$$\dot{x}_c(t) = A_c(\alpha)x_c(t) + B_c(\alpha)y(t), \tag{6.48a}$$

$$u(t) = C_c(\alpha)x_c(t), \tag{6.48b}$$

where $x_c(t) \in \mathbf{R}^n$ is the controller state vector; $A_c(\alpha)$, $B_c(\alpha)$, and $C_c(\alpha)$ are appropriately dimensioned constant matrices to be determined later.

Augmenting the model of (6.47a)–(6.47c) to include the states of the DOF controller in (6.48a)–(6.48b), we obtain the following closed-loop system:

$$\dot{\xi}(t) = \bar{A}(\alpha)\xi(t) + \bar{A}_d(\alpha)K\xi(t-d(t)) + \bar{A}_h(\alpha)K\dot{\xi}(t-h), \tag{6.49a}$$

$$\xi(t) = \varphi(t), \quad t \in \left[-\bar{d}, 0\right], \tag{6.49b}$$

where $\xi(t) \triangleq \begin{bmatrix} x(t) \\ x_c(t) \end{bmatrix}$, $K \triangleq \begin{bmatrix} I & 0 \end{bmatrix}$ and

$$\begin{cases} \bar{A}(\alpha) \triangleq \begin{bmatrix} A(\alpha) & B(\alpha)C_c(\alpha) \\ B_c(\alpha)C(\alpha) & A_c(\alpha) \end{bmatrix}, \\ \bar{A}_d(\alpha) \triangleq \begin{bmatrix} A_d(\alpha) \\ B_c(\alpha)C_d(\alpha) \end{bmatrix}, \quad \bar{A}_h(\alpha) \triangleq \begin{bmatrix} A_h(\alpha) \\ 0 \end{bmatrix}. \end{cases} \tag{6.50}$$

Associated with the closed-loop system (6.49a)–(6.49b) is the following cost function:

$$\mathcal{J} = \int_0^\infty (x^T(t)Ux(t) + u^T(t)Wu(t))dt, \tag{6.51}$$

where $U > 0$ and $W > 0$ are given matrices.

We now introduce the following definitions before presenting our main results in this section.

Definition 6.3.1 *Consider the switched neutral delay system in (6.47a)–(6.47c). If there exists a DOF controller $u(t)$ in the form of (6.48a)–(6.48b) and a positive scalar $\mathcal{J}^*$ such that the closed-loop system in (6.49a)–(6.49b) is stable and the cost function in (6.51) satisfies $\mathcal{J} \leq \mathcal{J}^*$, then $\mathcal{J}^*$ is said to be a guaranteed cost and $u(t)$ is said to be a guaranteed cost DOF controller for the switched hybrid system in (6.47a)–(6.47c).*

Definition 6.3.2 *The closed-loop system in (6.49a)–(6.49b) is said to be exponentially stable under α if its solution $\xi(t)$ satisfies*

$$\|\xi(t)\| \leq \eta \left\|\xi(t_0)\right\|_C e^{-\lambda(t-t_0)}, \quad \forall t \geq t_0,$$

where $\eta \geq 1$ and $\lambda > 0$ are two real constants, and

$$\|\xi(t_0)\|_C \triangleq \sup_{-\bar{d}\leq\theta\leq 0} \left\{\|\xi(t_0+\theta)\|, \|\dot{\xi}(t_0+\theta)\|\right\}.$$

Therefore, the problem to be addressed can be formulated as follows: for system (6.47a)–(6.47c), develop a procedure to design a DOF controller $u(t)$ in the form of (6.48a)–(6.48b) such that the closed-loop system in (6.49a)–(6.49b) is exponentially stable and the specified linear integral-quadratic cost function in (6.51) has an upper bound.

6.3.2 Main Results

We will first present a sufficient condition for the existence of the DOF controller in (6.48a)–(6.48b), then give a parameterized representation of the controller in terms of the feasible solutions to a certain set of LMIs.

Theorem 6.3.3 *Consider the closed-loop system in (6.49a)–(6.49b). For a given constant $\beta > 0$, suppose there exist matrices $P(i) > 0$, $Q(i) > 0$, and $R(i) > 0$ such that for $i \in \mathcal{N}$,*

$$\begin{bmatrix} \Pi_{11}(i) + \Psi(i) & P(i)\bar{A}_d(i) & P(i)\bar{A}_h(i) & \bar{A}^T(i)K^T R(i) \\ \star & -(1-\tau)e^{-\beta d}Q(i) & 0 & \bar{A}_d^T(i)K^T R(i) \\ \star & \star & -e^{-\beta h}R(i) & \bar{A}_h^T(i)K^T R(i) \\ \star & \star & \star & -R(i) \end{bmatrix} < 0, \tag{6.52}$$

where

$$\begin{cases} \Psi(i) \triangleq \begin{bmatrix} U & 0 \\ 0 & C_c^T(i)WC_c(i) \end{bmatrix}, \\ \Pi_{11}(i) \triangleq P(i)\bar{A}(i) + \bar{A}^T(i)P(i) + \beta P(i) + K^T Q(i)K. \end{cases}$$

Then the closed-loop system in (6.49a)–(6.49b) is exponentially stable and the cost function (6.51) has the bound of

$$\begin{aligned} \mathcal{J}^* = \xi^T(0)P(i_0)\xi(0) &+ \int_{-d}^{0} e^{\beta s} x^T(s)Q(i_0)x(s)ds \\ &+ \int_{-h}^{0} e^{\beta s} \dot{x}^T(s)R(i_0)\dot{x}(s)ds, \end{aligned} \tag{6.53}$$

for any switching signal with average dwell time satisfying $T_a > T_a^* = \frac{\ln \mu}{\beta}$, *where* $\mu \geq 1$ *satisfies*

$$P(i) \leq \mu P(j),\ Q(i) \leq \mu Q(j),\ R(i) \leq \mu R(j),\ \forall i,j \in \mathcal{N}. \tag{6.54}$$

Moreover, an estimate of the state decay is given by

$$\|\xi(t)\| \leq \eta \, \|\xi(0)\|_{\mathrm{C}} \, e^{-\lambda t}, \tag{6.55}$$

where

$$\begin{cases} \lambda \triangleq \frac{1}{2}\left(\beta - \frac{\ln \mu}{T_a}\right) > 0, \quad \eta \triangleq \sqrt{\frac{b}{a}} \geq 1, \\ a \triangleq \min_{\forall i \in \mathcal{N}} \lambda_{\min}(P(i)), \\ b \triangleq \max_{\forall i \in \mathcal{N}} \lambda_{\max}(P(i)) + d \max_{\forall i \in \mathcal{N}} \lambda_{\max}(Q(i)) + h \max_{\forall i \in \mathcal{N}} \lambda_{\max}(R(i)). \end{cases} \tag{6.56}$$

Proof. Choose a Lyapunov function of the following form:

$$\begin{aligned} V(\xi_t, \alpha) \triangleq \xi^T(t)P(\alpha)\xi(t) &+ \int_{t-d(t)}^{t} e^{\beta(s-t)} \xi^T(s)K^T Q(\alpha)K\xi(s)ds \\ &+ \int_{t-h}^{t} e^{\beta(s-t)} \dot{\xi}^T(s)K^T R(\alpha)K\dot{\xi}(s)ds, \end{aligned} \tag{6.57}$$

where $P(\alpha) > 0$, $Q(\alpha) > 0$, and $R(\alpha) > 0$ ($\alpha \in \mathcal{N}$) are to be determined. Then, as with the solution of the closed-loop system in (6.49a)–(6.49b) for a fixed α, we have

$$\begin{aligned} \dot{V}(\xi_t, \alpha) \leq \xi^T(t)\left[P(\alpha)\bar{A}(\alpha) + \bar{A}^T(\alpha)P(\alpha)\right]\xi(t) + 2\xi^T(t)P(\alpha)\bar{A}_d(\alpha)K\xi(t-d(t)) \\ + 2\xi^T(t)P(\alpha)\bar{A}_h(\alpha)K\dot{\xi}(t-h) + \xi^T(t)K^T Q(\alpha)K\xi(t) \end{aligned}$$

$$
\begin{aligned}
&-(1-\tau)e^{-\beta d}\xi^T(t-d(t))K^TQ(\alpha)K\xi(t-d(t))\\
&+\dot{\xi}^T(t)K^TR(\alpha)K\dot{\xi}(t)-e^{-\beta h}\dot{\xi}^T(t-h)K^TR(\alpha)K\dot{\xi}(t-h)\\
&-\beta\int_{t-d(t)}^{t}e^{\beta(s-t)}\xi^T(s)K^TQ(\alpha)K\xi(s)ds\\
&-\beta\int_{t-h}^{t}e^{\beta(s-t)}\dot{\xi}^T(s)K^TR(\alpha)K\dot{\xi}(s)ds.
\end{aligned}
$$

Thus,

$$\dot{V}(\xi_t,\alpha)+\beta V(\xi_t,\alpha)\leq \eta^T(t)\Pi(\alpha)\eta(t),$$

where $\eta(t)\triangleq\begin{bmatrix}\xi(t)\\ K\xi(t-d(t))\\ K\dot{\xi}(t-h)\end{bmatrix}$ and

$$
\Pi(\alpha)\triangleq\begin{bmatrix}\Pi_{11}(\alpha) & P(\alpha)\bar{A}_d(\alpha) & P(\alpha)\bar{A}_h(\alpha)\\ \star & -(1-\tau)e^{-\beta d}Q(\alpha) & 0\\ \star & \star & -e^{-\beta h}R(\alpha)\end{bmatrix}
+\begin{bmatrix}\bar{A}^T(\alpha)K^T\\ \bar{A}_d^T(\alpha)K^T\\ \bar{A}_h^T(\alpha)K^T\end{bmatrix}R(\alpha)\begin{bmatrix}\bar{A}^T(\alpha)K^T\\ \bar{A}_d^T(\alpha)K^T\\ \bar{A}_h^T(\alpha)K^T\end{bmatrix}^T,
$$

with $\Pi_{11}(\alpha)$ defined in (6.52). It can be seen from (6.52) that $\Pi(\alpha)<0$, which implies

$$\dot{V}(\xi_t,\alpha)+\beta V(\xi_t,\alpha)\leq 0. \tag{6.58}$$

Now, for an arbitrary piecewise constant switching signal α, and for any $t>0$, we let $0=t_0<t_1<\cdots<t_k<\cdots$, $k=1,\ldots,$ denote the switching points of α over the interval $(0,t)$. As mentioned earlier, the i_kth subsystem is activated when $t\in[t_k,t_{k+1})$. Integrating (6.58) from t_k to t gives

$$V(\xi_t,\alpha)\leq e^{-\beta(t-t_k)}V(\xi_{t_k},\alpha(t_k)). \tag{6.59}$$

Using (6.54) and (6.57), at switching instant t_k, we have

$$V(\xi_{t_k},\alpha(t_k))\leq \mu V(\xi_{t_k^-},\alpha(t_k^-)). \tag{6.60}$$

Therefore, it follows from (6.59)–(6.60) and the fact $\vartheta=N_\alpha(0,t)\leq(t-0)/T_a$ that

$$
\begin{aligned}
V(\xi_t,\alpha)&\leq e^{-\beta(t-t_k)}\mu V(\xi_{t_k^-},\alpha(t_k^-))\leq\cdots\\
&\leq e^{-\beta(t-0)}\mu^{\vartheta}V(\xi_0,\alpha(0))\\
&\leq e^{-(\beta-\ln\mu/T_a)t}V(\xi_0,\alpha(0)).
\end{aligned} \tag{6.61}
$$

It can be shown from (6.57) that there exist scalars $a > 0$ and $b > 0$ such that the following holds:

$$V(\xi_t, \alpha) \geq a \|\xi(t)\|^2, \quad V(\xi_0, \alpha(0)) \leq b \|\xi(0)\|_C^2, \tag{6.62}$$

where a and b are defined in (6.56). Combining (6.61) and (6.62) yields

$$\|\xi(t)\|^2 \leq \frac{1}{a} V(\xi_t, \alpha) \leq \frac{b}{a} e^{-(\beta - \ln \mu / T_a)t} \|\xi(0)\|_C^2,$$

which implies (6.55). By Definition 6.3.2 with $t_0 = 0$, the closed-loop system in (6.49a)–(6.49b) is exponentially stable.

However, it can be seen from (6.52) that

$$\dot{V}(\xi_t, \alpha) + \beta V(\xi_t, \alpha) + \xi^T(t)\Psi(\alpha)\xi(t) \leq 0,$$

where $\Psi(\alpha)$ is as defined previously. The above inequality implies

$$\xi^T(t)\Psi(\alpha)\xi(t) \leq -\dot{V}(\xi_t, \alpha) - \beta V(\xi_t, \alpha) \leq -\dot{V}(\xi_t, \alpha).$$

Thus, we have

$$\begin{aligned} x^T(t)Ux(t) + u^T(t)Wu(t) &= x^T(t)Ux(t) + x_c^T(t)C_c^T(\alpha)WC_c(\alpha)x_c(t) \\ &= \xi^T(t)\Psi(\alpha)\xi(t) \leq -\dot{V}(\xi_t, \alpha). \end{aligned} \tag{6.63}$$

Moreover, by integrating both sides of (6.63) from 0 to ∞ and using the initial condition, we obtain

$$\begin{aligned} \mathcal{J} &= \int_0^\infty \left[x^T(t)Ux(t) + u^T(t)Wu(t) \right] dt \leq V(\xi_0, \alpha(0)) \\ &\leq \xi^T(0)P(\alpha(0))\xi(0) + \int_{-d}^{0} e^{\beta s}\xi^T(s)K^T Q(\alpha(0))K\xi(s)ds \\ &\quad + \int_{-h}^{0} e^{\beta s}\dot{\xi}^T(s)K^T R(\alpha(0))K\dot{\xi}(s)ds \\ &= \xi^T(0)P(\alpha(0))\xi(0) + \int_{-d}^{0} e^{\beta s}x^T(s)Q(\alpha(0))x(s)ds \\ &\quad + \int_{-h}^{0} e^{\beta s}\dot{x}^T(s)R(\alpha(0))\dot{x}(s)ds. \end{aligned}$$

The desired result follows Definition 6.3.1, thus the proof is completed. ■

Now, we present a solution to the guaranteed cost DOF control problem for the switched neutral delay system in (6.47a)–(6.47c).

Theorem 6.3.4 *Consider the switched neutral delay system in (6.47a)–(6.47c). For a given constant $\beta > 0$, suppose there exist matrices $\mathcal{P}_1(i) > 0$, $\mathcal{P}_3(i) > 0$, $\mathcal{Q}_1(i) > 0$, $\mathcal{Q}_3(i) > 0$, $\mathcal{R}(i) > 0$, $R(i) > 0$, $\mathcal{P}_2(i)$, $\mathcal{Q}_2(i)$, $\mathcal{A}_c(i)$, $\mathcal{B}_c(i)$, $C_c(i)$, $\mathcal{F}$, $\mathcal{G}$, and $\mathcal{K}$ such that for $i \in \mathcal{N}$,*

$$\begin{bmatrix} \Sigma_{11}(i) & \Sigma_{12}(i) & \Sigma_{13}(i) & \Sigma_{14}(i) & \Sigma_{15}(i) & \mathcal{F}^T A_h(i) & A^T(i) & 0 & 0 \\ \star & \Sigma_{22}(i) & \Sigma_{23}(i) & \Sigma_{24}(i) & A_d(i) & A_h(i) & \Sigma_{27}^T(i) & \mathcal{G}^T & C_c^T(i) \\ \star & \star & \Sigma_{33}(i) & \Sigma_{34}(i) & \Sigma_{15}(i) & \mathcal{F}^T A_h(i) & 0 & 0 & 0 \\ \star & \star & \star & \Sigma_{44}(i) & A_d(i) & A_h(i) & 0 & 0 & 0 \\ \star & \star & \star & \star & \Sigma_{55}(i) & 0 & A_d^T(i) & 0 & 0 \\ \star & \star & \star & \star & \star & -e^{-\beta h} R(i) & A_h^T(i) & 0 & 0 \\ \star & \star & \star & \star & \star & \star & -\mathcal{R}(i) & 0 & 0 \\ \star & \star & \star & \star & \star & \star & \star & -U^{-1} & 0 \\ \star & \star & \star & \star & \star & \star & \star & \star & -W^{-1} \end{bmatrix} < 0, \tag{6.64a}$$

$$\mathcal{P}(i) \triangleq \begin{bmatrix} \mathcal{P}_1(i) & \mathcal{P}_2(i) \\ \star & \mathcal{P}_3(i) \end{bmatrix} > 0, \tag{6.64b}$$

$$\mathcal{Q}(i) \triangleq \begin{bmatrix} \mathcal{Q}_1(i) & \mathcal{Q}_2(i) \\ \star & \mathcal{Q}_3(i) \end{bmatrix} > 0, \tag{6.64c}$$

$$R(i)\mathcal{R}(i) = I, \tag{6.64d}$$

where

$$\begin{cases} \Sigma_{11}(i) \triangleq \mathcal{F}^T A(i) + A^T(i)\mathcal{F} + \mathcal{B}_c(i)C(i) + C^T(i)\mathcal{B}_c(i) + \beta\mathcal{P}_1(i) + \mathcal{Q}_1(i) + U, \\ \Sigma_{12}(i) \triangleq \mathcal{A}_c(i) + A^T(i) + \beta\mathcal{P}_2(i) + \mathcal{Q}_2(i) + U\mathcal{G}, \\ \Sigma_{22}(i) \triangleq A(i)\mathcal{G} + \mathcal{G}^T A^T(i) + B(i)C_c(i) + C_c^T(i)B^T(i) + \beta\mathcal{P}_3(i) + \mathcal{Q}_3(i), \\ \Sigma_{13}(i) \triangleq \mathcal{P}_1(i) - \mathcal{F}^T + A^T(i)\mathcal{F} + C^T(i)\mathcal{B}_c^T(i), \\ \Sigma_{23}(i) \triangleq \mathcal{P}_2^T(i) - I + \mathcal{A}_c^T(i), \\ \Sigma_{33}(i) \triangleq -\mathcal{F} - \mathcal{F}^T, \\ \Sigma_{14}(i) \triangleq \mathcal{P}_2(i) - \mathcal{K} + A^T(i), \\ \Sigma_{24}(i) \triangleq \mathcal{P}_3(i) - \mathcal{G} + \mathcal{G}^T A^T(i) + C_c^T(i)B^T(i), \\ \Sigma_{34}(i) \triangleq -\mathcal{K} - I, \\ \Sigma_{44}(i) \triangleq -\mathcal{G} - \mathcal{G}^T, \\ \Sigma_{15}(i) \triangleq \mathcal{F}^T A_d(i) + \mathcal{B}_c(i)C_d(i), \\ \Sigma_{55}(i) \triangleq -(1-\tau)e^{-\beta d}\mathcal{Q}_1(i), \\ \Sigma_{27}(i) \triangleq A(i)\mathcal{G} + B(i)C_c(i). \end{cases}$$

Then there exists a DOF controller in the form of (6.48a)–(6.48b), such that the closed-loop system in (6.49a)–(6.49b) is exponentially stable and the cost function (6.51) has the bound of

$$\mathcal{J}^* = x^T(0)\mathcal{P}_1(i_0)x(0) + \int_{-d}^{0} e^{\beta s} x^T(s)\mathcal{Q}_1(i_0)x(s)ds + \int_{-h}^{0} e^{\beta s} \dot{x}^T(s)R(i_0)\dot{x}(s)ds, \tag{6.65}$$

for any switching signal with average dwell time satisfying $T_a > T_a^* = \frac{\ln \mu}{\beta}$, *where* $\mu \geq 1$ *satisfies*

$$\mathcal{P}(i) \leq \mu \mathcal{P}(j),\ \mathcal{Q}(i) \leq \mu \mathcal{Q}(j),\ \mathcal{R}(i) \leq \mu \mathcal{R}(j), \quad \forall i, j \in \mathcal{N}. \tag{6.66}$$

Moreover, if the above conditions are feasible, then a desired DOF controller realization is given by

$$\begin{cases} \mathcal{A}_c(i) \triangleq \mathcal{F}^T A(i)\mathcal{G} + F_4^T \mathcal{B}_c(i)C(i)\mathcal{G} + \mathcal{F}^T B(i)\mathcal{C}_c(i)G_4 + F_4^T \mathcal{A}_c(i)G_4, \\ \mathcal{B}_c(i) \triangleq F_4^T \mathcal{B}_c(i), \\ \mathcal{C}_c(i) \triangleq \mathcal{C}_c(i)G_4. \end{cases} \tag{6.67}$$

Proof. By Theorem 6.3.3 and introducing a slack matrix F, it is not difficult to see that the conditions in Theorem 6.3.3 are satisfied if there exist matrices $P(i) > 0$, $Q(i) > 0$, $R(i) > 0$, and F such that the following LMI condition holds:

$$\begin{bmatrix} \bar{\Pi}_{11}(i) & \bar{\Pi}_{12}(i) & F^T \bar{A}_d(i) & F^T \bar{A}_h(i) & \bar{A}^T(i)K^T \\ \star & -F - F^T & F^T \bar{A}_d(i) & F^T \bar{A}_h(i) & 0 \\ \star & \star & \bar{\Pi}_{33}(i) & 0 & \bar{A}_d^T(i)K^T \\ \star & \star & \star & -e^{-\beta h}R(i) & \bar{A}_h^T(i)K^T \\ \star & \star & \star & \star & -R^{-1}(i) \end{bmatrix} < 0, \tag{6.68}$$

where

$$\begin{cases} \bar{\Pi}_{11}(i) \triangleq F^T \bar{A}(i) + \bar{A}^T(i)F + \beta P(i) + K^T Q(i)K + \Psi(i), \\ \bar{\Pi}_{12}(i) \triangleq P(i) - F^T + \bar{A}^T(i)F, \\ \bar{\Pi}_{33}(i) \triangleq -(1-\tau)e^{-\beta d}Q(i). \end{cases}$$

The above can be verified by performing a projection transformation on (6.68) by

$$\Lambda(i) \triangleq \begin{bmatrix} I & 0 & 0 & 0 \\ \bar{A}(i) & \bar{A}_d(i) & \bar{A}_h(i) & 0 \\ 0 & I & 0 & 0 \\ 0 & 0 & I & 0 \\ 0 & 0 & 0 & R(i) \end{bmatrix},$$

and we can obtained (6.52).

In what follows, we will use (6.68), rather than (6.52), to solve the DOF control problem because there is no product term between the parameter-dependent Lyapunov matrix and the system dynamic matrix in (6.68), which is crucial to solving the present problem.

Notice that if the condition in (6.68) holds, then matrix F is nonsingular. Let the matrix F and its inverse matrix be partitioned respectively as

$$F \triangleq \begin{bmatrix} F_1 & F_2 \\ F_4 & F_3 \end{bmatrix}, \quad G = F^{-1} \triangleq \begin{bmatrix} G_1 & G_2 \\ G_4 & G_3 \end{bmatrix}. \tag{6.69}$$

Without loss of generality, we assume that F_4 and G_4 are nonsingular – if not, F_4 and G_4 may be perturbed respectively by matrices ΔF_4 and ΔG_4 with sufficiently small norms such that $F_4 + \Delta F_4$ and $G_4 + \Delta G_4$ are nonsingular and satisfy (6.68).

Define the following matrices that are also nonsingular:

$$\mathcal{J}_F \triangleq \begin{bmatrix} F_1 & I \\ F_4 & 0 \end{bmatrix}, \quad \mathcal{J}_G \triangleq \begin{bmatrix} I & G_1 \\ 0 & G_4 \end{bmatrix}. \tag{6.70}$$

Noticing that $F\mathcal{J}_G = \mathcal{J}_F$, $G\mathcal{J}_F = \mathcal{J}_G$, and $F_1G_1 + F_2G_4 = I$, and performing a congruence transformation on (6.68) by diag $\{\mathcal{J}_G, \mathcal{J}_G, I, I, I\}$, we obtain

$$\begin{bmatrix} \tilde{\Pi}_{11}(i) & \tilde{\Pi}_{12}(i) & \mathcal{J}_G^T F^T \bar{A}_d(i) & \mathcal{J}_G^T F^T \bar{A}_h(i) & \mathcal{J}_G^T \bar{A}^T(i) K^T \\ \star & \tilde{\Pi}_{22}(i) & \mathcal{J}_G^T F^T \bar{A}_d(i) & \mathcal{J}_G^T F^T \bar{A}_h(i) & 0 \\ \star & \star & \bar{\Pi}_{33}(i) & 0 & \bar{A}_d^T(i) K^T \\ \star & \star & \star & -e^{-\beta h} R(i) & \bar{A}_h^T(i) K^T \\ \star & \star & \star & \star & -R^{-1}(i) \end{bmatrix} < 0, \tag{6.71}$$

where

$$\begin{cases} \tilde{\Pi}_{11}(i) \triangleq \mathcal{J}_G^T \left(F^T \bar{A}(i) + \bar{A}^T(i) F + \beta P(i) + K^T Q(i) K + \Psi(i) \right) \mathcal{J}_G, \\ \tilde{\Pi}_{12}(i) \triangleq \mathcal{J}_G^T \bar{\Pi}_{12}(i) \mathcal{J}_G, \\ \tilde{\Pi}_{22}(i) \triangleq -\mathcal{J}_G^T \left(F + F^T \right) \mathcal{J}_G. \end{cases}$$

Define the following matrices:

$$\begin{cases} \mathcal{R}(i) \triangleq R^{-1}(i), \ \mathcal{F} \triangleq F_1, \ \mathcal{G} \triangleq G_1, \ \mathcal{K} \triangleq F_1^T G_1 + F_4^T G_4, \\ P(i) \triangleq \begin{bmatrix} P_1(i) & P_2(i) \\ \star & P_3(i) \end{bmatrix}, \\ \mathcal{P}(i) \triangleq \mathcal{J}_G^T P(i) \mathcal{J}_G \triangleq \begin{bmatrix} \mathcal{P}_1(i) & \mathcal{P}_2(i) \\ \star & \mathcal{P}_3(i) \end{bmatrix} > 0, \\ \mathcal{Q}(i) \triangleq \mathcal{J}_G^T K^T Q(i) K \mathcal{J}_G \triangleq \begin{bmatrix} \mathcal{Q}_1(i) & \mathcal{Q}_2(i) \\ \star & \mathcal{Q}_3(i) \end{bmatrix} > 0. \end{cases} \tag{6.72}$$

It can be easily seen from (6.72) that $P_1(i) = \mathcal{P}_1(i)$ and $Q(i) = \mathcal{Q}_1(i)$. Also, we define

$$\begin{cases} \mathcal{A}_c(i) \triangleq F_1^T A(i) G_1 + F_4^T B_c(i) C(i) G_1 + F_1^T B(i) C_c(i) G_4 + F_4^T A_c(i) G_4, \\ \mathcal{B}_c(i) \triangleq F_4^T B_c(i), \\ \mathcal{C}_c(i) \triangleq C_c(i) G_4. \end{cases} \tag{6.73}$$

Thus, by considering (6.50), (6.69)–(6.70) and (6.72)–(6.73), we have

$$\begin{cases} \mathcal{J}_G^T F^T \bar{A}(i) \mathcal{J}_G \triangleq \begin{bmatrix} \mathcal{F}^T A(i) + \mathcal{B}_c(i) C(i) & \mathcal{A}_c(i) \\ A(i) & A(i)\mathcal{G} + B(i)\mathcal{C}_c(i) \end{bmatrix}, \\ \mathcal{J}_G^T F^T \bar{A}_d(i) \triangleq \begin{bmatrix} \mathcal{F}^T A_d(i) + \mathcal{B}_c(i) C_d(i) \\ A_d(i) \end{bmatrix}, \\ \mathcal{J}_G^T F^T \bar{A}_h(i) \triangleq \begin{bmatrix} \mathcal{F}^T A_h(i) \\ A_h(i) \end{bmatrix}, \\ \mathcal{J}_G^T F^T \mathcal{J}_G \triangleq \begin{bmatrix} \mathcal{F}^T & \mathcal{K} \\ I & \mathcal{G} \end{bmatrix}, \\ K\bar{A}(i)\mathcal{J}_G \triangleq [\, A(i) \quad A(i)\mathcal{G} + B(i)\mathcal{C}_c(i) \,], \\ \mathcal{J}_G^T \Psi(i) \mathcal{J}_G \triangleq \begin{bmatrix} U & U\mathcal{G} \\ \star & \mathcal{G}^T U \mathcal{G} + \mathcal{C}_c^T(i) W \mathcal{C}_c(i) \end{bmatrix}. \end{cases} \tag{6.74}$$

Substituting (6.74) into (6.71) implies (6.64a).

Moreover, considering (6.54) together with (6.72), gives (6.66), and considering (6.73) together with (6.72), yields (6.67). Finally, considering the zero initial condition of $x_c(t)$, that is, $x_c(0) = 0$, we have

$$\begin{aligned} \xi^T(0) P(i_0) \xi(0) &= x^T(0) K P(i_0) K^T x(0) \\ &= x^T(0) P_1(i_0) x(0) = x^T(0) \mathcal{P}_1(i_0) x(0). \end{aligned}$$

Therefore, replacing $\xi^T(0)P(i_0)\xi(0)$ and $Q(i_0)$ with $x^T(0)\mathcal{P}_1(i_0)x(0)$ and $\mathcal{Q}_1(i_0)$ in (6.53) of Theorem 6.3.3, respectively, supplies (6.65). This completes the proof. ■

Remark 6.3 *Notice that there exist product terms between the Lyapunov and system matrices in the LMI condition (6.52) of Theorem 6.3.3, which will make it difficult to solve the DOF controller synthesis problem. Hence, in the proof of Theorem 6.3.4, we have made a decoupling between the Lyapunov and system matrices by introducing a slack matrix variable F, and then obtained a new condition in (6.68). Although the new condition might have introduced some conservativeness, owing to the common matrix variable F, the introduced decoupling technique enables us to obtain a more easily tractable condition (6.68) for synthesis of the DOF controller.* ♦

Remark 6.4 *To solve the parameters of the DOF controller in (6.67), the matrices F_4 and G_4 should be available in advance, which can be obtained by taking any full rank factorization of $F_4^T G_4 = \mathcal{K} - \mathcal{F}^T \mathcal{G}$ (derived from $\mathcal{K} \triangleq F_1^T G_1 + F_4^T G_4$).* ♦

Remark 6.5 *Theorem 6.3.4 gives a set of DOF controllers characterized in terms of the solutions to (6.64a)–(6.64d) and (6.66), and (6.67) parameterizes the set of DOF controllers. Each controller ensures the exponential stability of the resulting closed-loop system and an upper bound on the cost function given by (6.65). In view of this, it is desirable to find one which minimizes the upper bound $\mathcal{J}^*$ in (6.65).* ♦

The following theorem presents a method of selecting an optimal DOF controller minimizing the upper bound of the guaranteed cost (6.65).

Theorem 6.3.5 *Consider the switched neutral delay system in (6.47a)–(6.47c) and cost function (6.51). For a given constant $\beta > 0$, suppose the following optimal problem:*

$$\mathcal{J}^\star \triangleq \min\{\varepsilon + \mathrm{trace}(\mathcal{S}) + \mathrm{trace}(\mathcal{T})\}, \tag{6.75}$$

subject to (6.64a)–(6.64d), (6.66) and

$$\begin{bmatrix} -\varepsilon & x^T(0)\mathcal{P}_1(i_0) \\ \star & -\mathcal{P}_1(i_0) \end{bmatrix} < 0, \tag{6.76a}$$

$$\begin{bmatrix} -\mathcal{S} & E^T \mathcal{Q}_1(i_0) \\ \star & -\mathcal{Q}_1(i_0) \end{bmatrix} < 0, \tag{6.76b}$$

$$\begin{bmatrix} -\mathcal{T} & F^T \\ \star & -\mathcal{R}(i_0) \end{bmatrix} < 0, \tag{6.76c}$$

has a feasible solution for $\mathcal{P}_1(i) > 0$, $\mathcal{P}_3(i) > 0$, $\mathcal{Q}_1(i) > 0$, $\mathcal{Q}_3(i) > 0$, $\mathcal{R}(i) > 0$, $R(i) > 0$, $\mathcal{S} > 0$, $\mathcal{T} > 0$, $\mathcal{P}_2(i)$, $\mathcal{Q}_2(i)$, $\mathcal{A}_c(i)$, $\mathcal{B}_c(i)$, $\mathcal{C}_c(i)$, $\mathcal{F}$, $\mathcal{G}$, $\mathcal{K}$ ($i \in \mathcal{N}$), and a scalar $\varepsilon > 0$, where

$$\begin{cases} \displaystyle\int_{-d}^{0} e^{\beta s} x(s) x^T(s) ds = EE^T, \\ \displaystyle\int_{-h}^{0} e^{\beta s} \dot{x}(s) \dot{x}^T(s) ds = FF^T, \end{cases} \tag{6.77}$$

with E and F being given constant matrices with appropriate dimensions. Then, the corresponding DOF controller in the form of (6.48a)–(6.48b) with (6.67) is an optimal DOF controller with a guaranteed cost in the sense that the upper bound on the closed-loop cost function (6.65) is minimized under this controller.

Proof. The proof of this theorem is along the same lines as that of Theorem 6.3.4. By Schur complement, LMI (6.76a) is equivalent to $x^T(0)\mathcal{P}_1(i_0)x(0) < \varepsilon$. However, noting (6.77), we have

$$\begin{aligned}
\int_{-d}^{0} e^{\beta s}x^T(s)\mathcal{Q}_1(i_0)x(s)ds &= \int_{-d}^{0} \text{trace}\left(e^{\beta s}x^T(s)\mathcal{Q}_1(i_0)x(s)\right) ds \\
&= \int_{-d}^{0} \text{trace}\left(e^{\beta s}\mathcal{Q}_1(i_0)x(s)x^T(s)\right) ds \\
&= \text{trace}\left(\mathcal{Q}_1(i_0)\right) \int_{-d}^{0} e^{\beta s}x(s)x^T(s)ds \\
&= \text{trace}\left(E^T\mathcal{Q}_1(i_0)E\right) \\
&< \text{trace}\,(\mathcal{S}).
\end{aligned}$$

Moreover, it follows from (6.77) and $R(i_0) = \mathcal{R}^{-1}(i_0)$ that

$$\begin{aligned}
\int_{-h}^{0} e^{\beta s}\dot{x}^T(s)R(i_0)\dot{x}(s)ds &= \int_{-h}^{0} \text{trace}\left(e^{\beta s}\dot{x}^T(s)\mathcal{R}^{-1}(i_0)\dot{x}(s)\right) ds \\
&= \int_{-h}^{0} \text{trace}\left(e^{\beta s}\mathcal{R}^{-1}(i_0)\dot{x}(s)\dot{x}^T(s)\right) ds \\
&= \text{trace}\left(\mathcal{R}^{-1}(i_0)\right) \int_{-h}^{0} e^{\beta s}\dot{x}(s)\dot{x}^T(s)ds \\
&= \text{trace}\left(F^T\mathcal{R}^{-1}(i_0)F\right) \\
&< \text{trace}(\mathcal{T}).
\end{aligned}$$

Thus, according to (6.65), we have $\mathcal{J}^* < \varepsilon + \text{trace}(\mathcal{S}) + \text{trace}(\mathcal{T})$, and the minimization of $\{\varepsilon + \text{trace}(\mathcal{S}) + \text{trace}(\mathcal{T})\}$ implies the minimization of the guaranteed cost for the switched neutral delay system in (6.47a)–(6.47c). The optimality of the solution to the optimization problem (6.75) follows from the convexity of the objective function as well as the constraints. This completes the proof. ∎

Notice that the conditions in Theorem 6.3.5 are not all of LMI form because of (6.64d), hence they can not be solved directly using LMI procedures. Now, by using the CCL method again, we suggest the following minimization problem involving LMI conditions instead of the original nonconvex feasibility problem formulated in Theorem 6.3.5.

Problem DOFC-SNDS-GC (DOF control of switched neutral delay systems with a guaranteed cost):

$$\min \text{trace}\left(\sum_{i\in\mathcal{N}} R(i)\mathcal{R}(i)\right)$$

subject to (6.64a)–(6.64c), (6.66), (6.76a)–(6.76c) and

$$\begin{bmatrix} R(i) & I \\ I & \mathcal{R}(i) \end{bmatrix} \geq 0, \quad \forall i \in \mathcal{N}. \tag{6.78}$$

The following algorithm is suggested to solve the Problem DOFC-SNDS-GC.

Algorithm DOFC-SNDS-GC

Step 1. Find a feasible set $(\mathcal{P}_1^{(0)}(i), \mathcal{P}_3^{(0)}(i), \mathcal{Q}_1^{(0)}(i), \mathcal{Q}_3^{(0)}(i), \mathcal{R}^{(0)}(i), R^{(0)}(i), \mathcal{P}_2^{(0)}(i), \mathcal{Q}_2^{(0)}(i), \mathcal{A}_c^{(0)}(i), \mathcal{B}_c^{(0)}(i), \mathcal{C}_c^{(0)}(i), \mathcal{S}^{(0)}, \mathcal{T}^{(0)}, \mathcal{F}^{(0)}, \mathcal{G}^{(0)}, \mathcal{K}^{(0)}, \varepsilon^{(0)})$ satisfying (6.64a)–(6.64c), (6.66), (6.76a)–(6.76c), and (6.78). Set $\kappa = 0$.
Step 2. Solve the following optimization problem:

$$\min \text{trace}\left(\sum_{i\in\mathcal{N}}\left[R^{(\kappa)}(i)\mathcal{R}(i) + R(i)\mathcal{R}^{(\kappa)}(i)\right]\right)$$

subject to (6.64a)–(6.64c), (6.66), (6.76a)–(6.76c), and (6.78)

and denote f^* as the optimized value.
Step 3. Substitute the obtained matrix variables $(\mathcal{P}_1(i), \mathcal{P}_3(i), \mathcal{Q}_1(i), \mathcal{Q}_3(i), \mathcal{R}(i), R(i), \mathcal{P}_2(i), \mathcal{Q}_2(i), \mathcal{A}_c(i), \mathcal{B}_c(i), \mathcal{C}_c(i), \mathcal{S}, \mathcal{T}, \mathcal{F}, \mathcal{G}, \mathcal{K}, \varepsilon)$ into (6.71). If (6.71) is satisfied, with

$$\left|f^* - 2Nn\right| < \delta,$$

for a sufficiently small scalar $\delta > 0$, then output the feasible solutions $(\mathcal{P}_1(i), \mathcal{P}_3(i), \mathcal{Q}_1(i), \mathcal{Q}_3(i), \mathcal{R}(i), R(i), \mathcal{P}_2(i), \mathcal{Q}_2(i), \mathcal{A}_c(i), \mathcal{B}_c(i), \mathcal{C}_c(i), \mathcal{S}, \mathcal{T}, \mathcal{F}, \mathcal{G}, \mathcal{K}, \varepsilon)$, so EXIT.
Step 4. If $\kappa > \mathbb{N}$ where $\mathbb{N}$ is the maximum number of iterations allowed, so EXIT.
Step 5. Set $\kappa = \kappa + 1$, $(\mathcal{P}_1^{(\kappa)}(i), \mathcal{P}_3^{(\kappa)}(i), \mathcal{Q}_1^{(\kappa)}(i), \mathcal{Q}_3^{(\kappa)}(i), \mathcal{R}^{(\kappa)}(i), R^{(\kappa)}(i), \mathcal{P}_2^{(\kappa)}(i), \mathcal{Q}_2^{(\kappa)}(i), \mathcal{A}_c^{(\kappa)}(i), \mathcal{B}_c^{(\kappa)}(i), \mathcal{C}_c^{(\kappa)}(i), \mathcal{S}^{(\kappa)}, \mathcal{T}^{(\kappa)}, \mathcal{F}^{(\kappa)}, \mathcal{G}^{(\kappa)}, \mathcal{K}^{(\kappa)}, \varepsilon^{(\kappa)}) = (\mathcal{P}_1(i), \mathcal{P}_3(i), \mathcal{Q}_1(i), \mathcal{Q}_3(i), \mathcal{R}(i), R(i), \mathcal{P}_2(i), \mathcal{Q}_2(i), \mathcal{A}_c(i), \mathcal{B}_c(i), \mathcal{C}_c(i), \mathcal{S}, \mathcal{T}, \mathcal{F}, \mathcal{G}, \mathcal{K}, \varepsilon)$, and go to Step 2.

6.3.3 *Illustrative Example*

Example 6.3.6 Consider system (6.47a)–(6.47c) with $N = 2$ (that is, there are two subsystems) and the related parameters are given as follows:

$$A(1) = \begin{bmatrix} -1.9 & 0.0 & 0.1 \\ 0.2 & -2.1 & 0.0 \\ 0.0 & 0.1 & 0.3 \end{bmatrix}, \quad A_d(1) = \begin{bmatrix} 0.2 & 0.0 & 0.1 \\ 0.1 & 0.1 & 0.1 \\ 0.0 & 0.1 & 0.2 \end{bmatrix}, \quad B(1) = \begin{bmatrix} 1.0 \\ 0.5 \\ 1.0 \end{bmatrix},$$

$$A(2) = \begin{bmatrix} -1.8 & -0.1 & 0.0 \\ 0.2 & -2.3 & 0.1 \\ 0.2 & -0.1 & -2.2 \end{bmatrix}, \quad A_d(2) = \begin{bmatrix} 0.2 & 0.1 & 0.0 \\ 0.1 & 0.2 & 0.1 \\ 0.0 & 0.1 & 0.1 \end{bmatrix}, \quad B(2) = \begin{bmatrix} 0.5 \\ 0.6 \\ 1.0 \end{bmatrix},$$

$$A_h(1) = \begin{bmatrix} 0.1 & 0.0 & 0.1 \\ 0.0 & 0.2 & 0.1 \\ 0.1 & 0.0 & 0.1 \end{bmatrix}, \quad A_h(2) = \begin{bmatrix} 0.2 & 0.1 & 0.0 \\ 0.1 & 0.1 & 0.1 \\ 0.1 & 0.0 & 0.1 \end{bmatrix},$$

$$C(1) = \begin{bmatrix} 1.2 & 1.0 & 1.4 \end{bmatrix}, \quad C_d(1) = \begin{bmatrix} 0.3 & 0.1 & 0.2 \end{bmatrix},$$

$$C(2) = \begin{bmatrix} 1.3 & 1.2 & 1.5 \end{bmatrix}, \quad C_d(2) = \begin{bmatrix} 0.1 & 0.3 & 0.2 \end{bmatrix},$$

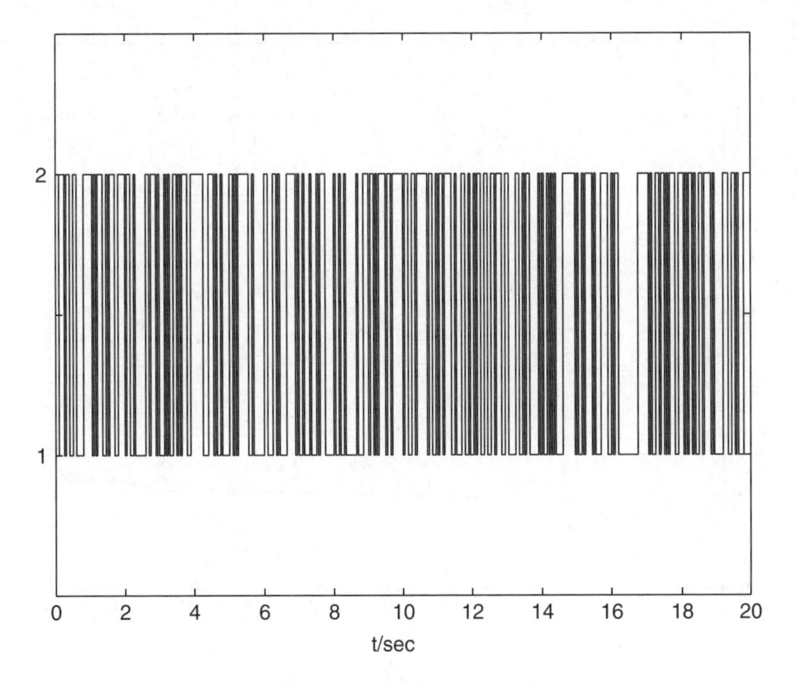

Figure 6.5 Switching signal

with $d(t) = 0.7 + 0.3\sin(t)$ and $\beta = 0.5$. A straightforward calculation gives $d = 1.0$ and $\tau = 0.3$. It can be checked that the system in (6.47a)–(6.47c) with $u(t) = 0$ and the above parameters is unstable for switching signal given in Figure 6.5 (which is generated randomly; here, '1' and '2' represent the first and second subsystems, respectively), and the states of open-loop system are shown in Figure 6.6 with the initial condition given by $x(\theta) = [\,0.1 \quad 0.3 \quad 0.2\,]^T$, $\theta \in [-1, 0]$. In view of this, our aim is to design a DOF controller $u(t)$ in the form of (6.48a)–(6.48b), such that the closed-loop system is exponentially stable and the specified linear integral-quadratic cost function in (6.51) has an upper bound.

Setting $\mu = 1.01$ (thus $T_a > \frac{\ln \mu}{\beta} = 0.0199$) and $U = \mathrm{diag}\{0.1, 0.1, 0.1\}$, $W = 0.5$, $i_0 = 1$, we solve Problem DOFC-SNDS-GC by Algorithm DOFC-SNDS-GC, and obtain

$$\mathcal{F} = \begin{bmatrix} 0.0824 & -0.0057 & -0.0330 \\ -0.0361 & 0.0965 & -0.0503 \\ -0.0206 & -0.0561 & 0.1413 \end{bmatrix},$$

$$\mathcal{K} = \begin{bmatrix} 0.4919 & -0.4490 & -0.2498 \\ -0.0094 & 0.4764 & -0.0552 \\ -0.3063 & -0.0147 & 0.3862 \end{bmatrix}, \quad \mathcal{G} = \begin{bmatrix} 12.5338 & -0.6635 & 0.0878 \\ 2.4689 & 16.0969 & 0.1650 \\ 0.0326 & 0.4439 & 17.5236 \end{bmatrix},$$

$$\mathcal{A}_c(1) = \begin{bmatrix} -1.2474 & 0.7324 & 0.2495 \\ -0.1116 & -1.5146 & -0.0846 \\ 0.1060 & -0.1952 & -1.3798 \end{bmatrix}, \quad \mathcal{B}_c(1) = \begin{bmatrix} -0.0692 \\ -0.0608 \\ -0.0661 \end{bmatrix},$$

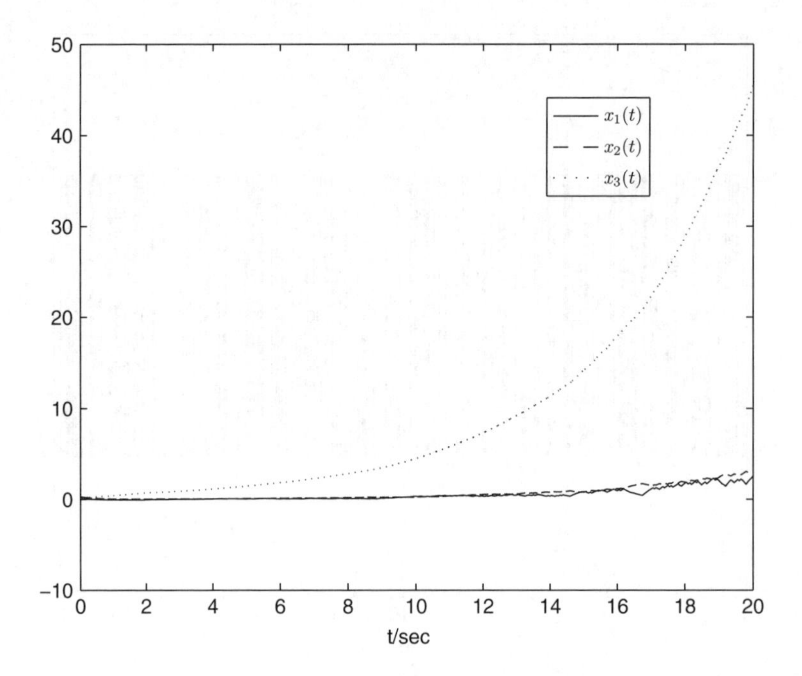

Figure 6.6 States of the open-loop system

$$\mathcal{A}_c(2) = \begin{bmatrix} -1.3014 & 0.4466 & 0.1132 \\ -0.2933 & -1.3092 & 0.0853 \\ 0.3635 & -0.2835 & -1.2655 \end{bmatrix}, \ \mathcal{B}_c(2) = \begin{bmatrix} -0.0704 \\ -0.0486 \\ -0.0562 \end{bmatrix},$$

$$\mathcal{C}_c(1) = \begin{bmatrix} 0.1977 & -0.3589 & -1.9555 \end{bmatrix},$$

$$\mathcal{C}_c(2) = \begin{bmatrix} -1.6509 & 0.7992 & -0.3083 \end{bmatrix}.$$

Setting $F_4 = I$, we have $G_4 = \mathcal{K} - \mathcal{F}^T\mathcal{G}$ by Remark 6.4. Therefore, it follows from (6.67) that

$$A_c(1) = \begin{bmatrix} -4.0081 & -2.4662 & -1.8121 \\ -1.6211 & -3.8452 & -1.3765 \\ -1.8396 & -2.2488 & -4.0423 \end{bmatrix}, \ B_c(1) = \begin{bmatrix} -0.0692 \\ -0.0608 \\ -0.0661 \end{bmatrix},$$

$$A_c(2) = \begin{bmatrix} -4.1412 & -2.4587 & -1.9010 \\ -1.2203 & -4.0549 & -1.4000 \\ -1.9281 & -2.2060 & -3.8444 \end{bmatrix}, \ B_c(2) = \begin{bmatrix} -0.0704 \\ -0.0486 \\ -0.0562 \end{bmatrix},$$

$$C_c(1) = \begin{bmatrix} -0.1919 & 1.3215 & 1.5099 \end{bmatrix},$$

$$C_c(2) = \begin{bmatrix} 3.7940 & 0.2577 & 0.4612 \end{bmatrix}.$$

Consequently, the optimal guaranteed cost of the closed-loop system is $\mathcal{J}^\star = 33.6685$. The states of the closed-loop system are given in Figure 6.7 and the DOF control input is shown in Figure 6.8.

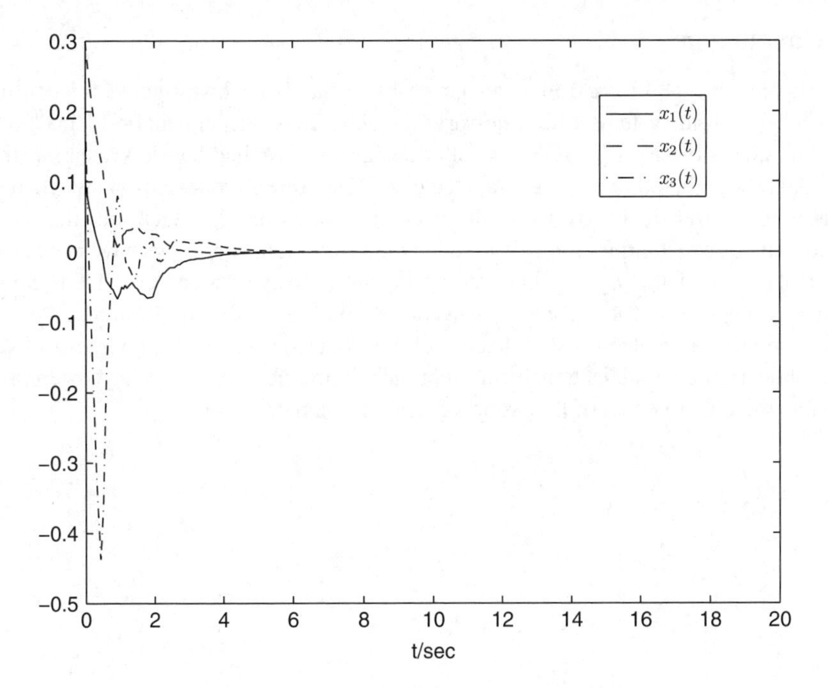

Figure 6.7 States of the closed-loop system

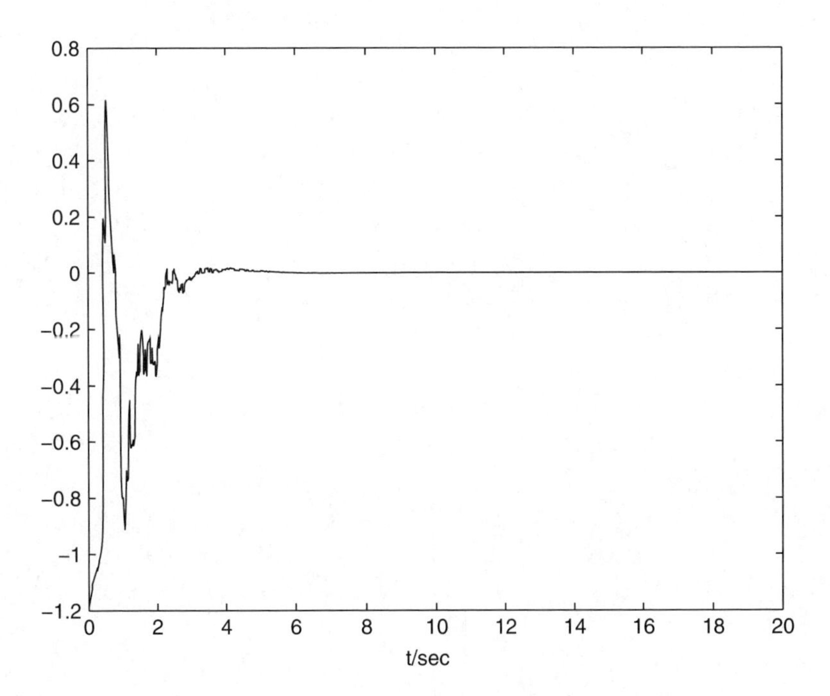

Figure 6.8 Control input

6.4 Conclusion

In this chapter, the DOF controller design problem has been investigated for continuous-time switched hybrid systems with time-varying delay. Two independent problems have been considered: one is the $\mathcal{L}_2$-$\mathcal{L}_\infty$ DOF controller design for switched hybrid systems with time-varying delay, and the other is the guaranteed cost DOF controller design for switched hybrid systems with neutral delay. By using the average dwell time approach and the piecewise Lyapunov function technique, some delay-dependent sufficient conditions have been proposed for the existences of the $\mathcal{L}_2$-$\mathcal{L}_\infty$ DOF controller and the guaranteed cost DOF controller, respectively. Then, the corresponding solvability conditions for the desired $\mathcal{L}_2$-$\mathcal{L}_\infty$ DOF controller and the guaranteed cost DOF controller have also been established, respectively, by using the linearizing variables transforms approach. Numerical examples have been provided to illustrate the effectiveness of the proposed design schemes.

7

SMC of Switched State-Delayed Hybrid Systems: Continuous-Time Case

7.1 Introduction

SMC theory and methodologies have been developed for many kinds of systems such as uncertain systems, time-delay systems, and stochastic systems. Unfortunately, little progress has been made toward solving SMC of switched hybrid systems. The research in this area has not been fully investigated and still remains challenging. In this chapter, we will investigate the SMC design problem for continuous-time switched hybrid systems with time-varying delay. First, the original system is transformed into a regular form through model transformation, and then by designing a linear sliding surface, the dynamical equation for the sliding mode dynamics is derived. By utilizing the average dwell time approach and the piecewise Lyapunov function technique, a delay-dependent sufficient condition for the existence of a desired sliding mode is proposed, and an explicit parametrization of the desired sliding surface is also given. Since the obtained conditions are not all expressed in terms of strict LMIs (some matrix equality constraints exist), the CCL method is exploited to cast them into a sequential minimization problem subject to LMI constraints, which can be easily solved numerically. Then, a discontinuous SMC law is synthesized, by which the system state trajectories can be driven onto the prescribed sliding surface in a finite time and maintained there for all subsequent time. Since the designed SMC law contains state-delay terms, it requires the time-varying delay to be explicitly known *a priori* in the practical implementation of the controller. However, in some practical situations, the information about time delay is unavailable, or difficult to measure. In such a case, the designed SMC law is not applicable. To overcome this, we suppose the the state-delay terms in controller are unknown and unmeasurable, but

Sliding Mode Control of Uncertain Parameter-Switching Hybrid Systems, First Edition. Ligang Wu, Peng Shi and Xiaojie Su.

that they are norm-bounded with an unknown upper bound. We will design an adaptive law to estimate the unknown upper bound, and thus an adaptive SMC law is synthesized, which can also guarantee that the system state trajectories reach onto the the prescribed sliding surface in a finite time.

7.2 System Description and Preliminaries

Consider the continuous-time state-delayed hybrid systems described by

$$\dot{x}(t) = A(\alpha(t))x(t) + A_d(\alpha(t))x(t - d(t)) + B\left[u(t) + F(\alpha(t))f(t)\right], \tag{7.1a}$$

$$x(t) = \phi(t), \quad t \in [-d, 0], \tag{7.1b}$$

where $x(t) \in \mathbf{R}^n$ is the system state vector; $u(t) \in \mathbf{R}^m$ is the control input; $f(t) \in \mathbf{R}^p$ is the nonlinearity representing the external disturbance or unmodeled dynamics; $\left\{\left(A(\alpha(t)), A_d(\alpha(t)), F(\alpha(t))\right) : \alpha(t) \in \mathcal{N}\right\}$ is a family of matrices parameterized by an index set $\mathcal{N} = \{1, 2, \ldots, N\}$; and $\alpha(t) : \mathbf{R} \to \mathcal{N}$ (denoted by α for simplicity) is a switching signal defined as the same in Chapter 5. Also, $\phi(t) \in \mathbf{C}_{n,d}$ is a differentiable vector-valued initial function on $[-d, 0]$ for a known constant $d > 0$, and $d(t)$ denotes the time-varying delays which satisfy $0 \leq d(t) \leq d$ and $\dot{d}(t) \leq \tau$.

For each possible value $\alpha(t) = i$, $i \in \mathcal{N}$, we denote the system matrices associated with mode i by $A(i) = A(\alpha)$, $A_d(i) = A_d(\alpha)$, and $F(i) = F(\alpha)$, where $A(i)$, $A_d(i)$, and $F(i)$ are constant matrices. In addition, B is assumed to be of full column rank, and for the nonlinearity $f(t)$, we suppose that

$$\|F(i)f(t)\| \leq \eta(i), \quad \forall i \in \mathcal{N},$$

where $\eta(i) > 0, i \in \mathcal{N}$ are real constants.

Introduce the following definitions for the autonomous system of (7.1a):

$$\dot{x}(t) = A(\alpha)x(t) + A_d(\alpha)x(t - d(t)). \tag{7.2}$$

Definition 7.2.1 *The switched state-delayed hybrid system in (7.2) is said to be exponentially stable under $\alpha(t)$ if the solution $x(t)$ of the system satisfies*

$$\|x(t)\| \leq \eta \left\|x(t_0)\right\|_{\mathrm{C}} e^{-\lambda(t-t_0)}, \quad \forall t \geq t_0,$$

where $\eta \geq 1$ and $\lambda > 0$ are two real constants, and

$$\left\|x(t_0)\right\|_{\mathrm{C}} \triangleq \sup_{-d \leq \theta \leq 0} \left\{\left\|x(t_0 + \theta)\right\|, \left\|\dot{x}(t_0 + \theta)\right\|\right\}.$$

7.3 Main Results

7.3.1 Sliding Mode Dynamics Analysis

In this section, we will consider the SMC problem for system (7.1a)–(7.1b). First of all, we design the switching function and analyze the stability of sliding mode dynamics. Since B is of full column rank by assumption, there exists a nonsingular matrix T such that

$$TB = \begin{bmatrix} 0_{(n-m)\times m} \\ B_1 \end{bmatrix}, \tag{7.3}$$

where $B_1 \in \mathbf{R}^{m\times m}$ is nonsingular. Taking a singular value decomposition of B, we have

$$B = U \begin{bmatrix} 0_{(n-m)\times m} \\ \Gamma \end{bmatrix} W^T, \tag{7.4}$$

where $U \triangleq \begin{bmatrix} U_1 & U_2 \end{bmatrix}$ and $W \in \mathbf{R}^{m\times m}$ are unitary matrices with $U_1 \in \mathbf{R}^{n\times(n-m)}$, $U_2 \in \mathbf{R}^{n\times m}$, and $\Gamma \in \mathbf{R}^{m\times m}$ is a diagonal positive-definite matrix. For convenience, choose $T = U^T$, then by the transformation $z(t) = Tx(t)$, system (7.1a)–(7.1b) becomes

$$\begin{aligned} \dot{z}(t) &= TA(\alpha)T^{-1}z(t) + TA_d(\alpha)T^{-1}z(t-d(t)) \\ &\quad + TB\,[u(t) + F(\alpha)f(t)]\,. \end{aligned} \tag{7.5}$$

Let $z(t) \triangleq \begin{bmatrix} z_1(t) \\ z_2(t) \end{bmatrix}$ with $z_1(t) \in \mathbf{R}^{n-m}$, $z_2(t) \in \mathbf{R}^m$, and

$$\begin{cases} \bar{A}(\alpha) \triangleq \begin{bmatrix} \bar{A}_{11}(\alpha) & \bar{A}_{12}(\alpha) \\ \bar{A}_{21}(\alpha) & \bar{A}_{22}(\alpha) \end{bmatrix} = TA(\alpha)T^{-1}, \\ \bar{A}_d(\alpha) \triangleq \begin{bmatrix} \bar{A}_{d11}(\alpha) & \bar{A}_{d12}(\alpha) \\ \bar{A}_{d21}(\alpha) & \bar{A}_{d22}(\alpha) \end{bmatrix} = TA_d(\alpha)T^{-1}, \end{cases}$$

then (7.5) can be written in the following regular form:

$$\begin{aligned} \begin{bmatrix} \dot{z}_1(t) \\ \dot{z}_2(t) \end{bmatrix} &= \begin{bmatrix} \bar{A}_{11}(\alpha) & \bar{A}_{12}(\alpha) \\ \bar{A}_{21}(\alpha) & \bar{A}_{22}(\alpha) \end{bmatrix} \begin{bmatrix} z_1(t) \\ z_2(t) \end{bmatrix} \\ &\quad + \begin{bmatrix} \bar{A}_{d11}(\alpha) & \bar{A}_{d12}(\alpha) \\ \bar{A}_{d21}(\alpha) & \bar{A}_{d22}(\alpha) \end{bmatrix} \begin{bmatrix} z_1(t-d(t)) \\ z_2(t-d(t)) \end{bmatrix} \\ &\quad + \begin{bmatrix} 0_{(n-m)\times m} \\ B_1 \end{bmatrix} (u(t) + F(\alpha)f(t))\,, \end{aligned} \tag{7.6}$$

where

$$
\begin{aligned}
&\bar{A}_{11}(\alpha) \triangleq U_1^T A(\alpha) U_1, \quad &&\bar{A}_{12}(\alpha) \triangleq U_1^T A(\alpha) U_2, \quad &&\bar{A}_{21}(\alpha) \triangleq U_2^T A(\alpha) U_1,\\
&\bar{A}_{22}(\alpha) \triangleq U_2^T A(\alpha) U_2, \quad &&\bar{A}_{d11}(\alpha) \triangleq U_1^T A_d(\alpha) U_1, \quad &&\bar{A}_{d12}(\alpha) \triangleq U_1^T A_d(\alpha) U_2,\\
&\bar{A}_{d21}(\alpha) \triangleq U_2^T A_d(\alpha) U_1, \quad &&\bar{A}_{d22}(\alpha) \triangleq U_2^T A_d(\alpha) U_2, \quad &&B_1 \triangleq \Gamma W^T.
\end{aligned}
$$

Obviously, the first subsystem of (7.6) represents the sliding mode dynamics. We design the following switching function:

$$
s(t) = Cz_1(t) + z_2(t), \tag{7.7}
$$

where $C \in \mathbf{R}^{m\times(n-m)}$ is the parametric matrix to be designed.

Remark 7.1 *Note that the switching function defined in (7.7) does not switch with the switching signal α (i.e. we design C not $C(\alpha)$ in (7.7)), that is, there is a unique non-switched sliding surface. The reason for this is to avoid repetitive jumps of the trajectories of the state components of the closed-loop system between sliding surfaces and hence the possible instability.* ♦

When the system state trajectories reach onto the sliding surface $s(t) = 0$, that is, $z_2(t) = -Cz_1(t)$, the sliding mode dynamics is attained. Substituting $z_2(t) = -Cz_1(t)$ into the first subsystem of (7.6) yields the sliding mode dynamics:

$$
\begin{aligned}
\dot{z}_1(t) = {}& \left(\bar{A}_{11}(\alpha) - \bar{A}_{12}(\alpha)C\right) z_1(t)\\
&+ \left(\bar{A}_{d11}(\alpha) - \bar{A}_{d12}(\alpha)C\right) z_1(t-d(t)).
\end{aligned} \tag{7.8}
$$

Now, we will analyze the stability of the sliding mode dynamics in (7.8) based on the result obtained in Theorem 5.2.3, and give the following theorem.

Theorem 7.3.1 *For a given constant $\beta > 0$, there exist matrices $P > 0$, $\mathcal{P} > 0$, $R(i) > 0$, $\mathcal{R}(i) > 0$, $\mathcal{Q}(i) > 0$, $\mathcal{S}(i) > 0$, $\mathcal{T}(i) > 0$, $\mathcal{X}(i)$, $\mathcal{Y}(i)$, and $\mathcal{K}$ such that for $i \in \mathcal{N}$,*

$$
\begin{bmatrix}
\breve{\Pi}_{11}(i) & \breve{\Pi}_{12}(i) & d\left(\bar{A}_{11}(i)\mathcal{P} - \bar{A}_{12}(i)\mathcal{K}\right)^T & d\mathcal{X}(i)\\
\star & \breve{\Pi}_{22}(i) & d\left(\bar{A}_{d11}(i)\mathcal{P} - \bar{A}_{d12}(i)\mathcal{K}\right)^T & d\mathcal{Y}(i)\\
\star & \star & -d\mathcal{R}(i) & 0\\
\star & \star & \star & -de^{-\beta d}\mathcal{S}(i)
\end{bmatrix} < 0, \tag{7.9a}
$$

$$
\begin{bmatrix}
-R(i) & P\\
\star & -\mathcal{T}(i)
\end{bmatrix} \leq 0, \tag{7.9b}
$$

$$
P\mathcal{P} = I,\ R(i)\mathcal{R}(i) = I,\ \mathcal{S}(i)\mathcal{T}(i) = I, \tag{7.9c}
$$

where

$$\begin{cases} \breve{\Pi}_{11}(i) \triangleq \bar{A}_{11}(i)\mathcal{P} + \mathcal{P}\bar{A}_{11}^T(i) - \bar{A}_{12}(i)\mathcal{K} - \mathcal{K}^T\bar{A}_{12}^T(i) \\ \qquad + \mathcal{X}(i) + \mathcal{X}^T(i) + \mathcal{Q}(i) + \beta\mathcal{P}, \\ \breve{\Pi}_{12}(i) \triangleq \bar{A}_{d11}(i)\mathcal{P} - \bar{A}_{d12}(i)\mathcal{K} + \mathcal{Y}^T(i) - \mathcal{X}(i), \\ \breve{\Pi}_{22}(i) \triangleq -(1-\tau)e^{-\beta d}\mathcal{Q}(i) - \mathcal{Y}(i) - \mathcal{Y}^T(i). \end{cases}$$

Then the sliding mode dynamics in (7.8) is exponentially stable for any switching signal with average dwell time satisfying $T_a > T_a^* = \frac{\ln \mu}{\beta}$, where $\mu \geq 1$ and satisfies

$$R(i) \leq \mu R(j),\ \mathcal{R}(i) \leq \mu\mathcal{R}(j),\ \mathcal{Q}(i) \leq \mu\mathcal{Q}(j),\quad \forall i,j \in \mathcal{N}. \tag{7.10}$$

Moreover, if the conditions above are feasible, the matrix C in (7.7) is given by $C = \mathcal{K}\mathcal{P}^{-1}$, that is, the switching function can be designed as

$$\begin{aligned} s(t) &= \mathcal{K}\mathcal{P}^{-1}z_1(t) + z_2(t) \\ &= \mathcal{K}Pz_1(t) + z_2(t). \end{aligned} \tag{7.11}$$

Proof. By Theorem 5.2.3, we know that if there exist matrices $P > 0$, $Q(i) > 0$, $R(i) > 0$, $X(i)$, and $Y(i)$ such that for $i \in \mathcal{N}$,

$$\begin{bmatrix} \hat{\Pi}_{11}(i) & \hat{\Pi}_{12}(i) & d\left(\bar{A}_{11}(i) - \bar{A}_{12}(i)C\right)^T R(i) & dX(i) \\ \star & \hat{\Pi}_{22}(i) & d\left(\bar{A}_{d11}(i) - \bar{A}_{d12}(i)C\right)^T R(i) & dY(i) \\ \star & \star & -dR(i) & 0 \\ \star & \star & \star & -de^{-\beta d}R(i) \end{bmatrix} < 0, \tag{7.12}$$

where

$$\begin{cases} \hat{\Pi}_{11}(i) \triangleq P\left(\bar{A}_{11}(i) - \bar{A}_{12}(i)C\right) + \left(\bar{A}_{11}(i) - \bar{A}_{12}(i)C\right)^T P \\ \qquad + X(i) + X^T(i) + Q(i) + \beta P, \\ \hat{\Pi}_{12}(i) \triangleq P\left(\bar{A}_{d11}(i) - \bar{A}_{d12}(i)C\right) + Y^T(i) - X(i), \\ \hat{\Pi}_{22}(i) \triangleq -(1-\tau)e^{-\beta d}Q(i) - Y(i) - Y^T(i), \end{cases}$$

then the sliding mode dynamics in (7.8) is exponentially stable for any switching signal with average dwell time satisfying $T_a > T_a^* = \frac{\ln \mu}{\beta}$, where $\mu \geq 1$ and satisfies

$$Q(i) \leq \mu Q(j),\quad R(i) \leq \mu R(j),\quad \forall i,j \in \mathcal{N}. \tag{7.13}$$

Define the following matrices:

$$\begin{cases} \mathcal{P} \triangleq P^{-1}, \quad \mathcal{K} \triangleq C\mathcal{P}, \\ \mathcal{R}(i) \triangleq R^{-1}(i), \quad \mathcal{Q}(i) \triangleq \mathcal{P}Q(i)\mathcal{P}, \\ \mathcal{X}(i) \triangleq \mathcal{P}X(i)\mathcal{P}, \quad \mathcal{Y}(i) \triangleq \mathcal{P}Y(i)\mathcal{P}, \quad \forall i \in \mathcal{N}. \end{cases} \tag{7.14}$$

Performing a congruence transformation on (7.12) with diag $\{\mathcal{P}, \mathcal{P}, \mathcal{R}(i), \mathcal{P}\}$, we have

$$\begin{bmatrix} \check{\Pi}_{11}(i) & \check{\Pi}_{12}(i) & d\left(\bar{A}_{11}(i)\mathcal{P} - \bar{A}_{12}(i)\mathcal{K}\right)^T & d\mathcal{X}(i) \\ \star & \check{\Pi}_{22}(i) & d\left(\bar{A}_{d11}(i)\mathcal{P} - \bar{A}_{d12}(i)\mathcal{K}\right)^T & d\mathcal{Y}(i) \\ \star & \star & -d\mathcal{R}(i) & 0 \\ \star & \star & \star & -de^{-\beta d}\mathcal{P}\mathcal{R}^{-1}(i)\mathcal{P} \end{bmatrix} < 0, \tag{7.15}$$

where $\check{\Pi}_{11}(i)$, $\check{\Pi}_{12}(i)$, and $\check{\Pi}_{22}(i)$ are defined in (7.9a).

Notice that (7.15) is not of LMI form because of the term of $\mathcal{P}\mathcal{R}^{-1}(i)\mathcal{P}$. Now, replacing $\mathcal{P}\mathcal{R}^{-1}(i)\mathcal{P}$ in (7.15) with $\mathcal{S}(i) > 0$, it follows that (7.15) holds if (7.9a) holds and for $i \in \mathcal{N}$,

$$\mathcal{P}\mathcal{R}^{-1}(i)\mathcal{P} \geq \mathcal{S}(i). \tag{7.16}$$

By Schur complement, (7.16) is equivalent to

$$\begin{bmatrix} -\mathcal{R}^{-1}(i) & \mathcal{P}^{-1} \\ \star & -\mathcal{S}^{-1}(i) \end{bmatrix} \leq 0, \tag{7.17}$$

which implies (7.9b) by (7.14) and letting $\mathcal{T}(i) = \mathcal{S}^{-1}(i)$.

Moreover, considering (7.13)–(7.14), we have (7.10). This completes the proof. ■

Remark 7.2 *It should be pointed out that the matrix variables P and $\mathcal{P}$ in Theorem 7.3.1 do not depend on the switching set and are fixed. As the designed switching function in (7.7) is a parameter-independent function, the parameter $C = \mathcal{K}\mathcal{P}^{-1}$ in (7.7) is guaranteed to be fixed if the matrix variable $\mathcal{P}$ is fixed.* ♦

Note that the conditions in Theorem 7.3.1 are not all of strict LMI form due to (7.9c), so we can not solve them by LMI procedures directly. Now, by using CCL method [66], we suggest the following minimization problem involving LMI conditions instead of the original nonconvex feasibility problem formulated in Theorem 7.3.1.

Problem SMDA (Sliding mode dynamics analysis)

$$\min \quad \text{trace}\left(P\mathcal{P} + \sum_{i\in\mathcal{N}} R(i)\mathcal{R}(i) + \sum_{i\in\mathcal{N}} \mathcal{S}(i)\mathcal{T}(i)\right)$$

subject to (7.9a)–(7.9b), (7.10) and for $i \in \mathcal{N}$,

$$\begin{bmatrix} P & I \\ I & \mathcal{P} \end{bmatrix} \geq 0, \quad \begin{bmatrix} R(i) & I \\ I & \mathcal{R}(i) \end{bmatrix} \geq 0, \quad \begin{bmatrix} \mathcal{S}(i) & I \\ I & \mathcal{T}(i) \end{bmatrix} \geq 0. \tag{7.18}$$

According to the CCL method [66], if the solution of the above minimization problem is $(1 + 2N)(n - m)$, then the conditions in Theorem 7.4.0 are solvable. We give the following algorithm to solve Problem SMDA.

Algorithm SMDA

Step 1. Find a feasible set $(P^{(0)}, \mathcal{P}^{(0)}, R^{(0)}(i), \mathcal{R}^{(0)}(i), \mathcal{Q}^{(0)}(i), \mathcal{S}^{(0)}(i), \mathcal{T}^{(0)}(i), \mathcal{X}^{(0)}(i), \mathcal{Y}^{(0)}(i), \mathcal{K}^{(0)})$ satisfying (7.9a)–(7.9b), (7.10), and (7.18). Set $\kappa = 0$.
Step 2. Solve the following optimization problem:

$$\min \quad \text{trace}\left(\begin{array}{c} P^{(\kappa)}\mathcal{P} + P\mathcal{P}^{(\kappa)} + \\ \sum_{i \in \mathcal{N}} \left(\begin{array}{c} R^{(\kappa)}(i)\mathcal{R}(i) + R(i)\mathcal{R}^{(\kappa)}(i) + \\ \mathcal{S}^{(\kappa)}(i)\mathcal{T}(i) + \mathcal{S}(i)\mathcal{T}^{(\kappa)}(i) \end{array} \right) \end{array} \right)$$

subject to (7.9a)–(7.9b), (7.10), and (7.18)

and denote f^* as the optimized value.
Step 3. Substitute the obtained matrices $(P, \mathcal{P}, R(i), \mathcal{R}(i), \mathcal{Q}(i), \mathcal{S}(i), \mathcal{T}(i), \mathcal{X}(i), \mathcal{Y}(i), \mathcal{K})$ into (7.17). If (7.17) is satisfied, with

$$\left| f^* - (2 + 4N)(n - m) \right| < \varepsilon,$$

for a sufficiently small scalar $\varepsilon > 0$, then output the feasible solutions $(P, \mathcal{P}, R(i), \mathcal{R}(i), \mathcal{Q}(i), \mathcal{S}(i), \mathcal{T}(i), \mathcal{X}(i), \mathcal{Y}(i), \mathcal{K})$, so EXIT.
Step 4. If $\kappa > \mathbb{N}$ where $\mathbb{N}$ is the maximum number of iterations allowed, so EXIT.
Step 5. Set $\kappa = \kappa + 1$, $(P^{(\kappa)}, \mathcal{P}^{(\kappa)}, R^{(\kappa)}(i), \mathcal{R}^{(\kappa)}(i), \mathcal{Q}^{(\kappa)}(i), \mathcal{S}^{(\kappa)}(i), \mathcal{T}^{(\kappa)}(i), \mathcal{X}^{(\kappa)}(i), \mathcal{Y}^{(\kappa)}(i), \mathcal{K}^{(\kappa)}) = (P, \mathcal{P}, R(i), \mathcal{R}(i), \mathcal{Q}(i), \mathcal{S}(i), \mathcal{T}(i), \mathcal{X}(i), \mathcal{Y}(i), \mathcal{K})$, and go to Step 2.

7.3.2 SMC Law Design

In the following, we are in a position to synthesize an SMC law to drive the system state trajectories onto the predefined sliding surface $s(t) = 0$, and give the following result.

Theorem 7.3.2 *Suppose that the conditions in (7.9a)–(7.10) have a set of feasible solutions* $P > 0$, $\mathcal{P} > 0$, $R(i) > 0$, $\mathcal{R}(i) > 0$, $\mathcal{Q}(i) > 0$, $\mathcal{S}(i) > 0$, $\mathcal{T}(i) > 0$, $\mathcal{X}(i)$, $\mathcal{Y}(i)$, *and* $\mathcal{K}$, *and the* *switching function is given by (7.11). Then the state trajectories of the closed-loop system (7.6) can be driven onto the sliding surface $s(t) = 0$ in a finite time by the control of*

$$\begin{aligned}u(t) = -B_1^{-1}\{&\mathcal{K}\mathcal{P}^{-1}[\bar{A}_{11}(i)z_1(t) + \bar{A}_{12}(i)z_2(t) + \bar{A}_{d11}(i)z_1(t-d(t))\\&+ \bar{A}_{d12}(i)z_2(t-d(t))] + \bar{A}_{21}(i)z_1(t) + \bar{A}_{22}(i)z_2(t)\\&+ \bar{A}_{d21}(i)z_1(t-d(t)) + \bar{A}_{d22}(i)z_2(t-d(t))\}\\&- (\rho(i) + \eta(i))\,\mathrm{sign}\left(B_1^T s(t)\right),\end{aligned} \tag{7.19}$$

where $\rho(i) > 0$, $i \in \mathcal{N}$ are adjustable parameters.

Proof. We will show that the control law (7.19) can not only drive the system state trajectories onto the sliding surface, but also keep it there for all subsequent time. Consider the switching function as

$$s(t) = \mathcal{K}\mathcal{P}^{-1}z_1(t) + z_2(t), \tag{7.20}$$

and choose the following Lyapunov function:

$$W(t) \triangleq \frac{1}{2}s^T(t)s(t). \tag{7.21}$$

Then as with the solution of the system in (7.6) for a fixed α, we have

$$\begin{aligned}\dot{W}(t) &= s^T(t)\dot{s}(t) = s^T(t)\left(\mathcal{K}\mathcal{P}^{-1}\dot{z}_1(t) + \dot{z}_2(t)\right)\\&= s^T(t)\{\mathcal{K}\mathcal{P}^{-1}[\bar{A}_{11}(\alpha)z_1(t) + \bar{A}_{12}(\alpha)z_2(t) + \bar{A}_{d11}(\alpha)z_1(t-d(t))\\&\quad + \bar{A}_{d12}(\alpha)z_2(t-d(t))] + \bar{A}_{21}(\alpha)z_1(t) + \bar{A}_{22}(\alpha)z_2(t) + \bar{A}_{d21}(\alpha)z_1(t-d(t))\\&\quad + \bar{A}_{d22}(\alpha)z_2(t-d(t)) + B_1u(t) + B_1F(\alpha)f(t)\}.\end{aligned} \tag{7.22}$$

Substituting the following control law into (7.22):

$$\begin{aligned}u(t) = &-B_1^{-1}\{\mathcal{K}\mathcal{P}^{-1}[\bar{A}_{11}(\alpha)z_1(t) + \bar{A}_{12}(\alpha)z_2(t) + \bar{A}_{d11}(\alpha)z_1(t-d(t))\\&+ \bar{A}_{d12}(\alpha)z_2(t-d(t))] + \bar{A}_{21}(\alpha)z_1(t) + \bar{A}_{22}(\alpha)z_2(t)\\&+ \bar{A}_{d21}(\alpha)z_1(t-d(t)) + \bar{A}_{d22}(\alpha)z_2(t-d(t))\}\\&- (\rho(\alpha) + \eta(\alpha))\,\mathrm{sign}\left(B_1^T s(t)\right),\end{aligned} \tag{7.23}$$

and noting that $\left\|s^T(t)B_1\right\| \leq \left|s^T(t)B_1\right|$, we have

$$\begin{aligned}
\dot{W}(t) &= s^T(t)\left[-B_1(\rho(\alpha)+\eta(\alpha))\,\mathrm{sign}\left(B_1^T s(t)\right) + B_1 F(\alpha)f(t)\right] \\
&= -s^T(t)B_1\left[(\rho(\alpha)+\eta(\alpha))\,\mathrm{sign}\left(B_1^T s(t)\right) - F(\alpha)f(t)\right] \\
&\leq -(\rho(\alpha)+\eta(\alpha))\left|s^T(t)B_1\right| + \eta(\alpha)\left\|s^T(t)B_1\right\| \\
&\leq -\rho(\alpha)\left\|s^T(t)B_1\right\| \\
&\leq -\sqrt{2\lambda_{\min}\left(B_1B_1^T\right)}\min_{\forall\alpha\in\mathcal{N}}(\rho(\alpha))\,W^{1/2}(t) \\
&\triangleq -\rho W^{1/2}(t) < 0,
\end{aligned}$$

where $\rho \triangleq \sqrt{2\lambda_{\min}\left(B_1B_1^T\right)}\min_{\forall\alpha\in\mathcal{N}}(\rho(\alpha)) > 0$.

As in the proof of Theorem 5.2.3, for an arbitrary piecewise constant switching signal α, and for any $t > 0$, we let $0 = t_0 < t_1 < \cdots < t_k < \cdots$, $k = 0, 1, \ldots$, denote the switching points of α over the interval $(0, t)$. The i_kth subsystem is activated when $t \in \left[t_k, t_{k+1}\right)$. Integrating $\dot{W}(t) \leq -\rho W^{1/2}(t)$ from t_k to t and t_{k-1} to t_k, $k = 1, 2, \ldots$, we have

$$\begin{cases}
W^{1/2}(t) - W^{1/2}(t_k) \leq -\frac{1}{2}\rho(t - t_k), \\
W^{1/2}(t_k) - W^{1/2}(t_{k-1}) \leq -\frac{1}{2}\rho(t_k - t_{k-1}), \\
\qquad\vdots \\
W^{1/2}(t_1) - W^{1/2}(0) \leq -\frac{1}{2}\rho(t_1 - 0).
\end{cases} \tag{7.24}$$

Summing the terms on both sides of (7.24) gives

$$W^{1/2}(t) - W^{1/2}(0) \leq -\frac{1}{2}\rho t. \tag{7.25}$$

It can be seen from (7.25) that there exists a time $t^* \leq 2W^{1/2}(0)/\rho$ such that $W(t) = 0$, and consequently $s(t) = 0$, for $t \geq t^*$, which means that the system state trajectories can reach onto the predefined sliding surface $s(t) = 0$ in a finite time. Since the reaching condition $s^T(t)\dot{s}(t) < 0$ holds, the system state trajectories can be driven onto the predefined sliding surface and maintained there for all subsequent time. This completes the proof. ■

Notice that the SMC law in (7.19) is applicable only when the time-varying delay $d(t)$ is explicitly known *a priori*, since there exist $z_1(t - d(t))$ and $z_2(t - d(t))$ in (7.19). However, in some practical situations, the information for delay $d(t)$ is unavailable, or difficult to measure. To overcome this, in what follows, we provide another kind of SMC law.

We assume that there exists a constant $r > 0$ such that

$$\|z(t+\theta)\| \leq r\,\|z(t)\|, \quad \theta \in [-d, 0], \tag{7.26}$$

where the constant r is not known *a priori*, which is often the case in practical situations. Therefore, to obtain the value of r, we should design an adaptive law first to estimate it, and thus give an adaptive SMC law for system (7.6). Let $r(t)$ represent the estimate of r. The corresponding estimation error is $\tilde{r}(t) = r(t) - r$.

Theorem 7.3.3 *Suppose the conditions in (7.9a)–(7.10) have a set of feasible solutions $P > 0$, $\mathcal{P} > 0$, $R(i) > 0$, $\mathcal{R}(i) > 0$, $\mathcal{Q}(i) > 0$, $\mathcal{S}(i) > 0$, $\mathcal{T}(i) > 0$, $\mathcal{X}(i)$, $\mathcal{Y}(i)$, and $\mathcal{K}$, and the switching function is given by (7.11). Then the state trajectories of the closed-loop system (7.6) can be driven onto the sliding surface $s(t) = 0$ with the control of*

$$
\begin{aligned}
u(t) = & \frac{-1}{\sqrt{\lambda_{\min}\left(B_1 B_1^T\right)}} \left\{ \delta(i) + \|B_1\| \eta(i) + \left\|\mathcal{K}\mathcal{P}^{-1}\right\| \left(\left\| \begin{bmatrix} \bar{A}_{11}(i) & \bar{A}_{12}(i) \end{bmatrix} \right\| \right.\right. \\
& + r(t) \left\| \begin{bmatrix} \bar{A}_{d11}(i) & \bar{A}_{d12}(i) \end{bmatrix} \right\| \Big) \|z(t)\| + \Big(\left\| \begin{bmatrix} \bar{A}_{21}(i) & \bar{A}_{22}(i) \end{bmatrix} \right\| \\
& \left.\left. + r(t) \left\| \begin{bmatrix} \bar{A}_{d21}(i) & \bar{A}_{d22}(i) \end{bmatrix} \right\| \right) \|z(t)\| \right\} \operatorname{sign}\left(B_1^T s(t)\right),
\end{aligned} \tag{7.27}
$$

where $\delta(i) > 0$, $i \in \mathcal{N}$ are constants, and the adaptive law is given as

$$
\begin{aligned}
\dot{r}(t) = \frac{1}{l} \min_{\forall i \in \mathcal{N}} & \left\{ \left\|\mathcal{K}\mathcal{P}^{-1}\right\| \left\| \begin{bmatrix} \bar{A}_{d11}(i) & \bar{A}_{d12}(i) \end{bmatrix} \right\| \right. \\
& \left. + \left\| \begin{bmatrix} \bar{A}_{d21}(i) & \bar{A}_{d22}(i) \end{bmatrix} \right\| \right\} \|z(t)\| \, \|s(t)\|,
\end{aligned} \tag{7.28}
$$

with $r(0) = 0$, where $l > 0$ is a given scalar.

Proof. Choose a Lyapunov function of the following form:

$$
W_1(t) \triangleq \frac{1}{2} \left(s^T(t)s(t) + l\tilde{r}^2(t) \right).
$$

Then as with the solution of the system in (7.6) for a fixed α and by noting (7.26), we have

$$
\begin{aligned}
\dot{W}_1(t) &= s^T(t) \left(\mathcal{K}\mathcal{P}^{-1}\dot{z}_1(t) + \dot{z}_2(t) \right) + l\tilde{r}(t)\dot{\tilde{r}}(t) \\
&= s^T(t) \left\{ \mathcal{K}\mathcal{P}^{-1} \left[\bar{A}_{11}(\alpha)z_1(t) + \bar{A}_{12}(\alpha)z_2(t) + \bar{A}_{d11}(\alpha)z_1(t-d(t)) \right.\right. \\
&\quad \left. + \bar{A}_{d12}(\alpha)z_2(t-d(t)) \right] + \bar{A}_{21}(\alpha)z_1(t) + \bar{A}_{22}(\alpha)z_2(t) + \bar{A}_{d21}(\alpha)z_1(t-d(t)) \\
&\quad \left. + \bar{A}_{d22}(\alpha)z_2(t-d(t)) + B_1 u(t) + B_1 F(\alpha) f(t) \right\} + l\tilde{r}(t)\dot{\tilde{r}}(t) \\
&\leq \|s(t)\| \left\{ \left\|\mathcal{K}\mathcal{P}^{-1}\right\| \left[\left\| \begin{bmatrix} \bar{A}_{11}(\alpha) & \bar{A}_{12}(\alpha) \end{bmatrix} \right\| + r \left\| \begin{bmatrix} \bar{A}_{d11}(\alpha) & \bar{A}_{d12}(\alpha) \end{bmatrix} \right\| \right] \|z(t)\| \right. \\
&\quad \left. + \left[\left\| \begin{bmatrix} \bar{A}_{21}(\alpha) & \bar{A}_{22}(\alpha) \end{bmatrix} \right\| + r \left\| \begin{bmatrix} \bar{A}_{d21}(\alpha) & \bar{A}_{d22}(\alpha) \end{bmatrix} \right\| \right] \|z(t)\| \right\} \\
&\quad + s^T(t) B_1 u(t) + \|s(t)\| \, \|B_1\| \, \|F(\alpha) f(t)\| + l\tilde{r}(t)\dot{\tilde{r}}(t).
\end{aligned} \tag{7.29}
$$

Substituting the control law (7.27) into (7.29) yields

$$\dot{W}_1(t) \leq -\tilde{r}(t)\left(\left\|\mathcal{K}\mathcal{P}^{-1}\right\|\left\|\left[\bar{A}_{d11}(\alpha)\ \ \bar{A}_{d12}(\alpha)\right]\right\| + \left\|\left[\bar{A}_{d21}(\alpha)\ \ \bar{A}_{d22}(\alpha)\right]\right\|\right)$$
$$\times \|z(t)\|\ \|s(t)\| - \delta(\alpha)\ \|s(t)\| + l\tilde{r}(t)\dot{\tilde{r}}(t). \tag{7.30}$$

Note from (7.28) that $\dot{r}(t) > 0$, which implies $\dot{\tilde{r}}(t) > 0$. Therefore, there exists a time instant $t^{\#}$ such that $\tilde{r}(t) > 0$ for $t > t^{\#}$, and consequently $l\tilde{r}(t)\dot{\tilde{r}}(t) > 0$ for $t > t^{\#}$. Substituting the adaptive law (7.28) (with i replaced by α) into (7.30), when $t > t^{\#}$ we have

$$\begin{aligned}\dot{W}_1(t) \leq -\delta(\alpha)\ \|s(t)\| &\leq -\min_{\forall\alpha\in\mathcal{N}}(\delta(\alpha))\ \|s(t)\| \\ &= -\delta\ \|s(t)\| < 0,\end{aligned} \tag{7.31}$$

where $\delta \triangleq \min_{\alpha\in\mathcal{N}}(\delta(\alpha)) > 0$. By (7.31) and noting $l\tilde{r}(t)\dot{\tilde{r}}(t) > 0$ for $t > t^{\#}$, we have $s^T(t)\dot{s}(t) < 0$ for $t > t^{\#}$, thus the reaching condition is satisfied. This completes the proof. ■

7.4 Illustrative Example

Example 7.4.1 Consider the switched state-delayed hybrid system in (7.1a)–(7.1b) with $N = 2$ (that is, there are two subsystems) and the following parameters:

$$A(1) = \begin{bmatrix} -0.9 & 0.2 & -0.2 \\ 0.2 & -0.1 & 0.3 \\ -0.3 & 0.1 & 0.3 \end{bmatrix},\ A_d(1) = \begin{bmatrix} 0.2 & 0 & 0.1 \\ 0.1 & 0.3 & 0.1 \\ 0.3 & 0.1 & 0.2 \end{bmatrix},\ B = \begin{bmatrix} 0.0 \\ 0.0 \\ 2.0 \end{bmatrix},$$

$$A(2) = \begin{bmatrix} -0.8 & -0.1 & -0.2 \\ 0.2 & -0.1 & 0.3 \\ 0.2 & -0.1 & 0.2 \end{bmatrix},\ A_d(2) = \begin{bmatrix} 0.2 & 0.1 & 0.0 \\ 0.1 & 0.2 & 0.1 \\ 0.1 & 0.1 & 0.3 \end{bmatrix},\ \begin{aligned} F(1) &= 1.6, \\ F(2) &= 2.0, \end{aligned}$$

and $d = 2.0$, $\beta = 0.6$, $\tau = 0.5$, and $f(t) = 0.5\exp(-t)\sin(t)$. It can be verified that system (7.1a)–(7.1b) with $u(t) = 0$ and the above parameters is unstable for a switching signal given in Figure 7.1 (which is generated randomly; here, '1' and '2' represent the first and second subsystems, respectively), the states of the open-loop system are shown in Figure 7.2 with the initial condition given by $x(\theta) = \begin{bmatrix} -1.0 & 0.5 & 1.0 \end{bmatrix}^T$, $\theta \in [-2, 0]$. Therefore, our aim is to design an SMC law $u(t)$ such that the closed-loop system is stable with arbitrary switching. To check the stability of (7.8) with arbitrary switching, we solve conditions (7.9a)–(7.9c) in Theorem 7.3.1 with $R(i) = R(j) = R$, $\mathcal{R}(i) = \mathcal{R}(j) = \mathcal{R}$, $\mathcal{Q}(i) = \mathcal{Q}(j) = \mathcal{Q}$, $\forall i, j \in \mathcal{N}$ by Algorithm SMDA, which gives

$$\mathcal{P} = \begin{bmatrix} 1.3480 & -0.4896 \\ -0.4896 & 0.3396 \end{bmatrix},\quad \mathcal{K} = \begin{bmatrix} 0.4537 & 0.3438 \end{bmatrix}.$$

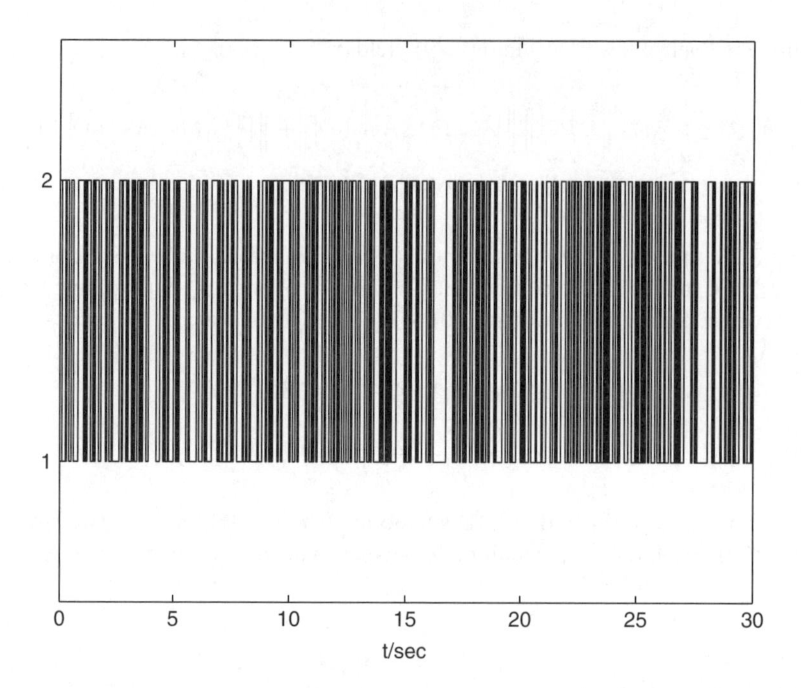

Figure 7.1 Switching signal

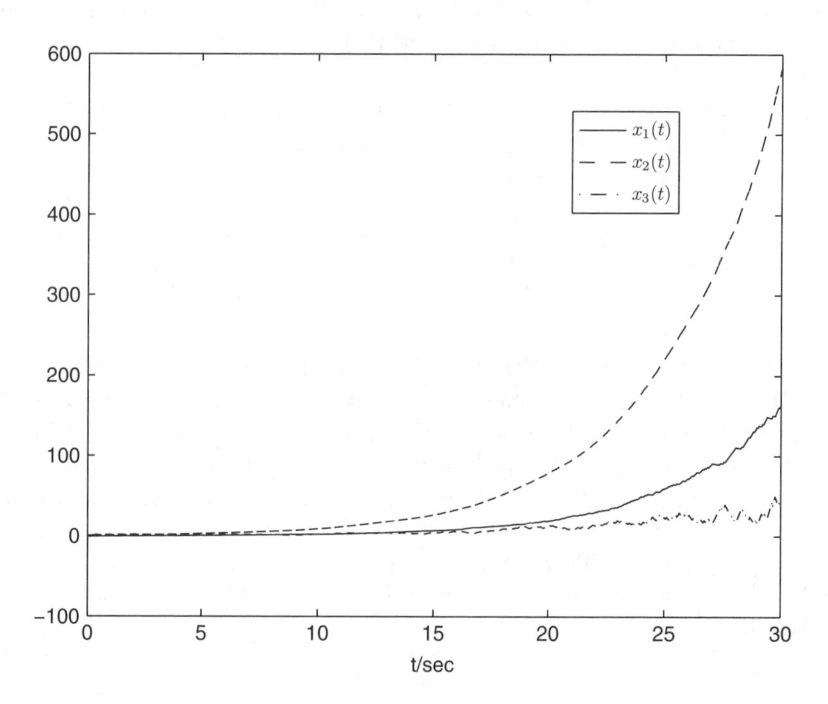

Figure 7.2 States of the open-loop system

According to (7.11), we have

$$s(t) = \mathcal{K}\mathcal{P}^{-1}z_1(t) + z_2(t) = \begin{bmatrix} 1.4786 & 3.1441 & 1.0000 \end{bmatrix} x(t). \tag{7.32}$$

The existence of a feasible solution shows that there exists a mode-independent Lyapunov function for checking the exponential stability of the sliding mode dynamics in (7.8), that is, we can find a desired switching function in (7.32) such that the resulting sliding mode dynamics in (7.8) is exponentially stable for arbitrary switching. The remaining task is to design an SMC law such that the system state trajectories can be driven onto the predefined sliding surface $s(t) = 0$ and maintained there for all subsequent time. When delay $d(t)$ in (7.1a)–(7.1b) is explicitly given as $d(t) = 1.5 + 0.5\sin t$, the SMC law in (7.19) can be computed as

$$u(t) = \begin{cases} u(t,1) = -\frac{1}{2}\left\{\begin{bmatrix} -1.0019 & 0.0813 & 0.9475 \end{bmatrix} x(t) \right. \\ \qquad \left. + \begin{bmatrix} 0.9101 & 1.0432 & 0.6623 \end{bmatrix} x(t-d(t))\right\} \\ \qquad - (\delta(1) + 0.8)\,\mathrm{sign}\,(s(t)), \quad i = 1, \\ u(t,2) = -\frac{1}{2}\left\{\begin{bmatrix} -0.3541 & -0.5623 & 0.8475 \end{bmatrix} x(t) \right. \\ \qquad \left. + \begin{bmatrix} 0.7101 & 0.8767 & 0.6144 \end{bmatrix} x(t-d(t))\right\} \\ \qquad - (\delta(2) + 1.0)\,\mathrm{sign}\,(s(t)), \quad i = 2. \end{cases} \tag{7.33}$$

When delay $d(t)$ in (7.1a)–(7.1b) is unknown, the SMC law designed in (7.27)–(7.28) can be applied, and given by

$$u(t) = \begin{cases} u(t,1) = -\frac{1}{2}\{(3.8603 + 1.5899r(t))\,\|x(t)\| \\ \qquad + \delta(1) + 1.6\}\,\mathrm{sign}\,(s(t)), \quad i = 1, \\ u(t,2) = -\frac{1}{2}\{(3.3311 + 1.4042r(t))\,\|x(t)\| \\ \qquad + \delta(2) + 2.0\}\,\mathrm{sign}\,(s(t)), \quad i = 2. \end{cases} \tag{7.34}$$

Set $\delta(1) = \delta(2) = 2$ and $l = 1$. The adaptive law in (7.28) is computed as

$$\dot{r}(t) = 1.4042\,\|x(t)\|\,\|s(t)\|.$$

To reduce the chattering, we replace $\mathrm{sign}\,(s(t))$ with $s(t)/\,(0.01 + \|s(t)\|)$. Figure 7.3 shows the state response of the closed-loop switched system with (7.33). The switching function and the control input are given in Figures 7.4 and 7.5, respectively. The corresponding simulation results with (7.34) are given in Figures 7.6–7.9.

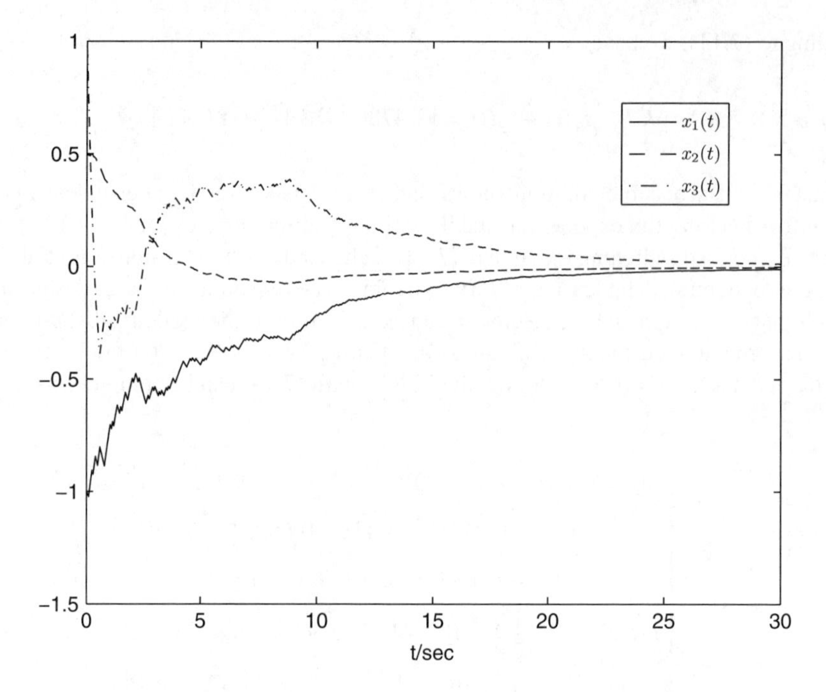

Figure 7.3 States of the closed-loop system with (7.33)

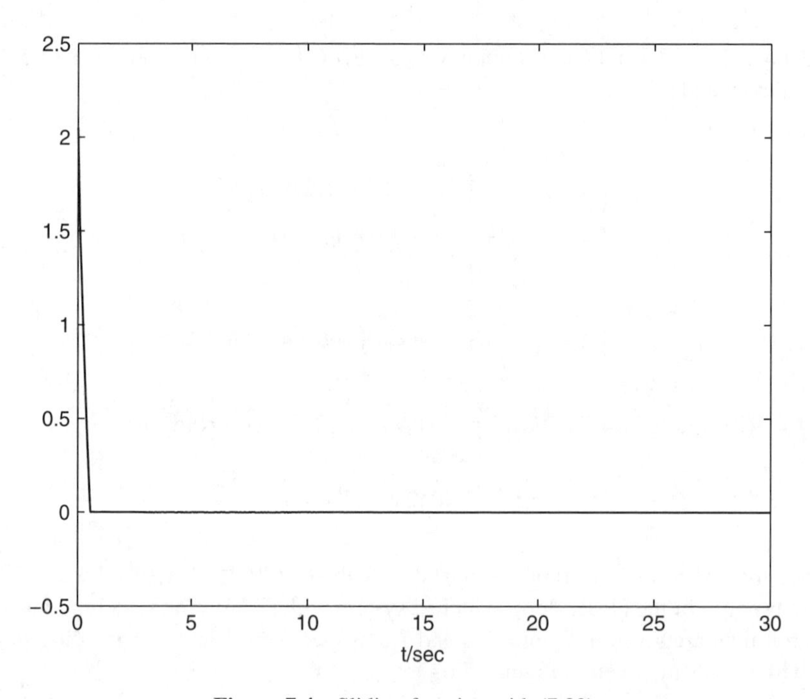

Figure 7.4 Sliding function with (7.33)

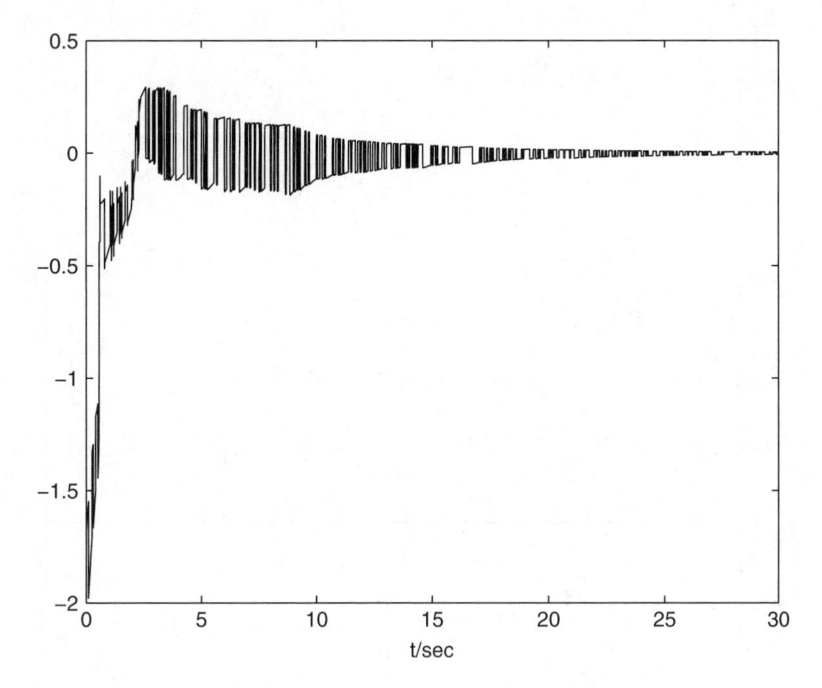

Figure 7.5 Control input (7.33)

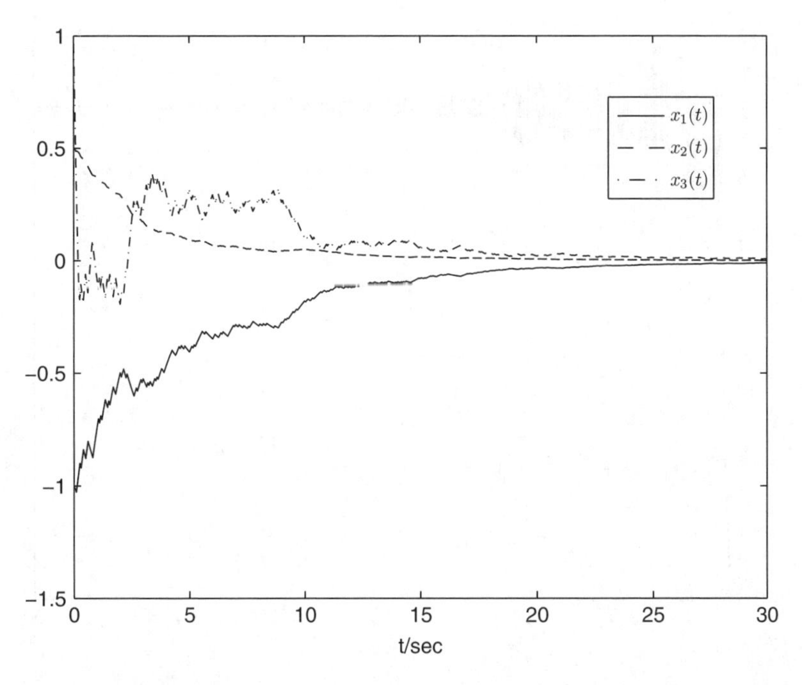

Figure 7.6 States of the closed-loop system with (7.34)

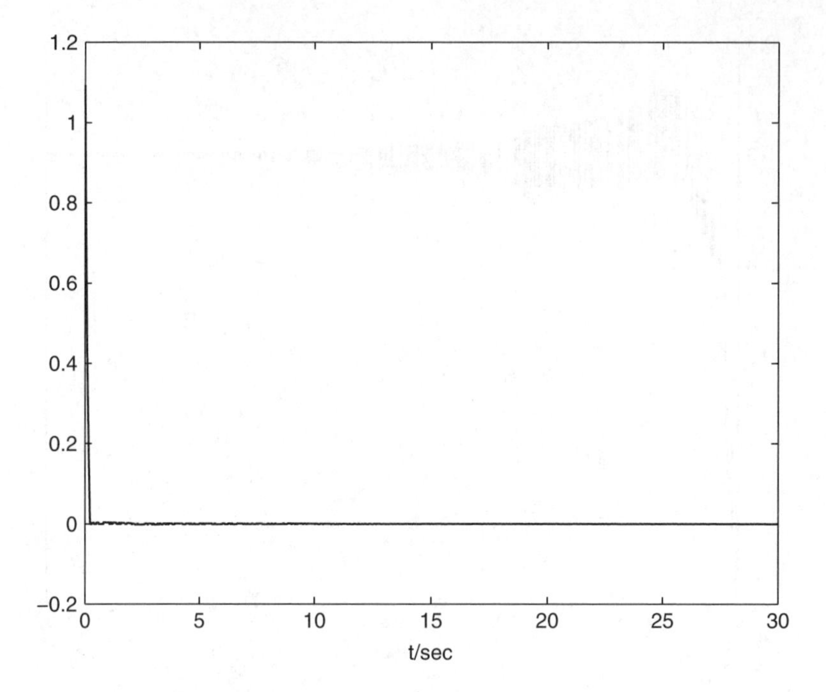

Figure 7.7 Sliding function with (7.34)

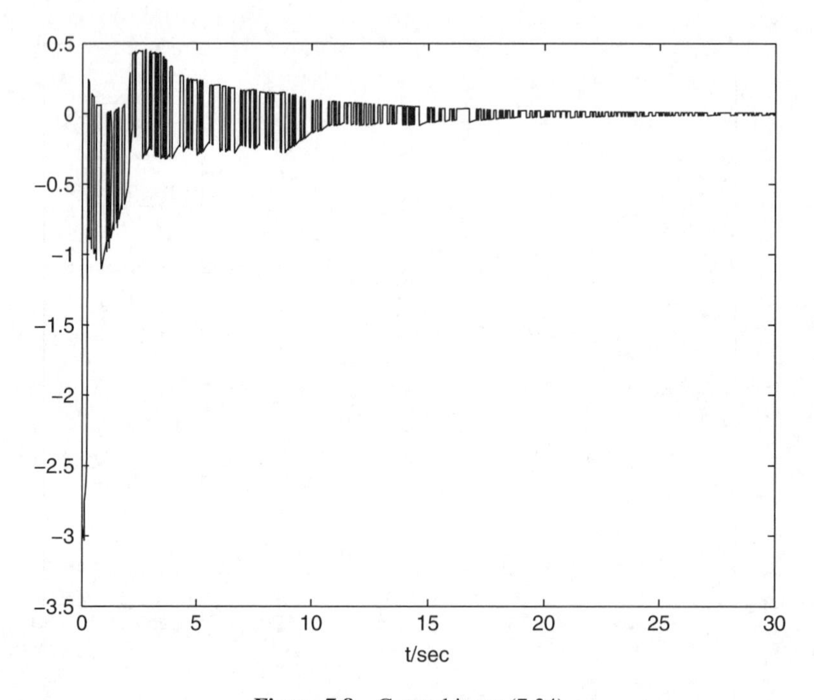

Figure 7.8 Control input (7.34)

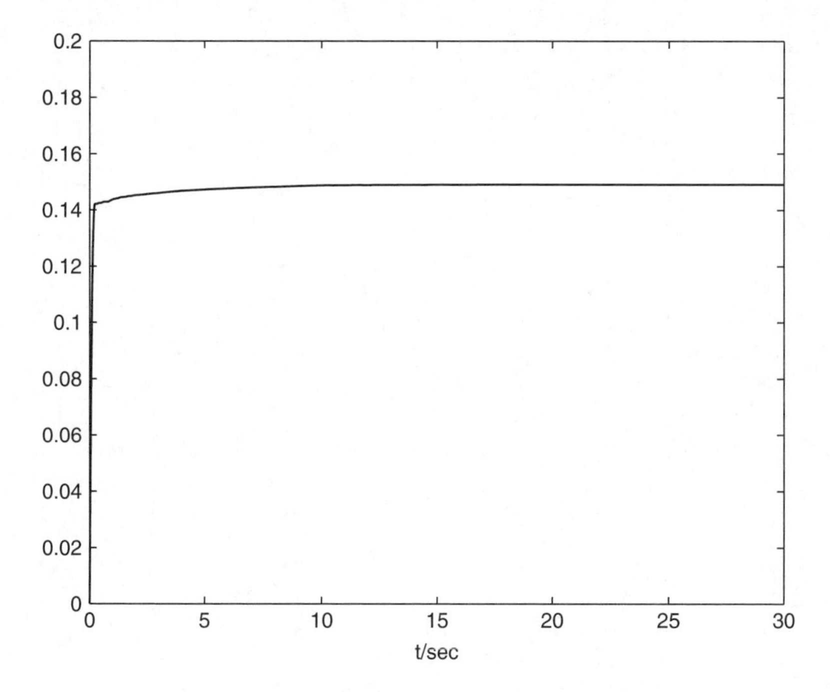

Figure 7.9 Adaptive estimate $r(t)$

7.5 Conclusion

In this chapter, the SMC design problem has been investigated for continuous-time switched systems with time-varying delay. By model transformation, the system has first been transformed into the regular form, and then the sliding mode dynamics has been derived by designing a linear switching function. The corresponding sufficient condition for the existence of resulting sliding mode dynamics has been derived, and an explicit parametrization of the desired sliding surface has also been given. In addition, an adaptive SMC law for reaching motion has been designed such that the system state trajectories can be driven onto the prescribed sliding surface in a finite time and maintained there for all subsequent time. Finally, a numerical example has been provided to illustrate the effectiveness of the design scheme.

8

SMC of Switched State-Delayed Hybrid Systems: Discrete-Time Case

8.1 Introduction

In this chapter, we will study the SMC design problem for discrete-time switched hybrid systems with time-varying delay. First, we transform the original system into a new one with regular form, and then by designing a linear switching function, a reduced-order sliding mode dynamics, described by a switched state-delayed hybrid system, is generated. By utilizing the average dwell time approach and the piecewise Lyapunov function technique, a delay-dependent sufficient condition for the existence of the desired sliding mode is proposed in terms of LMIs, and an explicit parametrization of the desired switching surface is also given. Here, to reduce the conservativeness induced by the time delay in the system, both the slack matrix technique and the delay partitioning method are employed, which makes the proposed existence condition less conservative and more practical. The delay partitioning – also called delay fractioning – has been considered as an effective approach to reduce the conservativeness of the stability condition for time-delay systems. This was initially proposed by Gouaisbaut and Peaucelle in [86], and was then developed in [58, 73, 220, 271]. The basic idea of the delay partitioning method is to evenly partition the time delay into several components (this generally means time-invariant delay), and then take each time-delay component into account individually when constructing a Lyapunov function. In this chapter, the time delay considered is a time-varying one with a known lower bound. In this case, combining with construction of an appropriate Lyapunov–Krasovskii function, the delay partitioning method is used by evenly partitioning the lower bound into several components. It is shown that the conservativeness of the obtained existence condition becomes less as the partitioning gets thinner. Finally, a discontinuous SMC law is designed to drive the state trajectories of the closed-loop system onto a prescribed sliding surface in a finite time and maintained there for all subsequent time.

Sliding Mode Control of Uncertain Parameter-Switching Hybrid Systems, First Edition. Ligang Wu, Peng Shi and Xiaojie Su.

8.2 System Description and Preliminaries

Consider the following discrete-time switched state-delayed hybrid system:

$$x(k+1) = A(\alpha(k))x(k) + A_d(\alpha(k))x(k-d(k)) + B(u(k) + F(\alpha(k))f(x,k)), \tag{8.1a}$$

$$x(k) = \phi(k), \quad k = -h_2, -h_2+1, -h_2+2, \ldots, 0, \tag{8.1b}$$

where $x(k) \in \mathbf{R}^n$ is the system state vector; $u(k) \in \mathbf{R}^m$ is the control input; $f(x,k) \in \mathbf{R}^p$ is the nonlinearity; $\{A(\alpha(k)), A_d(\alpha(k)), F(\alpha(k)) : \alpha(k) \in \mathcal{N}\}$ is a family of matrices parameterized by an index set $\mathcal{N} = \{1, 2, \ldots, N\}$; and $\alpha(k) : \mathbf{Z}^+ \rightarrow \mathcal{N}$ (denoted by α for simplicity) is the switching signal defined as the same in Chapter 5. Also, $x(k) = \phi(k)$, $k = -h_2, -h_2+1, \ldots, 0$ are the initial conditions, and $d(k)$ denotes the time-varying delays which satisfy $h_1 \leq d(k) \leq h_2$, where h_1 and h_2 are two positive constants representing its lower and upper bounds, respectively.

For each possible value $\alpha = i$, $i \in \mathcal{N}$, we denote the system matrices associated with mode i by $A(i) = A(\alpha)$, $A_d(i) = A_d(\alpha)$, and $F(i) = F(\alpha)$, where $A(i)$, $A_d(i)$, and $F(i)$ are constant matrices. Moreover, we assume that $(A(i), B)$ is controllable for each $i \in \mathcal{N}$, and matrix B is of full column rank. For the nonlinearity $f(x,k)$, we suppose that

$$\|F(i)f(x,k)\| \leq \eta(i), \quad i \in \mathcal{N},$$

where $\eta(i) > 0$ are scalars.

Since $(A(i), B)$ is controllable, there exists a nonsingular matrix T such that

$$TB = \begin{bmatrix} 0_{(n-m)\times m} \\ B_1 \end{bmatrix},$$

where $B_1 \in \mathbf{R}^{m\times m}$ is nonsingular. For convenience, choose

$$T = \begin{bmatrix} U_2^T \\ U_1^T \end{bmatrix},$$

where $U_1 \in \mathbf{R}^{n\times m}$ and $U_2 \in \mathbf{R}^{n\times(n-m)}$ are two sub-blocks of a unitary matrix resulting from the singular value decomposition of B, that is,

$$B = \begin{bmatrix} U_2 & U_1 \end{bmatrix} \begin{bmatrix} 0_{(n-m)\times m} \\ \Gamma \end{bmatrix} W^T,$$

where $\Gamma \in \mathbf{R}^{m\times m}$ is a diagonal positive-definite matrix and $W \in \mathbf{R}^{m\times m}$ is a unitary matrix.

By state transformation $z(k) = Tx(k)$, system (8.1a) takes the form

$$z(k+1) = \bar{A}(\alpha)z(k) + \bar{A}_d(\alpha)z(k-d(k)) + \begin{bmatrix} 0_{(n-p)\times p} \\ B_1 \end{bmatrix} (u(k) + F(\alpha)f(z,k)), \tag{8.2}$$

where $\bar{A}(\alpha) = TA(\alpha)T^{-1}$ and $\bar{A}_d(\alpha) = TA_d(\alpha)T^{-1}$. Let $z(k) \triangleq \begin{bmatrix} z_1(k) \\ z_2(k) \end{bmatrix}$ with $z_1(k) \in \mathbf{R}^{(n-m)}$, $z_2(k) \in \mathbf{R}^m$, and

$$\bar{A}(\alpha) \triangleq \begin{bmatrix} \bar{A}_{11}(\alpha) & \bar{A}_{12}(\alpha) \\ \bar{A}_{21}(\alpha) & \bar{A}_{22}(\alpha) \end{bmatrix}, \quad \bar{A}_d(\alpha) \triangleq \begin{bmatrix} \bar{A}_{d11}(\alpha) & \bar{A}_{d12}(\alpha) \\ \bar{A}_{d21}(\alpha) & \bar{A}_{d22}(\alpha) \end{bmatrix},$$

then (8.2) can be expressed in the following regular form:

$$\begin{aligned} \begin{bmatrix} z_1(k+1) \\ z_2(k+1) \end{bmatrix} &= \begin{bmatrix} \bar{A}_{11}(\alpha) & \bar{A}_{12}(\alpha) \\ \bar{A}_{21}(\alpha) & \bar{A}_{22}(\alpha) \end{bmatrix} \begin{bmatrix} z_1(k) \\ z_2(k) \end{bmatrix} \\ &+ \begin{bmatrix} \bar{A}_{d11}(\alpha) & \bar{A}_{d12}(\alpha) \\ \bar{A}_{d21}(\alpha) & \bar{A}_{d22}(\alpha) \end{bmatrix} \begin{bmatrix} z_1(k-d(k)) \\ z_2(k-d(k)) \end{bmatrix} \\ &+ \begin{bmatrix} 0_{(n-m)\times m} \\ B_1 \end{bmatrix} (u(k) + F(\alpha)f(z,k)), \end{aligned} \tag{8.3}$$

where $\bar{A}_{11}(\alpha) = U_1^T A(\alpha)U_1$, $\bar{A}_{12}(\alpha) = U_1^T A(\alpha)U_2$, $\bar{A}_{21}(\alpha) = U_2^T A(\alpha)U_1$, $\bar{A}_{22}(\alpha) = U_2^T A(\alpha)U_2$, $\bar{A}_{d11}(\alpha) = U_1^T A_d(\alpha)U_1$, $\bar{A}_{d12}(\alpha) = U_1^T A_d(\alpha)U_2$, $\bar{A}_{d21}(\alpha) = U_2^T A_d(\alpha)U_1$, $\bar{A}_{d22}(\alpha) = U_2^T A_d(\alpha)U_2$, and $B_1 = \Gamma W^T$.

It is obvious that the first equation of system (8.3) represents the sliding motion dynamics of system (8.2), hence the corresponding sliding surface can be chosen as:

$$s(k) = Cz_1(k) + z_2(k), \tag{8.4}$$

where $C \in \mathbf{R}^{m\times(n-m)}$ is the parameter to be designed.

When the system state trajectories reach onto the sliding surface $s(k) = 0$, that is, $z_2(k) = -Cz_1(k)$, the sliding mode dynamics is attained. Substituting $z_2(k) = -Cz_1(k)$ into the first equation of system (8.3) gives the sliding mode dynamics as

$$\begin{aligned} z_1(k+1) &= \left(\bar{A}_{11}(\alpha) - \bar{A}_{12}(\alpha)C\right) z_1(k) \\ &+ \left(\bar{A}_{d11}(\alpha) - \bar{A}_{d12}(\alpha)C\right) z_1(k-d(k)). \end{aligned} \tag{8.5}$$

Definition 8.2.1 *The sliding mode dynamics (8.5) is said to be exponentially stable under α if the solution $z_1(k)$ satisfies*

$$\|z_1(k)\| \le \eta \left\|z_1(k_0)\right\|_{\mathrm{C}} \rho^{k-k_0}, \quad \forall k \ge k_0,$$

where $\eta \ge 1$ and $0 < \rho < 1$ are two real constants, and

$$\|z_1(k_0)\|_{\mathrm{C}} \triangleq \sup_{\theta=-h_2,-h_2+1,\ldots,0} \left\{\|z_1(k_0+\theta)\|, \|\xi(k_0)\|\right\},$$

where $\xi(k) \triangleq z_1(k+1) - z_1(k)$.

8.3 Main Results

8.3.1 Sliding Mode Dynamics Analysis

In this section, we analyze the stability for the sliding mode dynamics in (8.5), and present the following result.

Theorem 8.3.1 *Given an integer m and a scalar $\beta > 0$, if there exist matrices $P(i) > 0$, $Q_{\varrho|\kappa}(i) > 0$, $R(i) > 0$, $S(i) > 0$, and $Z(i) > 0$, and matrices $L(i)$ such that for $i \in \mathcal{N}$,*

$$\begin{bmatrix} \Phi(i) & \hat{A}^T(i)P(i) & \beta^{h_2+1}h_2\tilde{A}^T(i)Z(i) & h_2L(i) \\ \star & -P(i) & 0 & 0 \\ \star & \star & -\beta^{h_2+1}h_2Z(i) & 0 \\ \star & \star & \star & -h_2Z(i) \end{bmatrix} < 0, \tag{8.6}$$

where

$$\begin{cases} \Phi(i) \triangleq \beta\text{diag}\left\{-P(i) + Q_{1|1}(i) + R(i) + S(i), -\beta^{h_2}Q_{1|(m-1)}(i) + Q_{2|m}(i), \right. \\ \quad \left. -\beta^{h_2}Q_{m|m}(i), -\beta^{h_2}R(i), -\beta^{h_2}S(i)\right\} \\ \quad + 2\beta^{h_2+1}L(i)\left[I_{n-p} \;\; 0_{(n-p)\times(m+2)(n-p)} \;\; -I_{n-p}\right], \\ \hat{A}(i) \triangleq \left[\bar{A}_{11}(i)-\bar{A}_{12}(i)C \;\; 0_{(n-p)\times m(n-p)} \;\; \bar{A}_{d11}(i)-\bar{A}_{d12}(i)C \;\; 0_{n-p}\right], \\ \tilde{A}(i) \triangleq \left[\bar{A}_{11}(i)-\bar{A}_{12}(i)C-I \;\; 0_{(n-p)\times m(n-p)} \;\; \bar{A}_{d11}(i)-\bar{A}_{d12}(i)C-I \;\; 0_{n-p}\right], \end{cases}$$

then the sliding mode dynamics in (8.5) is exponentially stable for any switching signal with average dwell time satisfying $T_a > T_a^* = \text{ceil}\left(-\frac{\ln \mu}{\ln \beta}\right)$, where $\mu \geq 1$ satisfies that $\forall i, j \in \mathcal{N}$,

$$\begin{aligned} &P(i) \leq \mu P(j), \quad Q_{\varrho|\kappa}(i) \leq \mu Q_{\varrho|\kappa}(j), \\ &R(i) \leq \mu R(j), \quad S(i) \leq \mu S(j), \quad Z(i) \leq \mu Z(j). \end{aligned} \tag{8.7}$$

Proof. Choose a Lyapunov function of the following form:

$$V(k) \triangleq \sum_{j=1}^{5} V_j(k), \tag{8.8}$$

with

$$\begin{cases} V_1(k) \triangleq z_1^T(\alpha)P(\alpha)z_1(\alpha), \\ V_2(k) \triangleq \sum_{s=k-\frac{h_1}{m}}^{k-1} \beta^{k-s}\psi^T(s)Q_{1|m}(\alpha)\psi(s), \\ V_3(k) \triangleq \sum_{s=k-d(k)}^{k-1} \beta^{k-s}z_1^T(s)R(\alpha)z_1(s), \\ V_4(k) \triangleq \sum_{s=k-h_2}^{k-1} \beta^{k-s}z_1^T(s)S(\alpha)z_1(s), \\ V_5(k) \triangleq \sum_{l=-h_2}^{-1}\sum_{s=l+k}^{k-1} \beta^{k-s}\xi^T(s)Z(\alpha)\xi(s), \end{cases}$$

where $P(\alpha) > 0$, $Q_{1|m}(\alpha) > 0$, $R(\alpha) > 0$, $S(\alpha) > 0$, and $Z(\alpha) > 0$ are real matrices to be determined, and

$$\psi(k) = \begin{bmatrix} z_1(k) \\ z_1\left(k - \frac{h_1}{m}\right) \\ \vdots \\ z_1\left(k - \frac{m-1}{m}h_1\right) \end{bmatrix}.$$

Then, as with the solution of (8.5) for a fixed α, we have

$$\begin{aligned}
\Delta V_1(k) &= z_1^T(k+1)P(\alpha)z_1(k+1) - z_1^T(k)P(\alpha)z_1(k), \\
\Delta V_2(k) &= -(1-\beta)\sum_{s=k-\frac{h_1}{m}}^{k-1} \beta^{k-s}\psi^T(s)Q_{1|m}(\alpha)\psi(s) + \beta\psi^T(k)Q_{1|m}(\alpha)\psi(k) \\
&\quad - \beta^{\frac{h_1}{m}+1}\psi^T\left(k - \frac{h_1}{m}\right)Q_{1|m}(\alpha)\psi\left(k - \frac{h_1}{m}\right), \\
\Delta V_3(k) &= -(1-\beta)\sum_{s=k-d(k)}^{k-1} \beta^{k-s}z_1^T(s)R(\alpha)z_1(s) + \beta z_1^T(k)R(\alpha)z_1(k) \\
&\quad - \beta^{d(k)+1}z_1^T(k-d(k))R(\alpha)z_1(k-d(k)), \\
\Delta V_4(k) &= -(1-\beta)\sum_{s=k-h_2}^{k-1} \beta^{k-s}z_1^T(s)S(\alpha)z_1(s) + \beta z_1^T(k)S(\alpha)z_1(k) \\
&\quad - \beta^{h_2+1}z_1^T(k-h_2)S(\alpha)z_1(k-h_2), \\
\Delta V_5(k) &= -(1-\beta)\sum_{l=-h_2}^{-1}\sum_{s=l+k}^{k-1} \beta^{k-s}\xi^T(s)Z(\alpha)\xi(s) + \beta h_2\xi^T(k)Z(\alpha)\xi(k) \\
&\quad - \sum_{s=k-h_2}^{k-1} \beta^{k-s+1}\xi^T(s)Z(\alpha)\xi(s).
\end{aligned} \tag{8.9}$$

Moreover, for any matrix

$$\Upsilon(k) \triangleq \begin{bmatrix} \psi^T(k) & z_1^T(k-h_1) & z_1^T(k-d(k)) & z_1^T(k-h_2) & \xi^T(k) \end{bmatrix}^T$$

and any matrices $L(\alpha)$ and $Z(\alpha)$, the following equations are true:

$$\begin{aligned}
2\Upsilon^T(k)L(\alpha)\left[z_1(k) - z_1\left(k-h_2\right) - \sum_{s=k-h_2}^{k-1}\xi(s)\right] &= 0, \\
h_2\Upsilon^T(k)L(\alpha)Z^{-1}(\alpha)L^T(\alpha)\Upsilon(k) - \sum_{s=k-h_2}^{k-1}\Upsilon^T(k)L(\alpha)Z^{-1}(\alpha)L^T(\alpha)\Upsilon(k) &= 0.
\end{aligned} \tag{8.10}$$

Considering (8.9)–(8.10) and denoting

$$\bar{\Upsilon}(k) \triangleq \left[\psi^T(k) \quad z_1^T(k-h_1) \quad z_1^T(k-d(k)) \quad z_1^T(k-h_2)\right]^T,$$

we have

$$\begin{aligned}
&\Delta V(k) + (1-\beta)V(k) \\
&\quad < \left[\bar{A}_{11}(\alpha) - \bar{A}_{12}(\alpha)Cz_1(k) + \bar{A}_{d11}(\alpha) - \bar{A}_{d12}(\alpha)Cz_1(k-d(k))\right]^T P(\alpha) \\
&\quad \times \left[\bar{A}_{11}(\alpha) - \bar{A}_{12}(\alpha)Cz_1(k) + \bar{A}_{d11}(\alpha) - \bar{A}_{d12}(\alpha)Cz_1(k-d(k))\right] \\
&\quad - \beta z_1^T(k)P(\alpha)z_1(k) + \beta\psi^T(k)Q_{1|m}(\alpha)\psi(k) \\
&\quad - \beta^{h_2+1}\psi^T\left(k - \frac{h_1}{m}\right)Q_{1|m}(\alpha)\psi\left(k - \frac{h_1}{m}\right) \\
&\quad + \beta z_1^T(k)R(\alpha)z_1(k) + \beta z_1^T(k)S(\alpha)z_1(k) + \beta h_2\xi^T(k)Z(\alpha)\xi(k) \\
&\quad - \beta^{h_2+1}z_1^T(k-d(k))R(\alpha)z_1(k-d(k)) - \beta^{h_2+1}z_1^T(k-h_2)S(\alpha)z_1(k-h_2) \\
&\quad + h_2\Upsilon^T(k)L(\alpha)Z^{-1}(\alpha)L(\alpha)\Upsilon(k) + 2\beta^{h_2+1}\Upsilon^T(k)L(\alpha)\left[z_1(k) - z_1\left(k-h_2\right)\right] \\
&\quad - \beta^{h_2+1}\sum_{s=k-h_2}^{k-1}\left[\xi^T(s)Z(\alpha) + \Upsilon^T(k)L(\alpha)\right]Z^{-1}(\alpha)\left[Z(\alpha)\xi(s) + L^T(\alpha)\Upsilon(k)\right] \\
&\quad = \bar{\Upsilon}^T(k)\left[\Phi(\alpha) + \hat{A}^T(\alpha)P(\alpha)\hat{A}(\alpha) + \beta^{h_2+1}h_2\tilde{A}^T(\alpha)Z(\alpha)\tilde{A}(\alpha)\right. \\
&\quad \left. + h_2L(\alpha)Z^{-1}(\alpha)L^T(\alpha)\right]\bar{\Upsilon}(k) \\
&\quad - \beta^{h_2+1}\sum_{s=k-h_2}^{k-1}\left[\xi^T(s)Z(\alpha) + \Upsilon^T(k)L(\alpha)\right]Z^{-1}(\alpha)\left[Z(\alpha)\xi(s) + L^T(\alpha)\Upsilon(k)\right],
\end{aligned}$$

where $\Phi(\alpha)$ is defined in (8.6).

Moreover, from (8.6), it follows that

$$\Phi(\alpha) + \hat{A}^T(\alpha)P(\alpha)\hat{A}(\alpha) + \beta^{h_2+1}h_2\tilde{A}^T(\alpha)Z(\alpha)\tilde{A}(\alpha) + h_2L(\alpha)Z(\alpha)^{-1}L^T(\alpha) < 0.$$

Then it can be easily seen that

$$\Delta V(z_1(k), \alpha(k)) + (1-\beta)V(z_1(k), \alpha(k)) < 0. \tag{8.11}$$

Now, for an arbitrary piecewise constant switching signal α, and for and $k > 0$, we let $k_0 < k_1 < \cdots < k_l < \cdots$, $l = 1, 2, \ldots$, denote the switching point of α over the interval $(0, k)$. Therefore, for $k \in [k_l, k_{l+1})$, it holds from (8.11) that

$$V(z_1(k), \alpha(k)) < \beta^{k-k_l}V(z_1(k_l), \alpha(k_l)). \tag{8.12}$$

Using (8.7) and (8.8), at switching instant t_k, we have

$$V(z_1(k_l), \alpha(k_l)) \leq \mu V(z_1(k_l), \alpha(k_{l-1})). \tag{8.13}$$

Therefore, it follows from (8.12)–(8.13) and the relationship $\vartheta = N_\alpha(0,k) \leq (k-k_0)/K_\alpha$ that

$$\begin{aligned} V(z_1(k), \alpha(k)) &\leq \beta^{k-k_l} \mu V(z_1(k_l), \alpha(k_{l-1})) \\ &\leq \cdots \\ &\leq \beta^{k-k_0} \mu^{\vartheta} V(z_1(k_0), \alpha(k_0)) \\ &\leq (\beta \mu^{1/T_a})^{k-k_0} V(z_1(k_0), \alpha(k_0)). \end{aligned} \tag{8.14}$$

Notice from (8.8) that there exist two positive constants a and b $(a \leq b)$ such that

$$V(z_1(k), \alpha(k)) \geq a\|z_1(k)\|^2, \quad V(z_1(k_0), \alpha(k_0) \leq b\|z_1(k_0)\|_C^2. \tag{8.15}$$

Combining (8.14) and (8.15) yields

$$\begin{aligned} \|z_1(k)\|^2 &\leq \frac{1}{a} V(z_1(k), \alpha(k)) \\ &\leq \frac{b}{a} \left(\beta \mu^{1/T_a}\right)^{k-k_0} \|z_1(k_0)\|_C^2. \end{aligned} \tag{8.16}$$

Furthermore, letting $\rho \triangleq \sqrt{\beta \mu^{1/T_a}}$, it follows that

$$\|z_1(k)\| \leq \sqrt{\frac{b}{a}} \rho^{k-k_0} \|z_1(k_0)\|_C. \tag{8.17}$$

By Definition 8.2.1, we know that if $0 < \rho < 1$, that is, $T_a > T_a^* = \text{ceil}(-\frac{\ln \mu}{\ln \beta})$, the switched system (8.5) is exponentially stable, where function ceil(a) represents rounding real number a to the nearest integer greater than or equal to a. The proof is completed. ■

Remark 8.1 *The matrices $Q_{\varrho|\kappa}(i)$ used in the above proof have two advantages: 1) the matrix for each part of the time partition can be chosen respectively according to the constraints of LMI, which decrease the conservativeness of our approach; and 2) it is simple to show the series of matrices from $Q_\varrho(i)$ to $Q_\kappa(i)$, which makes the result simpler and clearer.* ♦

Remark 8.2 *It should be pointed out that the switching function defined in (8.4) does not switch with the switching signal α. That is, we design C not $C(\alpha)$ in (8.4). In this way, we can avoid repetitive jumps of the state trajectories of the state components of the closed-loop system between sliding surfaces and hence the possible instability.* ♦

Now, we are on the path to solve the parameter matrices in (8.6). Considering the convenience of solving an LMI, more transformation has to be made to turn the inequality in (8.6) into an LMI, and the following theorem is obtained.

Theorem 8.3.2 *For a given constant $\beta > 0$, suppose that there exist matrices $\mathcal{P} > 0$, $\mathcal{Q}_{\varrho|\kappa}(i) > 0$, $\mathcal{R}(i) > 0$, $\mathcal{S}(i) > 0$, and $\mathcal{Z}(i) > 0$, and matrices $\mathcal{K}$, $\mathcal{L}(i)$, such that for $i \in \mathcal{N}$,*

$$\begin{bmatrix} \bar{\Phi}(i) & \hat{\mathcal{A}}^T(i) & \beta^{h_2+1}h_2\tilde{\mathcal{A}}^T(i) & h_2\mathcal{L}(i) \\ \star & -\mathcal{P} & 0 & 0 \\ \star & \star & -\beta^{h_2+1}h_2\mathcal{Z}(i) & 0 \\ \star & \star & \star & -h_2\left(\mathcal{Z}(i) - 2\mathcal{P}\right) \end{bmatrix} < 0, \tag{8.18}$$

where

$$\begin{cases} \bar{\Phi}(i) \triangleq \beta\text{diag}\left\{-\mathcal{P} + \mathcal{Q}_{1|1}(i) + \mathcal{R}(i) + \mathcal{S}(i),\ -\beta^{h_2}\mathcal{Q}_{1|m-1}(i) + \mathcal{Q}_{2|m}(i),\right. \\ \qquad \left. -\ \beta^{h_2}\mathcal{Q}_{m|m}(i),\ -\beta^{h_2}\mathcal{R}(i),\ -\beta^{h_2}\mathcal{S}(i)\right\} \\ \qquad +\ 2\beta^{h_2+1}\mathcal{L}(i)\left[I_{n-p} \quad 0_{(n-p)\times(m+2)(n-p)} \quad -I_{n-p}\right], \\ \hat{\mathcal{A}}(i) \triangleq \left[\bar{A}_{11}(i)\mathcal{P} - \bar{A}_{12}(i)\mathcal{K} \quad 0_{(n-p)\times m(n-p)} \quad \bar{A}_{d11}(i)\mathcal{P} - \bar{A}_{d12}(i)\mathcal{K} \quad 0_{n-p}\right], \\ \tilde{\mathcal{A}}(i) \triangleq \left[\bar{A}_{11}(i)\mathcal{P} - \bar{A}_{12}(i)\mathcal{K} - \mathcal{P} \quad 0_{(n-p)\times m(n-p)}\right. \\ \qquad \left. \bar{A}_{d11}(i)\mathcal{P} - \bar{A}_{d12}(i)\mathcal{K} - \mathcal{P} \quad 0_{n-p}\right]. \end{cases}$$

Then the sliding mode dynamics in (8.5) is exponentially stable for any switching signal with average dwell time satisfying $T_a > T_a^* = \text{ceil}\left(-\frac{\ln\mu}{\ln\beta}\right)$, where $\mu \geq 1$ and satisfies

$$\begin{aligned} \mathcal{Q}_{\varrho|\kappa}(i) &\leq \mu\mathcal{Q}_{\varrho|\kappa}(j), \quad \mathcal{R}(i) \leq \mu\mathcal{R}(j), \\ \mathcal{S}(i) &\leq \mu\mathcal{S}(j), \qquad\ \ \mathcal{Z}(i) \leq \mu\mathcal{Z}(j). \end{aligned} \tag{8.19}$$

Moreover, if the conditions above are feasible, the matrix C in (8.4) is given by $C = \mathcal{K}\mathcal{P}^{-1}$, that is, the switching function can be designed as

$$s(k) = \mathcal{K}\mathcal{P}^{-1}z_1(k) + z_2(k). \tag{8.20}$$

Proof. Defining the following matrices:

$$\begin{cases} \mathcal{P} \triangleq P^{-1}, \quad \mathcal{Q}_{\varrho|\kappa}(i) \triangleq \mathcal{P}^T Q_{\varrho|\kappa}(i)\mathcal{P}, \quad \mathcal{R}(i) \triangleq \mathcal{P}^T R(i)\mathcal{P}, \\ \mathcal{S}(i) \triangleq \mathcal{P}^T S(i)\mathcal{P}, \quad \mathcal{L}(i) \triangleq \mathcal{P}^T L(i)\mathcal{P}, \quad i \in \mathcal{N}, \end{cases}$$

and performing a congruence transformation on (8.6) with diag $\{I_{m+3} \otimes \mathcal{P}, \mathcal{P}, \mathcal{Z}(i), \mathcal{P}\}$, we have

$$\begin{bmatrix} \bar{\Phi}(i) & \bar{\mathcal{A}}^T(i) & \beta^{h_2+1}h_2\tilde{\mathcal{A}}^T(i) & h_2\mathcal{L}(i) \\ \star & -\mathcal{P} & 0 & 0 \\ \star & \star & -\beta^{h_2+1}h_2\mathcal{Z}(i) & 0 \\ \star & \star & \star & -h_2\mathcal{P}\mathcal{Z}^{-1}(i)\mathcal{P} \end{bmatrix} < 0. \tag{8.21}$$

Moreover, notice that

$$\begin{aligned} 0 &\leq (\mathcal{P} - \mathcal{Z}(i))\,\mathcal{Z}^{-1}(i)\,(\mathcal{P} - \mathcal{Z}(i)) \\ &= \mathcal{P}\mathcal{Z}^{-1}(i)\mathcal{P} - \mathcal{P} - \mathcal{P} + \mathcal{Z}(i), \end{aligned}$$

which implies

$$-\mathcal{P}\mathcal{Z}^{-1}(i)\mathcal{P} \leq \mathcal{Z}(i) - 2\mathcal{P}.$$

Thus, inequality in (8.21) holds if that in (8.18) holds. This completes the proof. ■

Remark 8.3 *It should be mentioned that the matrices P and $\mathcal{P}$ in Theorem 8.3.2 do not depend on the switching signal α and are fixed. Since the designed sliding surface in (8.4) is parameter-independent, the parameter $C = \mathcal{K}\mathcal{P}^{-1}$ in (8.4) is guaranteed to be fixed if P and $\mathcal{P}$ are fixed.* ♦

8.3.2 SMC Law Design

In this section, we design an SMC law to drive the system state trajectories onto the sliding surface $s(k) = 0$, and have the following result.

Theorem 8.3.3 *With the switching function given by (8.20), the state trajectories of the closed-loop system in (8.3) can be driven onto the sliding surface by the following control and finally converges into a residual set of the origin:*

$$\begin{cases} u(k) = -B_1^{-1}\left[\Pi s(k) + \bar{C}\bar{A}(\alpha)z(k) + \bar{C}\bar{A}_d(\alpha)z(k-d_k)\right] + u_N(k), \\ u_N(k) = \begin{cases} -\text{sign}\left(B_1^T s(k)\right)\eta, & \|B_1^T s(k)\| > \varepsilon, \\ -\dfrac{B_1^T s(k)}{\varepsilon}\eta^2, & \|B_1^T s(k)\| \leq \varepsilon, \end{cases} \end{cases} \tag{8.22}$$

where Π is a positive definite matrix.

Proof. We will complete the proof by showing that the control law (8.22) can not only drive the system state trajectories onto the liner sliding surface, but also keeps it there for all subsequent time. From the sliding surface (8.4), we have

$$
\begin{aligned}
s(k+1) &= \bar{C}z(k+1) \\
&= \bar{C}\bar{A}(\alpha)z(k) + \bar{C}\bar{A}_d(\alpha)z(k-d_k) + B_1 u(k,\alpha) + B_1 F(\alpha)f(z,k) \\
&= -\Pi s(k) + B_1 F(\alpha)f(z,k) + B_1 u_N(k),
\end{aligned}
$$

where $\bar{C} = \begin{bmatrix} C & I \end{bmatrix}$.

Consider the following Lyapunov function:

$$
V(k) = \frac{1}{2}s^T(k)s(k). \tag{8.23}
$$

Then the incremental $\Delta V(k)$ is

$$
\begin{aligned}
\Delta V(k) &= s^T(k)\Delta s(k) + \frac{1}{2}\Delta s^T(k)\Delta s(k) \\
&= s^T(k)\left[-(\Pi + I)s(k) + B_1 F(\alpha)f(z,k) + B_1 u_N(k)\right] + \frac{1}{2}\Delta s^T(k)\Delta s(k) \\
&\leq -\, s^T(k)(\Pi + I)s(k) + \|s^T(k)B_1\|\|F(\alpha)f(z,k)\| \\
&\quad + s^T(k)B_1 u_N(k) + \frac{1}{2}\Delta s^T(k)\Delta s(k).
\end{aligned}
$$

If $\|B_1^T s(k)\| > \varepsilon$, with the control law (8.22), holding that $\|B_1^T s(k)\| \leq |B_1^T s(k)|$, we have

$$
\begin{aligned}
\Delta V(k) &\leq -\, s^T(k)(\Pi + I)s(k) + \|s^T(k)B_1\|\eta \\
&\quad -\, s^T(k)B_1 \operatorname{sign}\left(B_1^T s(k)\right)\eta + \frac{1}{2}\Delta s^T(k)\Delta s(k) \\
&\leq -\, s^T(k)(\Pi + I)s(k) + \frac{1}{2}\Delta s^T(k)\Delta s(k).
\end{aligned}
$$

If $\|B_1^T s(k)\| \leq \varepsilon$, with the control law (8.22), we have

$$
\begin{aligned}
\Delta V(k) \leq &-\, s^T(k)(\Pi + I)s(k) + \frac{\eta}{\varepsilon}\left(\|B_1^T s(k)\|\varepsilon - \|B_1^T s(k)\|^2\eta\right) \\
&+ \frac{1}{2}\Delta s^T(k)\Delta s(k).
\end{aligned}
$$

Since $\Pi > 0$ is to be tuned, an appropriate Π can be selected large enough such that $\Delta V(k) < 0$ as long as $s(k)$ is within a certain bounded region which contains an equilibrium point. Then $\Delta s(k)$ is reasonably bounded, although it is not asymptotically convergent to zero, which shows that the state trajectories of (8.3) can be driven onto the sliding surface by the control law (8.22) and maintained there for all the subsequent time. This completes the proof. ∎

8.4 Illustrative Example

Consider system (8.1a)–(8.1b) with $N = 2$ and the following parameters:

$$A(1) = \begin{bmatrix} -0.25 & 0.1 \\ 0.1 & 0.3 \end{bmatrix}, \quad A_d(1) = \begin{bmatrix} -0.05 & 0 \\ 0.1 & -0.06 \end{bmatrix},$$

$$A(2) = \begin{bmatrix} 0.3 & -0.2 \\ -0.2 & 0.2 \end{bmatrix}, \quad A_d(2) = \begin{bmatrix} -0.04 & 0.01 \\ 0 & -0.03 \end{bmatrix},$$

$$B = \begin{bmatrix} 0 \\ 2.0 \end{bmatrix}, \quad F(1) = 1.6, \quad F(2) = 2.0,$$

and $d(k) = 4 + \text{round}(\sin(k))$, where round($a$) represents the nearest integer to number a. Some other parameters of the system are given as $f(k) = 0.5\exp(-k)\sin(\sqrt{x_1^2 + x_2^2})$, $\beta = 0.5$, $m = 3$, $\mu = 1$, $h_1 = 3$, and $h_2 = 5$. Using the LMI Toolbox in Matlab to solve conditions (8.18)–(8.19) in Theorem 8.3.2, we have $\mathcal{P} = 7.2129$ and $\mathcal{K} = 0.72373$, thus,

$$\begin{aligned} s(k) &= \mathcal{K}\mathcal{P}^{-1}x_1(k) + x_2(k) \\ &= \begin{bmatrix} 0.1003 & 1.0000 \end{bmatrix} x(k). \end{aligned}$$

Let the initial condition be $\phi(k) = [\,-0.8 \quad 1.0\,]^T$ $(k = -5, -4, \ldots, 0)$. The switching signal is shown in Figure 8.1, and the states of the closed-loop system are illustrated in Figure 8.2. Figure 8.3 depicts the control input, and the switching function is given in Figure 8.4 with $\Pi = 3$ and $\varepsilon = 0.2$.

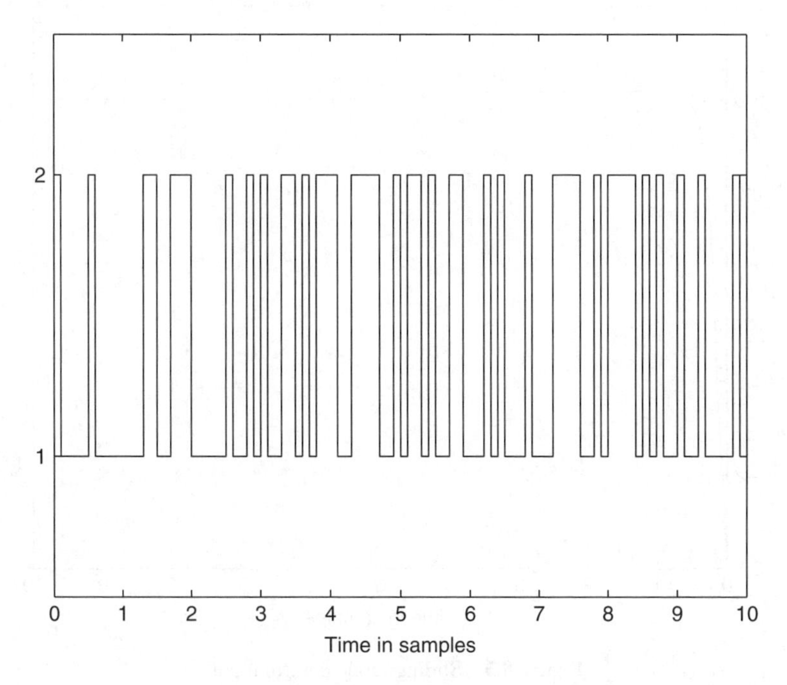

Figure 8.1 Switching signal

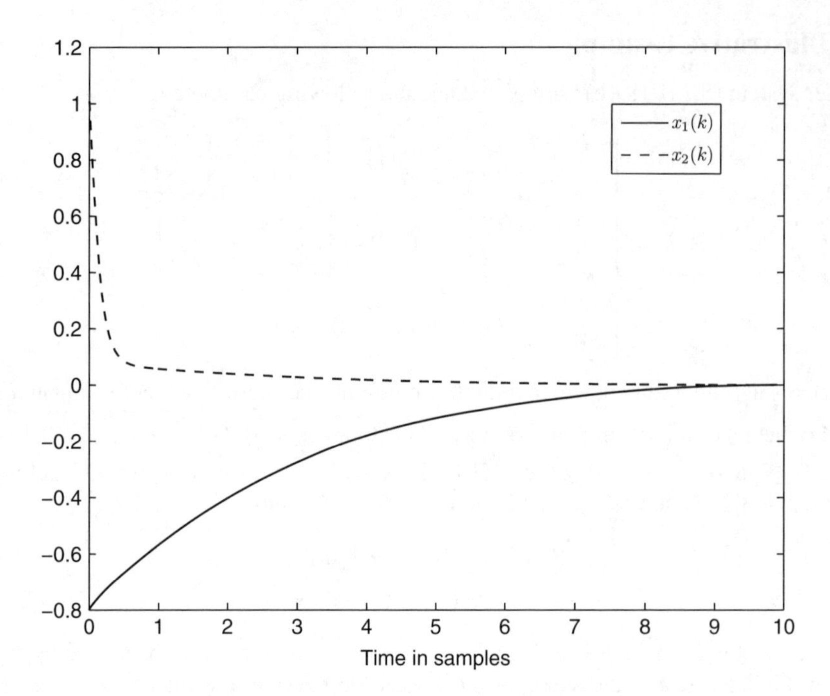

Figure 8.2 States of the closed-loop system

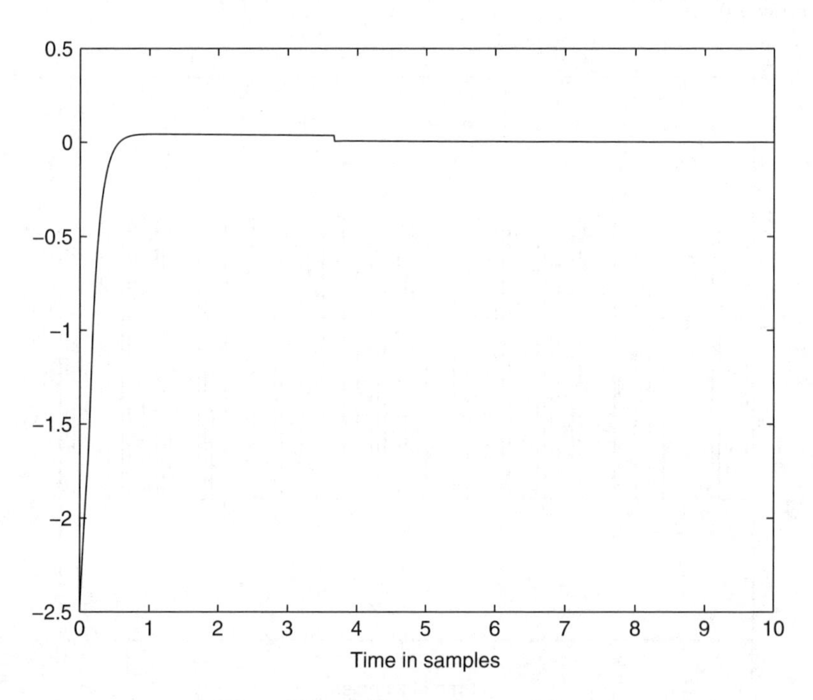

Figure 8.3 Sliding mode control input

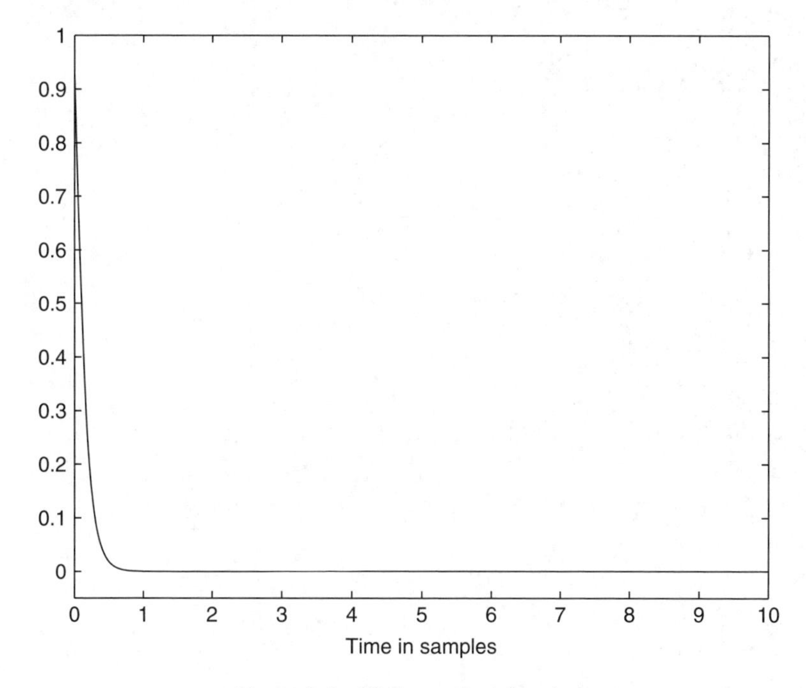

Figure 8.4 Sliding surface function

8.5 Conclusion

In this chapter, the problem of SMC of a discrete-time switched hybrid system with time-varying delay has been investigated. Within the LMIs framework, a sufficient condition, which is dependent on the maximum and minimum delay bounds, has been established to guarantee the existence of a linear sliding surface. The conservativeness of the obtained results has been reduced by employing the delay partitioning method and the slack matrix technique. An SMC law has been designed to force the closed-loop system to be driven onto a prescribed sliding surface and maintained there for all subsequent time. Finally, a numerical example has been included to demonstrate the usefulness of the developed new design techniques.

Part Three

SMC of Switched Stochastic Hybrid Systems

9

Control of Switched Stochastic Hybrid Systems: Continuous-Time Case

9.1 Introduction

Stochastic systems play an important role in many branches of science and engineering applications, thus they have received much attention during recent decades. Many results reported about stochastic systems can be found in the literature, for example, stochastic stability analysis, stabilization, optimal and robust control, filtering, and model reduction. Recently, a great deal of work has been reported on stochastic systems with Markovian switching. These results motivate us to study some interesting topics on stochastic systems whose parameters operate by a switching signal, that is, the switched stochastic hybrid systems. This work is interesting and challenging since this kind of hybrid system integrates the switched hybrid systems into that of the stochastic systems, and thus is theoretically significant.

In this chapter, we investigate the $\mathcal{H}_\infty$ control (including state feedback control and DOF control) problems for continuous-time switched stochastic hybrid systems. The average dwell time approach combined with the piecewise Lyapunov function technique are applied to derive the main results. There are two main advantages of using this approach to the switched system. First, this approach uses a mode-dependent Lyapunov function, which avoids some conservativeness caused by using a common Lyapunov function for all the subsystems. The other main advantage is that the obtained result is not just an asymptotic stability condition, but an exponential one. Therefore, by this approach, a sufficient conditions is first proposed, which guarantees the mean-square exponential stability of the unforced switched stochastic hybrid system. When system states are available, a state feedback controller is designed such that the closed-loop system is mean-square exponentially stable with an $\mathcal{H}_\infty$ performance. However, when system states are not all available, a DOF controller is designed, and the mean-square exponential stability with an $\mathcal{H}_\infty$ performance is also guaranteed. Sufficient solvability conditions for the desired controllers are proposed in terms of LMIs.

Sliding Mode Control of Uncertain Parameter-Switching Hybrid Systems, First Edition. Ligang Wu, Peng Shi and Xiaojie Su.

9.2 System Description and Preliminaries

Consider a class of switched stochastic hybrid systems of the form:

$$dx(t) = [A(\alpha(t))x(t) + B(\alpha(t))u(t) + D(\alpha(t))\omega(t)]\, dt + E(\alpha(t))x(t)d\varpi(t), \tag{9.1a}$$

$$z(t) = C(\alpha(t))x(t), \tag{9.1b}$$

where $x(t) \in \mathbf{R}^n$ is the state vector; $u(t) \in \mathbf{R}^m$ is the control input; $\omega(t) \in \mathbf{R}^p$ is the disturbance input which belongs to $\mathcal{L}_2\,[0, \infty)$; $z(t) \in \mathbf{R}^q$ is the controlled output; $\varpi(t)$ is a one-dimensional Brownian motion satisfying $\mathbf{E}\{d\varpi(t)\} = 0$; and $\mathbf{E}\left\{d\varpi^2(t)\right\} = dt$. Also, $\{(A(\alpha(t)), B(\alpha(t)), C(\alpha(t)), D(\alpha(t)), E(\alpha(t))) : \alpha(t) \in \mathcal{N}\}$ is a family of matrices parameterized by an index set $\mathcal{N} = \{1, 2, \ldots, N\}$ and $\alpha(t) : \mathbf{R} \rightarrow \mathcal{N}$ (denoted by α for simplicity) is a switching signal defined the same as in Chapter 5. In addition, we assume that the switch signal $\alpha(t)$ has an average dwell time.

Here, we design a stabilization controller and an $\mathcal{H}_\infty$ state feedback controller with the following general structure:

$$u(t) = K\,(\alpha)\, x(t), \tag{9.2}$$

where $K\,(\alpha) \in \mathbf{R}^{m\times n}$ are parametric matrices to be designed. Substituting the controller $u(t)$ into the system (9.1a)–(9.1b), we obtain the closed-loop stabilization system as

$$dx(t) = [A(\alpha) + B(\alpha)K(\alpha)]\, x(t)dt + E(\alpha)x(t)d\varpi(t), \tag{9.3}$$

and the closed-loop $\mathcal{H}_\infty$ control system as

$$dx(t) = \{[A(\alpha) + B(\alpha)K(\alpha)]\, x(t) + D(\alpha)\omega(t)\}\, dt + E(\alpha)x(t)d\varpi(t), \tag{9.4a}$$

$$z(t) = C(\alpha)x(t). \tag{9.4b}$$

The above state feedback controller requires that the system states are fully accessible. However, in practical applications, it is usually either not accessible or hard to access. In such a case, one option is to assume the availability of a measured output signal vector given by

$$dy(t) = [G(\alpha)x(t) + H(\alpha)\omega(t)]\, dt + F(\alpha)x(t)d\varpi(t), \tag{9.5}$$

where $y(t) \in \mathbf{R}^r$ is the measured output, and $G(\alpha)$, $H(\alpha)$, and $F(\alpha)$ are real constant matrices.

For each possible value $\alpha(t) = i, i \in \mathcal{N}$, we will denote the system matrices associated with mode i by $A(i) = A(\alpha)$, $B(i) = B(\alpha)$, $C(i) = C(\alpha)$, $D(i) = D(\alpha)$, $E(i) = E(\alpha)$, $F(i) = F(\alpha)$, $G(i) = G(\alpha)$, and $H(i) = H(\alpha)$, where $A(i), B(i), C(i), D(i), E(i), F(i), G(i)$, and $H(i)$ are constant matrices.

We are also interested in designing a DOF controller in the form of

$$d\hat{x}(t) = A_c(\alpha)\hat{x}(t)dt + B_c(\alpha)dy(t) \tag{9.6a}$$

$$u(t) = C_c(\alpha)\hat{x}(t), \tag{9.6b}$$

where $\hat{x}(t) \in \mathbf{R}^n$ is the controller state vector; $A_c(\alpha)$, $B_c(\alpha)$, and $C_c(\alpha)$ are matrices to be determined.

Augmenting the model of (9.1a)–(9.1b) to include the states of the DOF controller (9.6a)–(9.6b), we obtain the closed-loop system as

$$d\xi(t) = \left[\tilde{A}(\alpha)\xi(t) + \tilde{D}(\alpha)\omega(t)\right] dt + \tilde{E}(\alpha)M\xi(t)d\varpi(t), \tag{9.7a}$$

$$z(t) = \tilde{C}(\alpha)\xi(t), \tag{9.7a}$$

where $\xi(t) \triangleq \begin{bmatrix} x(t) \\ \hat{x}(t) \end{bmatrix}$ and

$$\begin{cases} \tilde{A}(\alpha) \triangleq \begin{bmatrix} A(\alpha) & B(\alpha)C_c(\alpha) \\ B_c(\alpha)G(\alpha) & A_c(\alpha) \end{bmatrix}, \ \tilde{D}(\alpha) \triangleq \begin{bmatrix} D(\alpha) \\ B_c(\alpha)H(\alpha) \end{bmatrix}, \\ \tilde{E}(\alpha) \triangleq \begin{bmatrix} E(\alpha) \\ B_c(\alpha)F(\alpha) \end{bmatrix}, \ \tilde{C}(\alpha) \triangleq \begin{bmatrix} C(\alpha) & 0 \end{bmatrix}, \ M \triangleq \begin{bmatrix} I & 0 \end{bmatrix}. \end{cases}$$

First we present the following definitions.

Definition 9.2.1 *The switched stochastic hybrid system in (9.1a)–(9.1b) with $u(t) = 0$ and $\omega(t) = 0$ is said to be mean-square exponentially stable under $\alpha(t)$ if its solution $x(t)$ satisfies*

$$\mathbf{E}\left\{\|x(t)\|\right\} \leq \eta \left\|x(t_0)\right\| e^{-\lambda(t-t_0)}, \quad \forall t \geq t_0,$$

for constants $\eta \geq 1$ and $\lambda > 0$.

Definition 9.2.2 *For $\beta > 0$ and $\gamma > 0$, the switched stochastic hybrid system in (9.1a)–(9.1b) with $u(t) = 0$ is said to be mean-square exponentially stable with a weighted $\mathcal{H}_\infty$ performance level γ under $\alpha(t)$, if it is mean-square exponentially stable with $\omega(t) = 0$, and under zero initial condition, that is, $x(0) = 0$, if it holds for all nonzero $\omega(t) \in \mathcal{L}_2[0, \infty)$ that*

$$\mathbf{E}\left\{\int_0^\infty e^{-\beta t} z^T(t)z(t)dt\right\} \leq \gamma^2 \int_0^\infty \omega^T(t)\omega(t)dt. \tag{9.8}$$

Therefore, the problems to be addressed in this chapter can be formulated as:

1. Stability analysis and stabilization. Propose a condition guaranteeing the mean-square exponential stability of the unforced switched stochastic hybrid system. Then, design of a stabilization controller such that the resulting closed-loop system is mean-square exponentially stable.
2. $\mathcal{H}_\infty$ control. Both the state feedback control and the DOF control are considered according to the availability of the system states. Specifically,

i) When system states are fully available, design a state feedback controller such that the closed-loop system is mean-square exponentially stable with a weighted $\mathcal{H}_\infty$ performance.
ii) When system states are not all available, design a DOF controller such that the closed-loop system is mean-square exponentially stable with a weighted $\mathcal{H}_\infty$ performance.

9.3 Stability Analysis and Stabilization

In this section, we apply the average dwell time approach combined with the piecewise Lyapunov function technique to investigate the mean-square exponential stability and stabilization problems for system (9.1a)–(9.1b) with $\omega(t) = 0$.

Before proceeding, we cite the following result of Itô's formula, which plays an important role in the stability analysis for stochastic systems (see [144] for a detailed account of Itô stochastic systems).

Lemma 9.3.1 [144] (Itô's formula) *Let $x(t)$ be an n-dimensional Itô's process on $t \geq 0$ with the stochastic differential*

$$dx(t) = f(t)dt + g(t)d\varpi(t),$$

where $f(t) \in \mathbf{R}^n$ and $g(t) \in \mathbf{R}^{n\times m}$. Let $V(x(t), t) \in \mathbf{C}^{2,1}(\mathbf{R}^n \times \mathbf{R}^+; \mathbf{R}^+)$. Then, $V(x(t), t)$ is a real-valued Itô process with its stochastic differential given by

$$\begin{cases} dV(x(t), t) = \mathcal{L}V(x(t), t)dt + V_x(x(t), t)g(t)d\varpi(t), \\ \mathcal{L}V(x(t), t) = V_t(x(t), t) + V_x(x(t), t)f(t) + \dfrac{1}{2}\text{trace}(g^T(t)V_{xx}(x(t), t)g(t)), \end{cases}$$

where $\mathbf{C}^{2,1}(\mathbf{R}^n \times \mathbf{R}^+; \mathbf{R}^+)$ denotes the family of all real-valued functions $V(x(t), t)$ such that they are continuously twice differentiable in x and t. If $V(x(t), t) \in \mathbf{C}^{2,1}(\mathbf{R}^n \times \mathbf{R}^+; \mathbf{R}^+)$, we set

$$\begin{cases} V_t(x(t), t) = \dfrac{\partial V(x(t), t)}{\partial t}, \\ V_x(x(t), t) = \left(\dfrac{\partial V(x(t), t)}{\partial x_1}, \dots, \dfrac{\partial V(x(t), t)}{\partial x_n}\right), \\ V_{xx}(x(t), t) = \left(\dfrac{\partial^2 V(x(t), t)}{\partial x_i x_j}\right)_{n\times n}. \end{cases}$$

Firstly, we present the following stability analysis result for the switched stochastic hybrid system in (9.1a)–(9.1b) with $u(t) = 0$ and $\omega(t) = 0$.

Theorem 9.3.2 *Given a scalar $\beta > 0$, suppose there exist matrices $P(i) > 0$ such that for $i \in \mathcal{N}$,*

$$\begin{bmatrix} P(i)A(i) + A^T(i)P(i) + \beta P(i) & E^T(i)P(i) \\ \star & -P(i) \end{bmatrix} < 0. \tag{9.9}$$

Then the switched stochastic hybrid system in (9.1a)–(9.1b) with $u(t) = 0$ *and* $\omega(t) = 0$ *is mean-square exponentially stable for any switching signal with average dwell time satisfying* $T_a > \frac{\ln \mu}{\beta}$, *where* $\mu \geq 1$ *and satisfies*

$$P(i) \leq \mu P(j), \quad \forall i, j \in \mathcal{N}. \tag{9.10}$$

Moreover, an estimate of the state decay is given by

$$\mathbf{E}\{\|x(t)\|\} \leq \eta \|x(0)\| e^{-\lambda t}, \tag{9.11}$$

where

$$\begin{cases} \lambda \triangleq \frac{1}{2}\left(\beta - \frac{\ln \mu}{T_a}\right) > 0, \quad \eta \triangleq \sqrt{\frac{b}{a}} \geq 1, \\ a \triangleq \min_{i \in \mathcal{N}} \lambda_{\min}(P(i)), \quad b \triangleq \max_{i \in \mathcal{N}} \lambda_{\max}(P(i)). \end{cases} \tag{9.12}$$

Proof. Choose a Lyapunov function as

$$V(x, \alpha) \triangleq x^T(t) P(\alpha) x(t), \tag{9.13}$$

where $P(\alpha) > 0$, $\alpha \in \mathcal{N}$ are to be determined. Then, as with the solution of the system (9.1a)–(9.1b) with $u(t) = 0$ and $\omega(t) = 0$ for a fixed α, by Itô's formula, we have

$$dV(x, \alpha) = \mathcal{L}V(x_t, \alpha)dt + 2x^T(t)P(\alpha)E(\alpha)x(t)d\varpi(t),$$

where

$$\begin{aligned} \mathcal{L}V(x, \alpha) &= 2x^T(t)P(\alpha)A(\alpha)x(t) + x^T(t)E^T(\alpha)P(\alpha)E(\alpha)x(t) \\ &= x^T(t)\left[P(\alpha)A(\alpha) + A^T(\alpha)P(\alpha) + E^T(\alpha)P(\alpha)E(\alpha)\right]x(t). \end{aligned} \tag{9.14}$$

By Schur complement, LMI (9.9) implies

$$P(\alpha)A(\alpha) + A^T(\alpha)P(\alpha) + E^T(\alpha)P(\alpha)E(\alpha) < -\beta P(\alpha),$$

which implies from (9.14) that

$$\mathcal{L}V(x, \alpha) < -\beta x^T(t)P(\alpha)x(t) = -\beta V(x, \alpha).$$

Thus, we have

$$dV(x, \alpha) < -\beta V(x, \alpha)dt + 2x^T(t)P(\alpha)E(\alpha)x(t)d\varpi(t).$$

Observe that

$$\begin{aligned} d\left[e^{\beta t}V(x,\alpha)\right] &= \beta e^{\beta t}V(x,\alpha)dt + e^{\beta t}dV(x,\alpha) \\ &< e^{\beta t}\left[\beta V(x,\alpha)dt - \beta V(x,\alpha)dt + 2x^T(t)P(\alpha)E(\alpha)x(t)d\varpi(t)\right] \\ &= 2e^{\beta t}x^T(t)P(\alpha)E(\alpha)x(t)d\varpi(t). \end{aligned} \tag{9.15}$$

Integrate both sides of (9.15) from $T > 0$ to t and take expectations. Then, by some mathematical operations, we have

$$\mathbf{E}\{V(x,\alpha)\} < e^{-\beta(t-T)}\mathbf{E}\{V(x(T),\alpha(T))\}. \tag{9.16}$$

Now, for an arbitrary piecewise constant switching signal α, and for any $t > 0$, we let $0 = t_0 < t_1 < \cdots < t_k < \cdots$, $k = 0, 1, \ldots$, denote the switching points of α over the interval $(0, t)$. As mentioned earlier, the i_kth subsystem is activated when $t \in \left[t_k, t_{k+1}\right)$. Letting $T = t_k$ in (9.16) gives

$$\mathbf{E}\{V(x,\alpha)\} < e^{-\beta(t-t_k)}\mathbf{E}\left\{V(x(t_k),\alpha(t_k))\right\}. \tag{9.17}$$

Using (9.10) and (9.13), at switching instant t_k, we have

$$\mathbf{E}\left\{V(x(t_k),\alpha(t_k))\right\} \le \mu\mathbf{E}\left\{V(x(t_k^-),\alpha(t_k^-))\right\}, \tag{9.18}$$

where t_k^- denotes the left limit of t_k.

Therefore, it follows from (9.17)–(9.18) and the relationship $\vartheta = N_\alpha(0,t) \le (t-0)/T_a$ that

$$\begin{aligned} \mathbf{E}\{V(x,\alpha)\} &\le e^{-\beta(t-t_k)}\mu\mathbf{E}\left\{V(x(t_k^-),\alpha(t_k^-))\right\} \\ &\le \cdots \\ &\le e^{-\beta(t-0)}\mu^{\vartheta}\mathbf{E}\{V(x(0),\alpha(0))\} \\ &\le e^{-(\beta-\ln\mu/T_a)t}\mathbf{E}\{V(x(0),\alpha(0))\} \\ &= e^{-(\beta-\ln\mu/T_a)t}V(x(0),\alpha(0)). \end{aligned} \tag{9.19}$$

Notice from (9.13) that

$$\mathbf{E}\{V(x,\alpha)\} \ge a\mathbf{E}\{\|x(t)\|^2\}, \quad V(x(0),\alpha(0)) \le b\|x(0)\|^2, \tag{9.20}$$

where a and b are defined in (9.12). Combining (9.19) and (9.20) yields

$$\begin{aligned} \mathbf{E}\left\{\|x(t)\|^2\right\} &\le \frac{1}{a}\mathbf{E}\{V(x,\alpha)\} \\ &\le \frac{b}{a}e^{-(\beta-\ln\mu/T_a)t}\|x(0)\|^2, \end{aligned} \tag{9.21}$$

which implies (9.11). By Definition 9.2.1 with $t_0 = 0$, the system in (9.1a)–(9.1b) with $u(t) = 0$ and $\omega(t) = 0$ is mean-square exponentially stable. This completes the proof. ■

Remark 9.1 *Note that the scalar β is introduced in the stability analysis of Theorem 9.3.2, this is the characteristic of the exponential stability for the switched system by using the average dwell time approach. Here, β plays a key role in controlling the low bound of the average dwell time due to $T_a > \frac{\ln \mu}{\beta}$. From $T_a > \frac{\ln \mu}{\beta}$ we can see that when β is given a bigger value, the lower bound of the average dwell time becomes smaller with a fixed μ, which may result in the instability of the system.* ♦

Remark 9.2 *When $\mu = 1$ in $T_a > \frac{\ln \mu}{\beta}$ we have $T_a > T_a^* = 0$, which means that the switching signal α can be arbitrary. In this case, (9.10) turns out to be $P(i) \leq P(j)$, $\forall i, j \in \mathcal{N}$. The only possibility for that is $P(i) = P(j) = P$, $\forall i, j \in \mathcal{N}$, and this implies that it requires a common (that is, mode-independent) Lyapunov function for all subsystems. However, when $\mu > 1$ and $\beta \to 0$ in $T_a > \frac{\ln \mu}{\beta}$, we have $T_a \to \infty$, that is, there is no switching. In such a case, switched stochastic hybrid system (9.1a)–(9.1b) is effectively operating at one of the subsystems all the time. We have the following result.* ♦

Corollary 9.3.3 *Suppose there is no switching in system (9.1a)–(9.1b) with $u(t) = 0$ and $\omega(t) = 0$ (when $\beta \to 0$ as discussed in Remark 9.2), that is, system (9.1a)–(9.1b) with $u(t) = 0$ and $\omega(t) = 0$ is transformed to a common stochastic system (thus, the parameters become as (A, E)). If there exists a matrix $P > 0$ such that*

$$\begin{bmatrix} PA + A^T P & E^T P \\ \star & -P \end{bmatrix} < 0, \tag{9.22}$$

then the common stochastic system is mean-square asymptotically stable.

Remark 9.3 *The mean-square asymptotic stability for the common stochastic system in Corollary 9.3.3 is consistent with the result in [240], which shows that Theorem 9.3.2 has extended some results in [240] to the switched hybrid systems.* ♦

Now, we present a solution to the stabilization problem, and give the following result.

Theorem 9.3.4 *Given a scalar $\beta > 0$, suppose there exist matrices $R(i) > 0$ and $L(i) > 0$ such that for $i \in \mathcal{N}$,*

$$\begin{bmatrix} \Sigma_{11}(i) & R(i)E^T(i) \\ \star & -R(i) \end{bmatrix} < 0, \tag{9.23}$$

where

$$\Sigma_{11}(i) \triangleq A(i)R(i) + R(i)A^T(i) + B(i)L(i) + L^T(i)B^T(i) + \beta R(i).$$

Then, the closed-loop system in (9.3) is mean-square exponentially stable for any switching signal with average dwell time satisfying $T_a > \frac{\ln \mu}{\beta}$*, where* $\mu \geq 1$ *and satisfies*

$$R(i) \leq \mu R(j), \quad \forall i,j \in \mathcal{N}. \tag{9.24}$$

Moreover, the gain matrices $K(i)$ *of the stabilization controller in (9.2) can be chosen by*

$$K(i) = L(i)R^{-1}(i), \quad i \in \mathcal{N}. \tag{9.25}$$

Proof. Replacing $A(i)$ in (9.9) with $A(i) + B(i)K(i)$ in Theorem 9.3.2, we have that the closed-loop stabilization system in (9.3) is mean-square exponentially stable if there exist matrices $P(i) > 0$ such that for $i \in \mathcal{N}$,

$$\begin{bmatrix} \bar{\Sigma}_{11}(i) & E^T(i)P(i) \\ \star & -P(i) \end{bmatrix} < 0, \tag{9.26}$$

where

$$\bar{\Sigma}_{11}(i) \triangleq P(i)\,(A(i) + B(i)K(i)) + (A(i) + B(i)K(i))^T\, P(i) + \beta P(i).$$

Letting $R(i) \triangleq P^{-1}(i)$ and performing a congruence transformation on (9.26) by $\text{diag}\,(R(i), R(i))$ yields

$$\begin{bmatrix} \tilde{\Sigma}_{11}(i) & R(i)E^T(i) \\ \star & -R(i) \end{bmatrix} < 0, \tag{9.27}$$

where

$$\tilde{\Sigma}_{11}(i) \triangleq (A(i) + B(i)K(i))\,R(i) + R(i)\,(A(i) + B(i)K(i))^T + \beta R(i).$$

Set $L(i) = K(i)R(i)$, thus (9.27) is equal to (9.23). This completes the proof. ■

9.4 $\mathcal{H}_\infty$ Control

9.4.1 $\mathcal{H}_\infty$ Performance Analysis

First, we will investigate the weighted $\mathcal{H}_\infty$ performance for the switched stochastic hybrid system in (9.1a)–(9.1b) with $u(t) = 0$.

Theorem 9.4.1 *Given scalars* $\beta > 0$ *and* $\gamma > 0$*, suppose there exist matrices* $P(i) > 0$ *such that for* $i \in \mathcal{N}$*,*

$$\begin{bmatrix} \bar{\Pi}_{11}(i) & P(i)D(i) & E^T(i)P(i) & C^T(i) \\ \star & -\gamma^2 I & 0 & 0 \\ \star & \star & -P(i) & 0 \\ \star & \star & \star & -I \end{bmatrix} < 0, \tag{9.28}$$

where

$$\bar{\Pi}_{11}(i) \triangleq P(i)A(i) + A^T(i)P(i) + \beta P(i).$$

Then the switched stochastic hybrid system in (9.1a)–(9.1b) with $u(t) = 0$ is mean-square exponentially stable with a weighted $\mathcal{H}_\infty$ performance level γ for any switching signal with average dwell time satisfying $T_a > \frac{\ln \mu}{\beta}$, where $\mu \geq 1$ satisfies (9.10).

Proof. The proof of mean-square exponential stability can be carried out along the same lines as that in the proof of Theorem 9.3.2. Now, we will establish the weighted $\mathcal{H}_\infty$ performance defined in (9.8); to this end, we introduce the following index:

$$\mathcal{J} \triangleq \mathcal{L}V(x,\alpha) + \beta V(x,\alpha) + z^T(t)z(t) - \gamma^2\omega^T(t)\omega(t),$$

where the Lyapunov function $V(x,\alpha)$ is given in (9.13) and

$$\begin{aligned}\mathcal{L}V(x,\alpha) &= 2x^T(t)P(\alpha)\left[A(\alpha)x(t) + D(\alpha)\omega(t)\right] + x^T(t)E^T(\alpha)P(\alpha)E(\alpha)x(t)\\ &= x^T(t)\left[P(\alpha)A(\alpha) + A^T(\alpha)P(\alpha) + E^T(\alpha)P(\alpha)E(\alpha)\right]x(t)\\ &\quad + 2x^T(t)P(\alpha)D(\alpha)\omega(t).\end{aligned}$$

Thus,

$$\mathcal{J} \triangleq \psi^T(t)\Pi(\alpha)\psi(t),$$

where $\psi(t) \triangleq \begin{bmatrix} x(t) \\ \omega(t) \end{bmatrix}$ and

$$\begin{cases} \Pi(\alpha) \triangleq \begin{bmatrix} \Pi_{11}(\alpha) & P(\alpha)D(\alpha) \\ \star & -\gamma^2 I \end{bmatrix}, \\ \Pi_{11}(\alpha) \triangleq P(\alpha)A(\alpha) + A^T(\alpha)P(\alpha) + \beta P(\alpha) + E^T(\alpha)P(\alpha)E(\alpha) + C^T(\alpha)C(\alpha). \end{cases}$$

By Schur complement, LMI (9.28) is equal to $\Pi(\alpha) < 0$, thus $\mathcal{J} < 0$. Let $\Gamma(t) \triangleq z^T(t)z(t) - \gamma^2\omega^T(t)\omega(t)$, then we have

$$\mathcal{L}V(x,\alpha) < -\beta V(x,\alpha) - \Gamma(t).$$

Thus, we have

$$\begin{aligned}dV(x,\alpha) &= \mathcal{L}V(x_t,\alpha)dt + 2x^T(t)P(\alpha)E(\alpha)x(t)d\varpi(t)\\ &< -\beta V(x,\alpha)dt - \Gamma(t)dt + 2x^T(t)P(\alpha)E(\alpha)x(t)d\varpi(t).\end{aligned}$$

Observe that

$$
\begin{aligned}
d\left[e^{\beta t}V(x,\alpha)\right] &= \beta e^{\beta t}V(x,\alpha)dt + e^{\beta t}dV(x,\alpha) \\
&< e^{\beta t}\left[\beta V(x,\alpha)dt - \beta V(x,\alpha)dt - \Gamma(t)dt + 2x^T(t)P(\alpha)E(\alpha)x(t)d\varpi(t)\right] \\
&= -e^{\beta t}\Gamma(t)dt + 2e^{\beta t}x^T(t)P(\alpha)E(\alpha)x(t)d\varpi(t). \qquad (9.29)
\end{aligned}
$$

Integrate both sides of (9.29) from $T > 0$ to t and take expectations. Then, by some mathematical operations, we have

$$
\mathbf{E}\{V(x,\alpha)\} < e^{-\beta(t-T)}\mathbf{E}\{V(x(T),\alpha(T))\} - \mathbf{E}\left\{\int_T^t e^{-\beta(t-s)}\Gamma(s)ds\right\}. \qquad (9.30)
$$

Let $0 = t_0 < t_1 < \cdots < t_k < \cdots$, $k = 1, \ldots,$ denote the switching points of α over the interval $(0, t)$, and suppose that the i_kth subsystem is activated when $t \in [t_k, t_{k+1})$. Setting $T = t_k$ in (9.30), we have

$$
\mathbf{E}\{V(x,\alpha)\} < e^{-\beta(t-t_k)}\mathbf{E}\left\{V(x(t_k),\alpha(t_k))\right\} - \mathbf{E}\left\{\int_{t_k}^t e^{-\beta(t-s)}\Gamma(s)ds\right\}. \qquad (9.31)
$$

Using (9.10) and (9.13), at switching instant t_k, we have

$$
\mathbf{E}\left\{V(x(t_k),\alpha(t_k))\right\} \le \mu\mathbf{E}\left\{V(x(t_k^-),\alpha(t_k^-))\right\}. \qquad (9.32)
$$

Therefore, it follows from (9.31)–(9.32) and the relationship $\vartheta = N_\alpha(0,t) \le (t-0)/T_a$ that

$$
\begin{aligned}
\mathbf{E}\{V(x,\alpha)\} &\le \mu e^{-\beta(t-t_k)}\mathbf{E}\left\{V(x(t_k^-),\alpha(t_k^-))\right\} - \mathbf{E}\left\{\int_{t_k}^t e^{-\beta(t-s)}\Gamma(s)ds\right\} \\
&\le \mu^\vartheta e^{-\beta t}\mathbf{E}\{V(x(0),\alpha(0))\} - \mu^\vartheta\mathbf{E}\left\{\int_0^{t_1} e^{-\beta(t-s)}\Gamma(s)ds\right\} \\
&\quad - \mu^{\vartheta-1}\mathbf{E}\left\{\int_{t_1}^{t_2} e^{-\beta(t-s)}\Gamma(s)ds\right\} - \cdots - \mu^0\mathbf{E}\left\{\int_{t_k}^{t} e^{-\beta(t-s)}\Gamma(s)ds\right\} \\
&= -\mathbf{E}\left\{\int_0^t e^{-\beta(t-s)+N_\alpha(s,t)\ln\mu}\Gamma(s)ds\right\} \\
&\quad + e^{-\beta t+N_\alpha(0,t)\ln\mu}V(x(0),\alpha(0)). \qquad (9.33)
\end{aligned}
$$

Under zero initial condition, that is, $x(0) = 0$, (9.33) implies

$$
\begin{aligned}
&\mathbf{E}\left\{\int_0^t e^{-\beta(t-s)+N_\alpha(s,t)\ln\mu}z^T(s)z(s)ds\right\} \\
&\le \gamma^2\mathbf{E}\left\{\int_0^t e^{-\beta(t-s)+N_\alpha(s,t)\ln\mu}\omega^T(s)\omega(s)ds\right\}. \qquad (9.34)
\end{aligned}
$$

Multiplying both sides of (9.34) by $e^{-N_\alpha(0,t)\ln\mu}$ yields

$$\mathbf{E}\left\{\int_0^t e^{-\beta(t-s)-N_\alpha(0,s)\ln\mu} z^T(s)z(s)ds\right\}$$
$$\leq \gamma^2\mathbf{E}\left\{\int_0^t e^{-\beta(t-s)-N_\alpha(0,s)\ln\mu}\omega^T(s)\omega(s)ds\right\}. \tag{9.35}$$

Notice that as $N_\alpha(0,s) \leq s/T_a$ and $T_a > T_a^* = \ln\mu/\beta$, we have $N_\alpha(0,s)\ln\mu \leq \beta s$. Thus, (9.35) implies

$$\mathbf{E}\left\{\int_0^t e^{-\beta(t-s)-\beta s} z^T(s)z(s)ds\right\} \leq \gamma^2\mathbf{E}\left\{\int_0^t e^{-\beta(t-s)}\omega^T(s)\omega(s)ds\right\}$$
$$= \gamma^2\int_0^t e^{-\beta(t-s)}\omega^T(s)\omega(s)ds.$$

Integrating the above inequality from $t = 0$ to ∞ yields (9.8). This completes the proof. ∎

Remark 9.4 *Note that Theorem 9.4.1 presents a weighted $\mathcal{H}_\infty$ performance for the switched stochastic hybrid system in (9.1a)–(9.1b) with $u(t) = 0$. The term 'weighted' refers to the weighting function $e^{-\beta t}$ in the left-hand side of (9.8). This is also the characteristic of the exponential stability result to the switched system by using the average dwell time approach combining with the piecewise Lyapunov function technique. When $\beta = 0$, we know from Remark 9.2 that there is no switching. Thus, the result in Theorem 9.4.1 becomes an asymptotic stability condition with an $\mathcal{H}_\infty$ performance for the deterministic stochastic system, which is also consistent with the results in [240].* ♦

9.4.2 State Feedback Control

In this sequel, we will present a solution to the $\mathcal{H}_\infty$ state feedback control problem.

Theorem 9.4.2 *Given scalars $\beta > 0$ and $\gamma > 0$, suppose there exist matrices $R(i) > 0$ and $L(i) > 0$, such that for $i \in \mathcal{N}$,*

$$\begin{bmatrix} \hat{\Pi}_{11}(i) & D(i) & R(i)E^T(i) & R(i)C^T(i) \\ \star & -\gamma^2 I & 0 & 0 \\ \star & \star & -R(i) & 0 \\ \star & \star & \star & -I \end{bmatrix} < 0, \tag{9.36}$$

where

$$\hat{\Pi}_{11}(i) \triangleq A(i)R(i) + R(i)A^T(i) + B(i)L(i) + L^T(i)B^T(i) + \beta R(i).$$

Then the closed-loop system in (9.4a)–(9.4b) is mean-square exponentially stable with a weighted $\mathcal{H}_\infty$ performance level γ for any switching signal with average dwell time satisfying $T_a > \frac{\ln \mu}{\beta}$, where $\mu \geq 1$ and satisfies

$$R(i) \leq \mu R(j), \quad \forall i,j \in \mathcal{N}. \tag{9.37}$$

Moreover, if the above LMIs have feasible solutions, then the gain matrices $K(i)$ of the $\mathcal{H}_\infty$ controller in (9.2) can be chosen by

$$K(i) = L(i)R^{-1}(i), \quad i \in \mathcal{N}. \tag{9.38}$$

The result can be carried out by employing the same techniques as used with Theorems 9.3.4 and 9.4.1.

9.4.3 $\mathcal{H}_\infty$ DOF Controller Design

In the following, we will study the $\mathcal{H}_\infty$ DOF control problem for the switched stochastic hybrid system (9.1a)–(9.1b). First, we present the following results, and its proof can be worked out along the same line of reasoning as in the derivation of Theorems 9.3.2 and 9.4.1.

Theorem 9.4.3 *Given scalars $\beta > 0$ and $\gamma > 0$, suppose there exist matrices $P(i) > 0$ such that for $i \in \mathcal{N}$,*

$$\begin{bmatrix} \tilde{\Pi}_{11}(i) & P(i)\tilde{D}(i) & M^T\tilde{E}^T(i)P(i) & \tilde{C}^T(i) \\ \star & -\gamma^2 I & 0 & 0 \\ \star & \star & -P(i) & 0 \\ \star & \star & \star & -I \end{bmatrix} < 0, \tag{9.39}$$

where

$$\tilde{\Pi}_{11}(i) \triangleq P(i)\tilde{A}(i) + \tilde{A}^T(i)P(i) + \beta P(i).$$

Then the closed-loop switched stochastic hybrid system in (9.7a)–(9.7b) is mean-square exponentially stable with a weighted $\mathcal{H}_\infty$ performance level γ for any switching signal with average dwell time satisfying $T_a > \frac{\ln \mu}{\beta}$, where $\mu \geq 1$ and satisfies

$$P(i) \leq \mu P(j), \quad \forall i,j \in \mathcal{N}. \tag{9.40}$$

Moreover, an estimate of the mean-square of the state decay is given by

$$\mathbf{E}\left\{\|\xi(t)\|\right\} \leq \eta \|\xi(0)\| e^{-\lambda t},$$

where

$$\begin{cases} \lambda = \dfrac{1}{2}\left(\beta - \dfrac{\ln\mu}{T_a}\right) > 0, \quad \eta = \sqrt{\dfrac{b}{a}} \geq 1, \\ a = \min\limits_{\forall i\in\mathcal{N}} \lambda_{\min}(P(i)), \quad b = \max\limits_{\forall i\in\mathcal{N}} \lambda_{\max}(P(i)). \end{cases}$$

Now, we present a solution to the $\mathcal{H}_\infty$ DOF control problem.

Theorem 9.4.4 *Consider the switched stochastic hybrid system in (9.1a)–(9.1b). For given constants $\beta > 0$ and $\gamma > 0$, suppose there exist matrices $\mathcal{P}(i) > 0$, $\mathcal{Q}(i) > 0$, $\mathcal{A}_c(i)$, $\mathcal{B}_c(i)$, and $\mathcal{C}_c(i)$ and a scalar $\varepsilon > 0$ such that for $i \in \mathcal{N}$,*

$$\begin{bmatrix} \Psi_{11}(i) & \Psi_{12}(i) & \Psi_{13}(i) & C^T(i) & 0 \\ \star & \Psi_{22}(i) & D(i) & \mathcal{Q}(i)C^T(i) & \varepsilon\mathcal{Q}(i) \\ \star & \star & -\gamma^2 I & 0 & 0 \\ \star & \star & \star & -I & 0 \\ \star & \star & \star & \star & -\varepsilon I \end{bmatrix} < 0, \tag{9.41a}$$

$$\begin{bmatrix} -\varepsilon I & E^T(i)\mathcal{P}(i) + F^T(i)\mathcal{B}_c^T(i) & E^T(i) \\ \star & -\mathcal{P}(i) & -I \\ \star & \star & -\mathcal{Q}(i) \end{bmatrix} < 0, \tag{9.41b}$$

where

$$\begin{cases} \Psi_{11}(i) \triangleq \mathcal{P}(i)A(i) + A^T(i)\mathcal{P}(i) + \mathcal{B}_c(i)G(i) + G^T(i)\mathcal{B}_c^T(i) + \beta\mathcal{P}(i) + \varepsilon I, \\ \Psi_{12}(i) \triangleq \mathcal{A}_c(i) + A(i) + I + \varepsilon\mathcal{Q}(i), \\ \Psi_{13}(i) \triangleq \mathcal{P}(i)D(i) + \mathcal{B}_c(i)H(i), \\ \Psi_{22}(i) \triangleq A(i)\mathcal{Q}(i) + \mathcal{Q}(i)A^T(i) + B(i)\mathcal{C}_c(i) + \mathcal{C}_c^T(i)B^T(i) + \beta\mathcal{Q}(i). \end{cases}$$

Then the closed-loop system in (9.7a)–(9.7b) is mean-square exponentially stable with a weighted $\mathcal{H}_\infty$ performance level γ for any switching signal with average dwell time satisfying $T_a > T_a^* = \frac{\ln\mu}{\beta}$, where $\mu \geq 1$ and satisfies

$$\begin{bmatrix} \mathcal{P}(i) & I \\ I & \mathcal{Q}(i) \end{bmatrix} \leq \mu \begin{bmatrix} \mathcal{P}(j) & I \\ I & \mathcal{Q}(j) \end{bmatrix}, \quad \forall i,j \in \mathcal{N}. \tag{9.42}$$

Moreover, if the above conditions have feasible solutions, then a desired weighted $\mathcal{H}_\infty$ DOF controller realization is given by

$$\begin{cases} A_c(i) \triangleq P_2^{-1}(i)[\mathcal{A}_c(i) - \mathcal{P}(i)A(i)Q(i) - P_2(i)B_c(i)G(i)Q(i) - \mathcal{P}(i)B(i)C_c(i)Q_2^T(i)]Q_2^{-T}(i), \\ B_c(i) \triangleq P_2^{-1}(i)\mathcal{B}_c(i), \\ C_c(i) \triangleq \mathcal{C}_c(i)Q_2^{-T}(i). \end{cases} \tag{9.43}$$

Proof. From Theorem 9.4.3, we know that the closed-loop system in (9.7a)–(9.7b) is mean-square exponentially stable with a weighted $\mathcal{H}_\infty$ performance level $\gamma > 0$, if there exist matrices $P(i) > 0$ satisfying (9.39). It is not difficult to see that these conditions are satisfied if there exist matrices $P(i) > 0$ and a scalar $\epsilon > 0$ such that for $i \in \mathcal{N}$,

$$\begin{bmatrix} \Phi_{11}(i) & P(i)\tilde{D}(i) & \tilde{C}^T(i) \\ \star & -\gamma^2 I & 0 \\ \star & \star & -I \end{bmatrix} < 0, \tag{9.44}$$

$$\begin{bmatrix} -\epsilon I & \tilde{E}^T(i)P(i) \\ \star & -P(i) \end{bmatrix} < 0, \tag{9.45}$$

where $\Phi_{11}(i) \triangleq P(i)\tilde{A}(i) + \tilde{A}^T(i)P(i) + \beta P(i) + \epsilon M^T M$.

Let $P(i)$ be partitioned as

$$\left.\begin{aligned} P(i) &\triangleq \begin{bmatrix} P_1(i) & P_2(i) \\ \star & P_3(i) \end{bmatrix} \\ Q(i) = P^{-1}(i) &\triangleq \begin{bmatrix} Q_1(i) & Q_2(i) \\ \star & Q_3(i) \end{bmatrix} \end{aligned}\right\}. \tag{9.46}$$

Without loss of generality, we assume $P_2(i)$ and $Q_2(i)$ are nonsingular (if not, $P_2(i)$ and $Q_2(i)$ may be perturbed by matrices $\Delta P_2(i)$ and $\Delta Q_2(i)$ with sufficiently small norms respectively such that $P_2(i) + \Delta P_2(i)$ and $Q_2(i) + \Delta Q_2(i)$ are nonsingular and satisfy (9.44)–(9.45)). Then, define the following nonsingular matrices:

$$\mathcal{J}_P(i) \triangleq \begin{bmatrix} P_1(i) & I \\ P_2^T(i) & 0 \end{bmatrix}, \quad \mathcal{J}_Q(i) \triangleq \begin{bmatrix} I & Q_1(i) \\ 0 & Q_2^T(i) \end{bmatrix}. \tag{9.47}$$

Notice that $P(i)\mathcal{J}_Q(i) = \mathcal{J}_P(i)$, $Q(i)\mathcal{J}_P(i) = \mathcal{J}_Q(i)$, and $P_1(i)Q_1(i) + P_2(i)Q_2^T(i) = I$, $i \in \mathcal{N}$.

Performing congruence transformations on (9.44) and (9.45) by matrices diag $\{\mathcal{J}_Q(i), I, I\}$ and diag $\{I, \mathcal{J}_Q(i)\}$, respectively, we obtain

$$\begin{bmatrix} \mathcal{J}_Q^T(i)\Phi_{11}(i)\mathcal{J}_Q(i) & \mathcal{J}_Q^T(i)P(i)\tilde{D}(i) & \mathcal{J}_Q^T(i)\tilde{C}^T(i) \\ \star & -\gamma^2 I & 0 \\ \star & \star & -I \end{bmatrix} < 0, \quad (9.48)$$

$$\begin{bmatrix} -\varepsilon I & \tilde{E}^T(i)P(i)\mathcal{J}_Q(i) \\ \star & -\mathcal{J}_Q^T(i)P(i)\mathcal{J}_Q(i) \end{bmatrix} < 0. \quad (9.49)$$

Define $\mathcal{P}(i) \triangleq P_1(i)$, $\mathcal{Q}(i) \triangleq Q_1(i)$ and

$$\begin{cases} \mathcal{A}_c(i) \triangleq P_1(i)A(i)Q_1(i) + P_2(i)B_c(i)G(i)Q_1(i) + P_1(i)B(i)C_c(i)Q_2^T(i) + P_2(i)A_c(i)Q_2^T(i), \\ \mathcal{B}_c(i) \triangleq P_2(i)B_c(i), \\ \mathcal{C}_c(i) \triangleq C_c(i)Q_2^T(i). \end{cases} \quad (9.50)$$

Then by (9.46)–(9.47) and (9.50), it follows from (9.48) and (9.49) that

$$\begin{bmatrix} \Psi_{11}(i) & \tilde{\Psi}_{12}(i) & \tilde{\Psi}_{13}(i) & C^T(i) \\ \star & \tilde{\Psi}_{22}(i) & D(i) & \mathcal{Q}(i)C^T(i) \\ \star & \star & -\gamma^2 I & 0 \\ \star & \star & \star & -I \end{bmatrix} < 0, \quad (9.51)$$

and (9.41), respectively, where

$$\begin{cases} \tilde{\Psi}_{12}(i) \triangleq \mathcal{A}_c(i) + A(i) + I + \varepsilon\mathcal{Q}(i), \\ \tilde{\Psi}_{13}(i) \triangleq \mathcal{P}(i)D(i) + \mathcal{B}_c(i)H(i), \\ \tilde{\Psi}_{22}(i) \triangleq A(i)\mathcal{Q}(i) + \mathcal{Q}(i)A^T(i) + B(i)\mathcal{C}_c(i) + \mathcal{C}_c^T(i)B^T(i) + \beta\mathcal{Q}(i) + \varepsilon\mathcal{Q}(i)\mathcal{Q}(i), \end{cases}$$

and $\Psi_{11}(i)$ are defined in (9.41a). By Schur complement, (9.41a) is equivalent to (9.51). Moreover, considering the conditions in (9.40) yields (9.42). This completes the proof. ■

Remark 9.5 *It should be pointed out that to solve the parameters of output feedback controller in (9.43), matrices $P_2(i)$ and $Q_2(i)$ should be available in advance, which can be obtained by taking any full rank factorization of $P_2(i)Q_2^T(i) = I - \mathcal{P}(i)\mathcal{Q}(i)$ (derived from $\mathcal{P}(i)\mathcal{Q}(i) + P_2(i)Q_2^T(i) = I$).* ♦

Remark 9.6 *Note that Theorem 9.4.4 provides a sufficient condition for solvability of the weighted $\mathcal{H}_\infty$ DOF control problem and, since the resulting condition is in LMI form, a desired $\mathcal{H}_\infty$ DOF controller which minimizes the weighted $\mathcal{H}_\infty$ performance level (i.e. maximize* *the level of noise removal) can be determined by solving the following convex optimization problem:*

$$\min \sigma \quad \textit{subject to } (9.41a) - (9.43), \textit{ where } \sigma \triangleq \gamma^2,$$

with matrix variables $\mathcal{P}(i) > 0$, $\mathcal{Q}(i) > 0$, $\mathcal{A}_c(i)$, $\mathcal{B}_c(i)$, and $\mathcal{C}_c(i)$ and a scalar $\varepsilon > 0$. ♦

9.5 Illustrative Example

Example 9.5.1 (Stabilization problem) Consider the switched stochastic hybrid system in (9.1a)–(9.1b) with $N = 2$ and the following parameters:

$$A(1) = \begin{bmatrix} -0.8 & 0.1 \\ -0.3 & -0.5 \end{bmatrix}, \; E(1) = \begin{bmatrix} 1.0 & 0.2 \\ 0.3 & 0.5 \end{bmatrix}, \; B(1) = \begin{bmatrix} 1.2 \\ 0.8 \end{bmatrix},$$

$$A(2) = \begin{bmatrix} -0.5 & 0.2 \\ -0.2 & -0.4 \end{bmatrix}, \; E(2) = \begin{bmatrix} 0.7 & 0.5 \\ 0.1 & 0.3 \end{bmatrix}, \; B(2) = \begin{bmatrix} 0.4 \\ 1.2 \end{bmatrix}.$$

Given $\beta = 0.5$. By simulation, when setting $\mu = 1.78$, thus $T_a > T_a^* = \frac{\ln \mu}{\beta} = 1.1532$, the conditions in (9.9) hold with

$$P(1) = 10^3 \times \begin{bmatrix} 3.4640 & 1.9001 \\ 1.9001 & 3.8329 \end{bmatrix},$$

$$P(2) = 10^3 \times \begin{bmatrix} 1.9687 & 0.8284 \\ 0.8284 & 5.2596 \end{bmatrix}.$$

As analyzed above, the open-loop system is mean-square exponentially stable for $T_a > T_a^* = 1.1532$. Moreover, taking $T_a = 1.2 > T_a^* = 1.1532$ and according to (9.11)–(9.12) we have $a = 1739.4$, $b = 5557.4$, $\eta = \frac{b}{a} = 3.1950$, and $\lambda = \beta - \frac{\ln \mu}{T_a} = 0.0195$, thus, an estimate of the mean-square of the state decay is given by

$$\mathbf{E}\left\{\|x(t)\|^2\right\} \leq 3.1950 e^{-0.0195t} \|x(0)\|^2 .$$

Now, we further simulate the stabilization problem. As analyzed above, the open-loop system is mean-square exponentially stable when $T_a \geq T_a^* = 1.1532$. Here, to show the effectiveness, we will design an appropriate stabilization controller in (9.2) such that the closed-loop system in (9.3) is mean-square exponentially stable for $T_a \geq T_a^* = 0.1$ (in this case, the allowable minimum of μ is $\mu_{\min} = 1.0513$). By solving conditions (9.23) in Theorem 9.3.4, we have

$$R(1) = \begin{bmatrix} 0.4747 & -0.1532 \\ -0.1532 & 0.6804 \end{bmatrix}, \; R(2) = \begin{bmatrix} 0.4695 & -0.1551 \\ -0.1551 & 0.6732 \end{bmatrix},$$

$$L(1) = \begin{bmatrix} -0.1685 & -0.1425 \end{bmatrix}, \; L(2) = \begin{bmatrix} -0.2031 & -0.2704 \end{bmatrix}.$$

Thus, by (9.25) we have

$$K(1) = \begin{bmatrix} -0.4556 & -0.3121 \end{bmatrix}, \quad K(2) = \begin{bmatrix} -0.6118 & -0.5425 \end{bmatrix}.$$

Therefore, by the stabilization controller in (9.2) with the above control gains, the closed-loop system is mean-square exponentially stable for $T_a \geq T_a^* = 0.1$.

Example 9.5.2 ($\mathcal{H}_\infty$ DOF control problem) Consider the switched stochastic hybrid system in (9.1a)–(9.1b) with $N = 2$ and

$$A(1) = \begin{bmatrix} 0.3 & 0.2 \\ -0.3 & -0.8 \end{bmatrix}, \; B(1) = \begin{bmatrix} 1.2 \\ 0.8 \end{bmatrix}, \; C(1) = \begin{bmatrix} 0.6 & 1.0 \end{bmatrix}, \; H(1) = 0.1,$$

$$A(2) = \begin{bmatrix} -0.8 & -0.1 \\ 0.3 & 0.1 \end{bmatrix}, \; B(2) = \begin{bmatrix} 1.4 \\ 1.2 \end{bmatrix}, \; C(2) = \begin{bmatrix} 1.4 & 0.9 \end{bmatrix}, \; H(2) = 0.2,$$

$$D(1) = \begin{bmatrix} 0.1 \\ 0.2 \end{bmatrix}, \; E(1) = \begin{bmatrix} 0.2 & 0.1 \\ 0.2 & 0.3 \end{bmatrix}, \; F(1) = \begin{bmatrix} 0.1 & 0.2 \end{bmatrix}, \; G(1) = \begin{bmatrix} 0.6 & 0.4 \end{bmatrix},$$

$$D(2) = \begin{bmatrix} 0.3 \\ 0.1 \end{bmatrix}, \; E(2) = \begin{bmatrix} 0.2 & 0.1 \\ 0.1 & 0.3 \end{bmatrix}, \; F(2) = \begin{bmatrix} 0.2 & 0.1 \end{bmatrix}, \; G(2) = \begin{bmatrix} 0.6 & 0.8 \end{bmatrix}.$$

Given $\beta = 0.5$, we checked that the considered switched stochastic hybrid system with the above two subsystems is not stable for a switching signal given in Figure 9.1 (which is generated randomly; here, '1' and '2' represent the first and second subsystems, respectively). Here, our aim is to design a DOF controller such that the resulting closed-loop system is mean-square exponentially stable with a weighted $\mathcal{H}_\infty$ performance level $\gamma > 0$ for $T_a > T_a^*$. Here, for example, we set $T_a^* = \frac{\ln \mu_{\min}}{\beta} = 0.0976$ (in this case, the allowable minimum of μ is $\mu_{\min} = 1.05$). Letting $\varepsilon = 0.18$ and solving (9.41a)–(9.41b) in Theorem 9.4.4, then setting $Q_2(i) = I$ and according to (9.43), we have $\gamma = 3.3166$ and

$$A_c(1) = \begin{bmatrix} -28.0435 & -19.1500 \\ 0.2633 & 0.1884 \end{bmatrix}, \; B_c(1) = \begin{bmatrix} 0.9100 \\ -0.6834 \end{bmatrix},$$

$$A_c(2) = \begin{bmatrix} -9.3112 & -7.8631 \\ -19.7947 & -16.2218 \end{bmatrix}, \; B_c(2) = \begin{bmatrix} -0.3707 \\ 0.8223 \end{bmatrix},$$

$$C_c(1) = \begin{bmatrix} -85.5350 & -57.8919 \end{bmatrix}, \; C_c(2) = \begin{bmatrix} -60.0887 & -49.8330 \end{bmatrix}.$$

Given the initial conditions as $x(0) = \begin{bmatrix} -1.0 & 1.0 \end{bmatrix}^T$ and $\hat{x}(0) = \begin{bmatrix} 0 & 0 \end{bmatrix}^T$, suppose the disturbance input $\omega(t)$ be $\omega(t) = 0.5e^{-t}\sin(t)$. By using the discretization approach [96], we simulate standard Brownian motion. Some initial parameters are given as follows: the simulation time $t \in [0, T^\star]$ with $T^\star = 20$, the normally distributed variance $\delta t = \frac{T^\star}{N^\star}$ with $N^\star = 2^{11}$, step size $\Delta t = \rho\delta t$ with $\rho = 2$, and the number of discretized Brownian paths $p = 10$. The simulation results are given in Figures 9.2–9.6. Among them, Figures 9.2–9.4 are the simulation results along an individual discretized Brownian path. Figures 9.2 and 9.3 give respectively the states

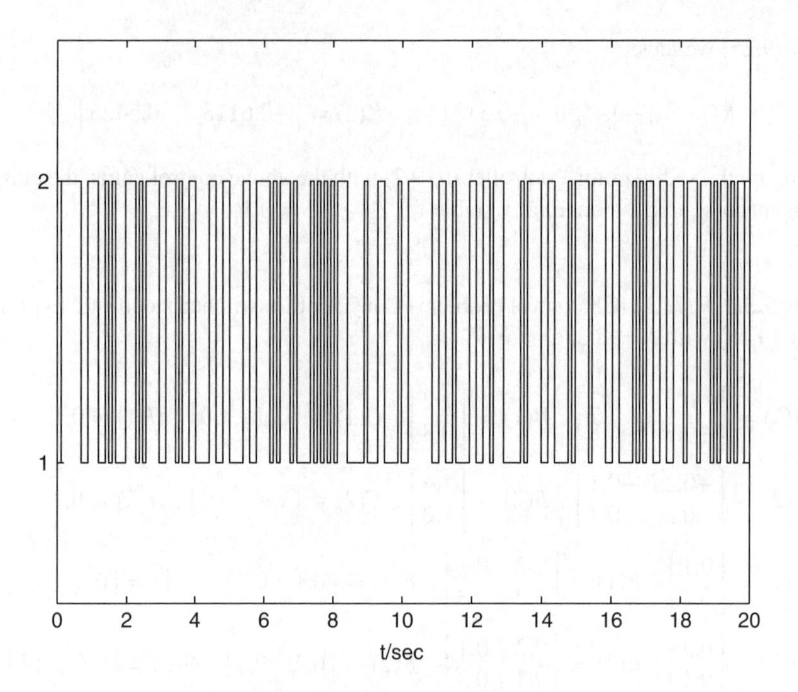

Figure 9.1 Switching signal

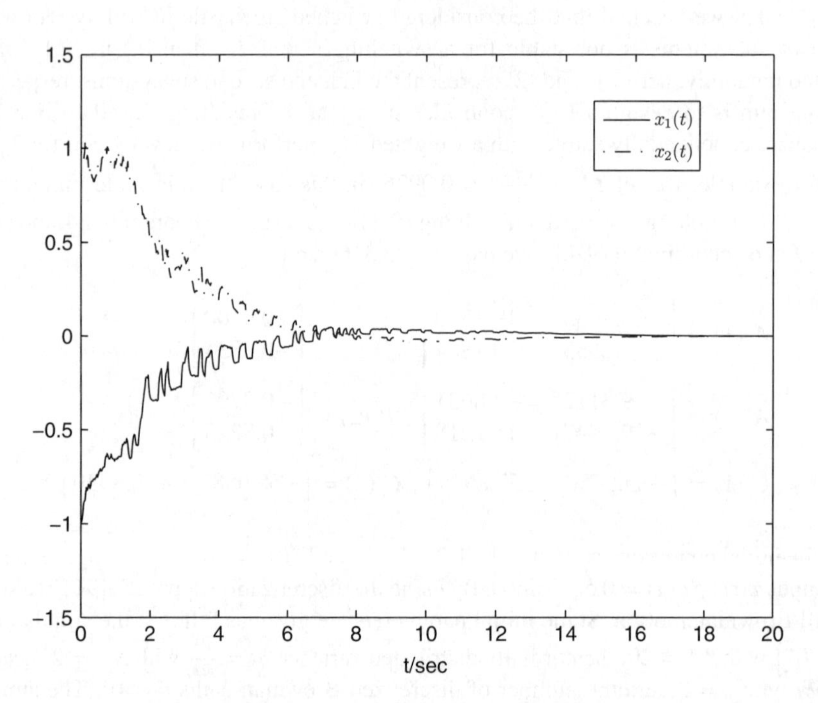

Figure 9.2 States of the closed-loop system

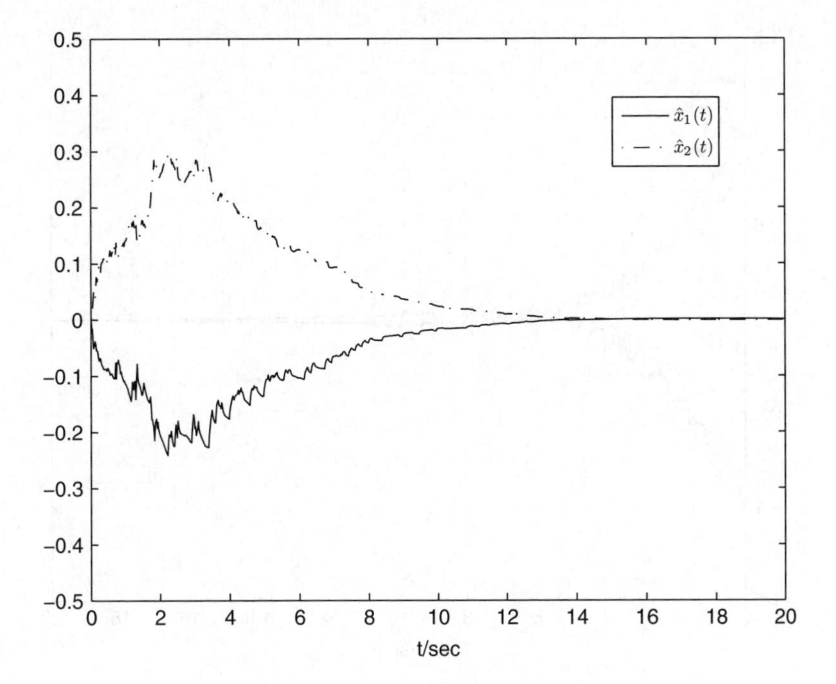

Figure 9.3 States of the DOF controller

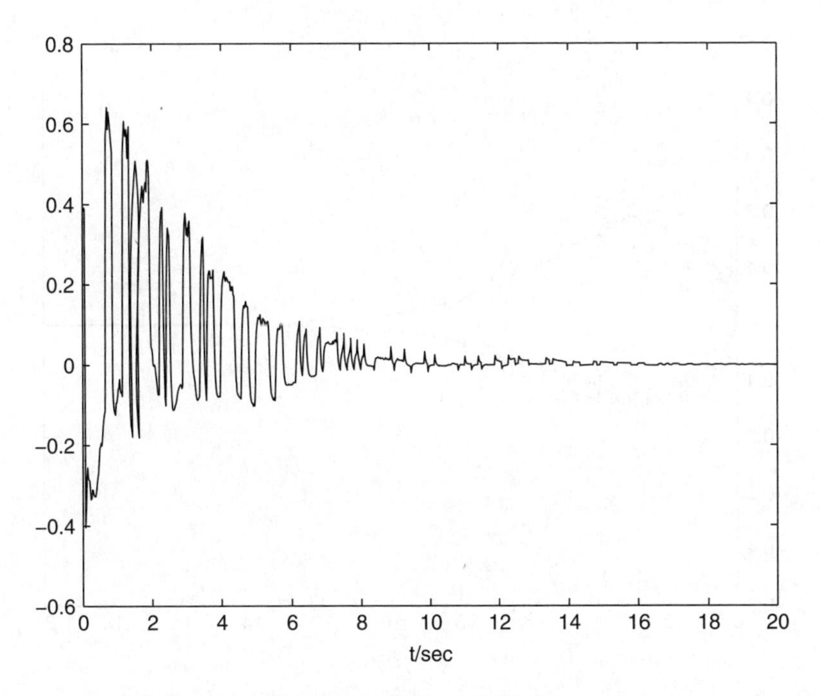

Figure 9.4 DOF control input

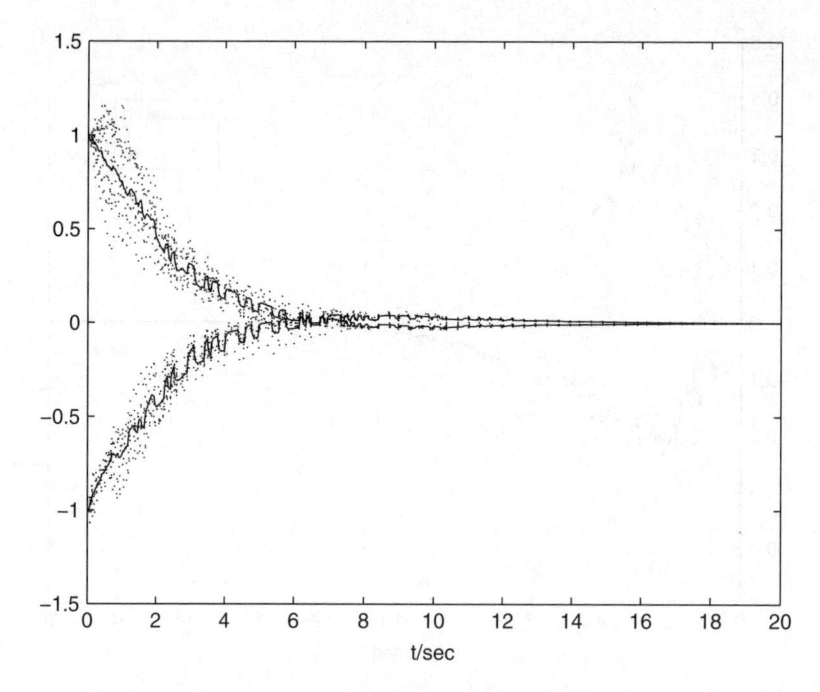

Figure 9.5 Individual paths and the average of the states of the closed-loop system

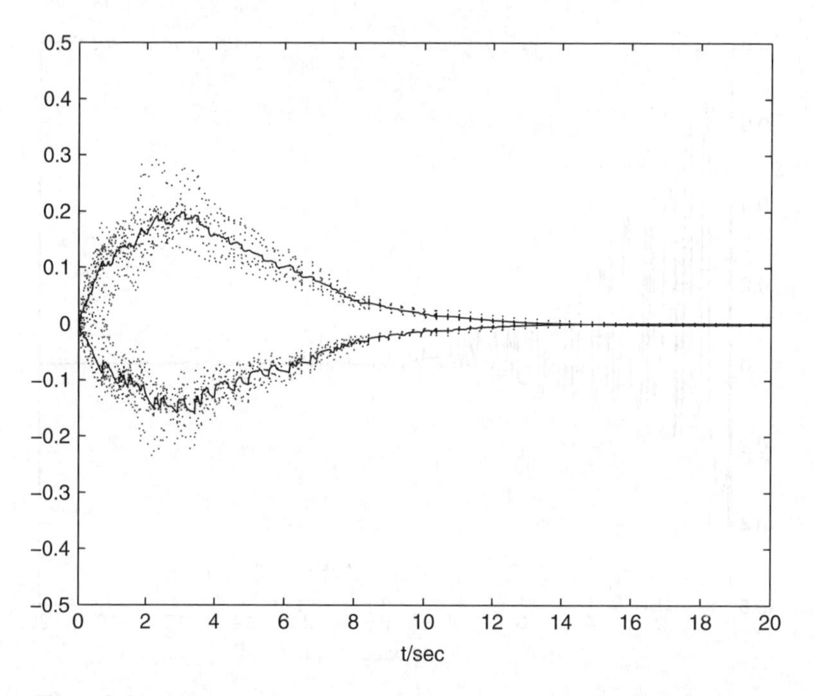

Figure 9.6 Individual paths and the average of the states of the DOF controller

of the closed-loop system and the DOF controller. The control input is shown in Figure 9.4. Figures 9.5 and 9.6 are the simulation results on $x(t)$ and $\hat{x}(t)$ along 10 individual paths (dotted lines) and the average over 10 paths (solid line), respectively.

9.6 Conclusion

In this chapter, the problems of stabilization and the $\mathcal{H}_\infty$ control have been investigated for continuous-time switched stochastic hybrid systems. By applying the average dwell time method and the piecewise Lyapunov function technique, sufficient conditions have been proposed for the mean-square exponential stability and the weighted $\mathcal{H}_\infty$ performance for the switched stochastic hybrid system. Then, the stabilization and the $\mathcal{H}_\infty$ control including the state feedback and DOF control problems have been solved. Finally, two numerical examples have been provided to illustrate the effectiveness of the proposed theories.

10

Control of Switched Stochastic Hybrid Systems: Discrete-Time Case

10.1 Introduction

In this chapter, we will study the problems of stability and stabilization, and the $\mathcal{H}_\infty$ control of discrete-time switched stochastic hybrid systems with time-varying delays. Similar to Chapter 9, the main results proposed in this chapter are obtained by employing the average dwell time approach and the piecewise Lyapunov function technique. However, the development of such methods in the continuous and discrete cases has certain technical differences. A sufficient condition, which guarantees the considered system mean-square exponentially stable, is first proposed in terms of LMIs, and by this the stabilization problem is then solved. The weighted $\mathcal{H}_\infty$ performance condition is then also established, and $\mathcal{H}_\infty$ control is designed. It is shown that the $\mathcal{H}_\infty$ control problem can be converted into a convex optimization problem with a set of LMI constraints which can be solved by applying interior-point algorithms.

10.2 System Description and Preliminaries

Consider the discrete-time switched stochastic hybrid system with time delays, which is described by the following dynamical equations:

$$
\begin{aligned}
x(k+1) &= A(\alpha(k))x(k) + A_d(\alpha(k))x(k-d(k)) + A_\tau(\alpha(k))f(x(k-\tau)) \\
&\quad + B_u(\alpha(k))u(k) + B_\omega(\alpha(k))\omega(k) \\
&\quad + \left[C(\alpha(k))x(k) + C_d(\alpha(k))x(k-d(k)) + D_\omega(\alpha(k))\omega(k)\right]\varpi(k), \qquad (10.1a) \\
z(k) &= L(\alpha(k))x(k), \qquad (10.1b) \\
x(k) &= \phi(k), \quad -\max\{\tau, d_2\} < k \leq 0, \qquad (10.1c)
\end{aligned}
$$

Sliding Mode Control of Uncertain Parameter-Switching Hybrid Systems, First Edition. Ligang Wu, Peng Shi and Xiaojie Su.

where $x(k) \in \mathbf{R}^n$ is the system state vector; $u(k) \in \mathbf{R}^m$ represents the control input; $\omega(k) \in \mathbf{R}^p$ is the noise signal that belongs to $\ell_2[0, +\infty)$; $z(k) \in \mathbf{R}^q$ is the controlled output; and $\varpi(k)$ is a zero-mean real scalar process on a probability space $(\varpi, \mathcal{F}, \mathcal{P})$ relative to an increasing family $(\mathcal{F}_k)_{k\in\mathbf{N}}$ of σ-algebras $\mathcal{F}_k \subset \mathcal{F}$ generated by $(\varpi(k))_{k\in\mathbf{N}}$. The stochastic process $\{\varpi(k)\}$ is independent, which is assumed to satisfy $\mathbf{E}\{\varpi(k)\} = 0$ and $\mathbf{E}\{\varpi^2(k)\} = \delta$, $k = 0, 1, \ldots$, where $\delta > 0$ is a known scalar. In addition, $\phi(k)$ denotes the initial conditions and $\alpha(k) : \mathbf{Z}^+ \to \mathcal{N} = \{1, 2, \ldots, N\}$ (denoted by α for simplicity) is a switching signal, which was defined in the same way in Chapter 5. Here, we assume that the switch signal $\alpha(k)$ has an average dwell time. The time-varying delay $d(k)$ satisfies $1 \le d_1 \le d(k) \le d_2$, where d_1 and d_2 are two constant positive scalars representing its lower and upper bounds, respectively.

Assumption 10.1 *For the nonlinear function $f(x) : \mathbf{R}^n \to \mathbf{R}^n$, there exist matrices Π_1 and Π_2 such that*

$$\left(f(x) - \Pi_1 x\right)^T \left(f(x) - \Pi_2 x\right) \le 0, \quad x \in \mathbf{R}^n. \tag{10.2}$$

We design a stabilization controller and an $\mathcal{H}_\infty$ state feedback controller with the following general structure:

$$u(k) = K(\alpha)x(k), \tag{10.3}$$

where $K(\alpha) \in \mathbf{R}^{m\times n}$ are parameter matrices to be designed. Substituting the controller $u(k)$ into system (10.1a)–(10.1c), we obtain the closed-loop stabilization system as

$$\begin{aligned} x(k+1) = {} & \left[A(\alpha) + B_u(\alpha)K(\alpha_k)\right] x(k) + A_d(\alpha)x(k - d(k)) + A_\tau(\alpha)f(x(k-\tau)) \\ & + \left[C(\alpha)x(k) + C_d(\alpha)x(k - d(k))\right] \varpi(k), \end{aligned} \tag{10.4}$$

and the closed-loop $\mathcal{H}_\infty$ control system as

$$\begin{aligned} x(k+1) = {} & \left[A(\alpha) + B_u(\alpha)K(\alpha)\right] x(k) + A_d(\alpha)x(k - d(k)) + A_\tau(\alpha)f(x(k-\tau)) \\ & + B_\omega(\alpha)\omega(k) + \left[C(\alpha)x(k) + C_d(\alpha)x(k - d(k)) + D_\omega(\alpha)\omega(k)\right] \varpi(k), \end{aligned} \tag{10.5a}$$

$$z(k) = L(\alpha)x(k). \tag{10.5b}$$

Remark 10.1 *For each possible value $\alpha = i$, $i \in \mathcal{N}$, we will denote the system matrices associated with mode i by $A(i) = A(\alpha)$, $A_d(i) = A_d(\alpha)$, $A_\tau(i) = A_\tau(\alpha)$, $C(i) = C(\alpha)$, $C_d(i) = C_d(\alpha)$, $B_u(i) = B_u(\alpha)$, $B_\omega(i) = B_\omega(\alpha)$, $D_\omega(i) = D_\omega(\alpha)$, $L(i) = L(\alpha)$, and $K(i) = K(\alpha)$, where $A(i)$, $A_d(i)$, $A_\tau(i)$, $C(i)$, $C_d(i)$, $B_u(i)$, $B_\omega(i)$, $D_\omega(i)$, $L(i)$, and $K(i)$ are constant matrices.* ♦

Definition 10.2.1 *The discrete-time switched stochastic hybrid system in (10.1a) with $u(k) = 0$ and $\omega(k) = 0$ is said to be mean-square exponentially stable under α if the solution $x(k)$*

satisfies

$$\mathbf{E}\left\{\|x(k)\|\right\} \leq \eta \left\|x(k_0)\right\|_C \rho^{(k-k_0)}, \quad \forall k \geq k_0,$$

for constants $\eta \geq 1$ *and* $0 < \rho < 1$, *and*

$$\left\|x(k_0)\right\|_C \triangleq \underbrace{\left\{\left\|x(k_0+\theta)\right\|, \left\|\varsigma(k_0+\theta)\right\|, \left\|f(\varsigma(k_0+\theta))\right\|\right\}}_{\sup_{-\max\{\tau,d_2\}<\theta\leq 0}},$$

where $\varsigma(\theta) \triangleq x(\theta+1) - x(\theta)$.

Definition 10.2.2 *For* $0 < \beta < 1$ *and* $\gamma > 0$, *the system in (10.1a)–(10.1c) with* $u(k) = 0$ *is said to be mean-square exponentially stable with a weighted* $\mathcal{H}_\infty$ *performance level* γ *under* α, *if it is mean-square exponentially stable with* $\omega(k) = 0$, *and under zero initial condition, it holds for all nonzero* $\omega(k) \in \ell_2[0, \infty)$ *that*

$$\mathbf{E}\left\{\sum_{s=k_0}^{\infty} \beta^s z^T(s) z(s)\right\} < \gamma^2 \sum_{s=k_0}^{\infty} \omega^T(s)\omega(s). \tag{10.6}$$

10.3 Stability Analysis and Stabilization

In this section, we apply the average dwell time approach combined with the piecewise Lyapunov function technique to investigate the mean-square exponential stability and stabilization problems for the system (10.1a).

Theorem 10.3.1 *Given a constant* $0 < \beta < 1$, *suppose that there exist matrices* $P(i) > 0$, $Q(i) > 0$, *and* $R(i) > 0$ *such that for* $i \in \mathcal{N}$,

$$\begin{bmatrix} \Phi_{11}(i) & 0 & \mathrm{H}_2 & 0 & A^T(i)P(i) & \delta C^T(i)P(i) \\ \star & -\beta^{d_2+1}Q(i) & 0 & 0 & A_d^T(i)P(i) & \delta C_d^T(i)P(i) \\ \star & \star & \beta R(i) - I & 0 & 0 & 0 \\ \star & \star & \star & -\beta^{\tau+1}R(i) & A_\tau^T(i)P(i) & 0 \\ \star & \star & \star & \star & -P(i) & 0 \\ \star & \star & \star & \star & \star & -\delta P(i) \end{bmatrix} < 0, \tag{10.7}$$

where

$$\begin{cases} \Phi_{11}(i) \triangleq -\beta P(i) + \beta(d_2 - d_1 + 1)Q(i) - \mathrm{H}_1, \\ \mathrm{H}_1 \triangleq \dfrac{\Pi_1^T \Pi_2 + \Pi_2^T \Pi_1}{2}, \\ \mathrm{H}_2 \triangleq \dfrac{\Pi_1^T + \Pi_2^T}{2}. \end{cases}$$

Then the discrete-time switched stochastic time-delay system in (10.1a) with $u(k) = 0$ and $\omega(k) = 0$ is mean-square exponentially stable for any switching signal with average dwell time satisfying $T_a > T_a^* = \text{ceil}\left(-\frac{\ln \mu}{\ln \beta}\right)$, where $\mu \geq 1$ satisfies

$$P(i) \leq \mu P(j), \quad Q(i) \leq \mu Q(j), \quad R(i) \leq \mu R(j). \tag{10.8}$$

Moreover, an estimate of the state decay is given by

$$\mathbf{E}\{\|x(k)\|\} \leq \eta \rho^{(k-k_0)} \|x(k_0)\|_C, \tag{10.9}$$

where $d \triangleq -\max\{\tau, d_2\}$ and

$$\begin{cases} \eta \triangleq \sqrt{\dfrac{b}{a}} \geq 1, \quad \rho \triangleq \sqrt{\beta \mu^{\frac{1}{T_a}}}, \\ a \triangleq \min\limits_{\forall i \in \mathcal{N}} \lambda_{\min}(P(i)), \\ b \triangleq \max\limits_{\forall i \in \mathcal{N}} \lambda_{\max}(P(i)) + (d + d^2) \max\limits_{\forall i \in \mathcal{N}} \lambda_{\max}(Q(i)) + d \max\limits_{\forall i \in \mathcal{N}} \lambda_{\max}(R(i)). \end{cases} \tag{10.10}$$

Proof. Choose a Lyapunov function of the following form:

$$\begin{cases} V(x, \alpha) \triangleq \sum\limits_{i=1}^{4} V_i(x, \alpha), \\ V_1(x, \alpha) \triangleq x^T(k)P(\alpha)x(k), \\ V_2(x, \alpha) \triangleq \sum\limits_{l=k-d(k)}^{k-1} \beta^{k-l} x^T(l)Q(\alpha)x(l), \\ V_3(x, \alpha) \triangleq \sum\limits_{s=-d_2+1}^{-d_1} \sum\limits_{l=k+s}^{k-1} \beta^{k-l} x^T(l)Q(\alpha)x(l), \\ V_4(x, \alpha) \triangleq \sum\limits_{l=k-\tau}^{k-1} \beta^{k-l} f^T(x(l))R(\alpha)f(x(l)), \end{cases} \tag{10.11}$$

where $P(\alpha) > 0$, $Q(\alpha) > 0$, and $R(\alpha) > 0$ are real matrices to be determined.

For $k \in [k_l, k_{l+1})$, define $\mathbf{E}\{\Delta V_j(x, \alpha)\} \triangleq \mathbf{E}\{V_j(x(k+1), \alpha) - V_j(x(k), \alpha)\}$, $j = 1, 2, 3, 4$, and thus we have $\mathbf{E}\{\Delta V(x, \alpha)\} = \sum_{i=1}^{4} \mathbf{E}\{\Delta V_i(x, \alpha)\}$ with

$$\begin{aligned} \mathbf{E}\{\Delta V_1(x, \alpha)\} &= \mathbf{E}\{x^T(k+1)P(\alpha)x(k+1) - x^T(k)P(\alpha)x(k)\} \\ &= \mathbf{E}\Big\{ [A(\alpha)x(k) + A_d(\alpha)x(k - d(k)) + A_\tau(\alpha)f(x(k-\tau))]^T P(\alpha) \\ &\quad \times [A(\alpha)x(k) + A_d(\alpha)x(k - d(k)) + A_\tau(\alpha)f(x(k-\tau))] \end{aligned}$$

$$+\delta\left[C(\alpha)x(k)+C_d(\alpha_k)x(k-d(k))\right]^T P(\alpha)$$
$$\times\left[C(\alpha)x(k)+C_d(\alpha)x(k-d(k))\right]-x^T(k)P(\alpha)x(k)\Big\}, \tag{10.12}$$

$$\mathbf{E}\left\{\Delta V_2(x,\alpha)\right\}\leq\mathbf{E}\Big\{-(1-\beta)\sum_{l=k-d(k)}^{k-1}\beta^{k-l}x^T(l)Q(\alpha)x(l)$$
$$+\sum_{l=k+1-d_2}^{k-d_1}\beta^{k+1-l}x^T(l)Q(\alpha)x(l)+\beta x^T(k)Q(\alpha)x(k)$$
$$-\beta^{d_2+1}x^T(k-d(k))Q(\alpha)x(k-d(k))\Big\}, \tag{10.13}$$

$$\mathbf{E}\left\{\Delta V_3(x,\alpha)\right\}=\mathbf{E}\Big\{-(1-\beta)\sum_{s=-d_2+1}^{-d_1}\sum_{l=k+s}^{k-1}\beta^{k-l}x^T(l)Q(\alpha)x(l)$$
$$+\beta(d_2-d_1)x^T(k)Q(\alpha)x(k)$$
$$-\sum_{l=k+1-d_2}^{k-d_1}\beta^{k+1-l}x^T(l)Q(\alpha)x(l)\Big\}, \tag{10.14}$$

$$\mathbf{E}\left\{\Delta V_4(x,\alpha)\right\}\leq\mathbf{E}\Big\{-(1-\beta)\sum_{l=k-\tau}^{k-1}\beta^{k-l}f^T(x(l))R(\alpha)f(x(l))$$
$$+\beta f^T(x(k))R(\alpha)f(x(k))$$
$$-\beta^{\tau+1}f^T(x(k-\tau))R(\alpha)f(x(k-\tau))\Big\}. \tag{10.15}$$

Moreover, Assumption 10.1 gives

$$\mathbf{E}\left\{\begin{bmatrix}x^T(k) & f^T(x(k))\end{bmatrix}\begin{bmatrix}\mathrm{H}_1 & -\mathrm{H}_2\\ \star & I\end{bmatrix}\begin{bmatrix}x(k)\\ f(x(k))\end{bmatrix}\right\}\leq 0, \tag{10.16}$$

where H_1 and H_2 are defined in (10.7) of Theorem 10.3.1.

Considering (10.12)–(10.15), and (10.16), we have

$$\mathbf{E}\left\{\Delta V(x,\alpha)\right\}+(1-\beta)\mathbf{E}\left\{V(x,\alpha)\right\}\triangleq\mathbf{E}\left\{\zeta^T(k)\Phi(\alpha)\zeta(k)\right\}, \tag{10.17}$$

where

$$\zeta(k)\triangleq\begin{bmatrix}x^T(k) & x^T(k-d(k)) & f^T(x(k)) & f^T(x(k-\tau))\end{bmatrix}^T,$$

and $\Phi(\alpha)$ is defined as

$$\Phi(\alpha) \triangleq \begin{bmatrix} \Phi_{11}(\alpha) & 0 & \mathrm{H}_2 & 0 \\ \star & -\beta^{d_2+1}Q(\alpha) & 0 & 0 \\ \star & \star & \beta R(\alpha) - I & 0 \\ \star & \star & \star & -\beta^{\tau+1}R(\alpha) \end{bmatrix}$$

$$+ \begin{bmatrix} A^T(\alpha) \\ A_d^T(\alpha) \\ 0 \\ A_\tau^T(\alpha) \end{bmatrix} P(\alpha) \begin{bmatrix} A^T(\alpha) \\ A_d^T(\alpha) \\ 0 \\ A_\tau^T(\alpha) \end{bmatrix}^T + \delta \begin{bmatrix} C^T(\alpha) \\ C_d^T(\alpha) \\ 0 \\ 0 \end{bmatrix} P(\alpha) \begin{bmatrix} C^T(\alpha) \\ C_d^T(\alpha) \\ 0 \\ 0 \end{bmatrix}^T,$$

where $\Phi_{11}(\alpha) \triangleq -\beta P(\alpha) + \beta(d_2 - d_1 + 1)Q(\alpha) - \mathrm{H}_1$.

Moreover, by Schur complement to (10.7) it follows that $\Phi(\alpha) < 0$, then one can easily obtain

$$\mathbf{E}\left\{\Delta V(x(k), \alpha(k)) + (1 - \beta)V(x(k), \alpha(k))\right\} < 0, \; \forall k \in [k_l, k_{l+1}). \tag{10.18}$$

Now, for an arbitrary piecewise constant switching signal α, and for any $k > 0$, we let $k_0 < k_1 < \cdots < k_l < \cdots, l = 1, \ldots,$ denote the switching points of α over the interval $(0, k)$. As mentioned earlier, the i_lth subsystem is activated when $k \in \left[k_l, k_{l+1}\right)$. Therefore, for $k \in [k_l, k_{l+1})$, it holds from (10.18) that

$$\mathbf{E}\left\{V(x(k), \alpha(k))\right\} < \beta^{k-k_l}\mathbf{E}\left\{V(x(k_l), \alpha(k_l))\right\}. \tag{10.19}$$

Using (10.8) and (10.11), we have

$$\mathbf{E}\left\{V(x(k_l), \alpha(k_l))\right\} \leq \mu\mathbf{E}\left\{V(x(k_l), \alpha(k_{l-1}))\right\}. \tag{10.20}$$

Therefore, it follows from (10.19)–(10.20) and the relationship $\vartheta = N_\alpha(k_0, k) \leq (k - k_0)/T_a$ that

$$\begin{aligned} \mathbf{E}\left\{V(x(k), \alpha(k))\right\} &\leq \beta^{k-k_l}\mu\mathbf{E}\left\{V(x(k_l), \alpha(k_{l-1}))\right\} \\ &\leq \cdots \\ &\leq \beta^{(k-k_0)}\mu^{\vartheta}\mathbf{E}\left\{V(x(k_0), \alpha(k_0))\right\} \\ &\leq (\beta\mu^{1/T_a})^{(k-k_0)}\mathbf{E}\left\{V(x(k_0), \alpha(k_0))\right\}. \end{aligned} \tag{10.21}$$

Notice from (10.11) that there exist two positive constants a and b ($a \leq b$) such that

$$\begin{aligned} \mathbf{E}\left\{V(x(k), \alpha(k))\right\} &\geq a\mathbf{E}\left\{\|x(k)\|^2\right\}, \\ \mathbf{E}\left\{V(x(k(0)), \alpha(k_0))\right\} &\leq b\left\|x(k_0)\right\|_{\mathrm{C}}^2. \end{aligned} \tag{10.22}$$

Combining (10.21) and (10.22) yields

$$\begin{aligned}\mathbf{E}\left\{\|x(k)\|^2\right\} &\leq \frac{1}{a}\mathbf{E}\left\{V(x(k),\alpha(k))\right\}\\ &\leq \frac{b}{a}(\beta\mu^{1/T_a})^{(k-k_0)}\left\|x(k_0)\right\|_{\mathrm{C}}^2. \end{aligned} \tag{10.23}$$

Furthermore, letting $\rho \triangleq \sqrt{\beta\mu^{1/T_a}}$, it follows that

$$\mathbf{E}\left\{\|x(k)\|\right\} \leq \sqrt{\frac{b}{a}}\left\|x(k_0)\right\|_{\mathrm{C}}\rho^{(k-k_0)}. \tag{10.24}$$

By Definition 10.2.1, we know that if $0 < \rho < 1$, that is, $T_a > T_a^* = \text{ceil}\left(-\frac{\ln\mu}{\ln\beta}\right)$, the discrete-time switched stochastic time-delay system in (10.1a) with $u(k) = 0$ and $\omega(k) = 0$ is mean-square exponentially stable, where function ceil(h) represents the rounding real number h to the nearest integer greater than or equal to h. The proof' is completed. ■

Remark 10.2 *In Theorem 10.3.1, we propose a sufficient condition for the mean-square exponential stability condition for the considered discrete-time switched stochastic time-delay system in (10.1a) with $u(k) = 0$ and $\omega(k) = 0$. Here, the parameter β plays a key role in controlling the lower bound of the average dwell time, which can be seen from $T_a > T_a^* = \text{ceil}\left(-\frac{\ln\mu}{\ln\beta}\right)$; specifically, if β is a smaller value, the lower bound of the average dwell time becomes smaller with a fixed μ, which may result in the instability of the system.* ♦

Remark 10.3 *Note that when $\mu = 1$ in $T_a > T_a^* = \text{ceil}\left(-\frac{\ln\mu}{\ln\beta}\right)$ we have $T_a > T_a^* = 0$, which means that the switching signal α_k can be arbitrary. In this case, (10.8) turns out to be $P(i) = P(j) = P$, $Q(i) = Q(j) = P$, $R(i) = R(j) = P$, $\forall i,j \in \mathcal{N}$, and the proposed approach becomes a quadratic one thus conservative. In this case, the system in (10.1a) with $u(k) = 0$ and $\omega(k) = 0$ turns out to be a discrete-time stochastic system with time delays. However, when $\beta = 1$ in $T_a > T_a^* = \text{ceil}\left(-\frac{\ln\mu}{\ln\beta}\right)$, we have $T_a = \infty$, that is, there is no switching.* ♦

Theorem 10.3.2 *Given a constant $0 < \beta < 1$, suppose that there exist matrices $X(i) > 0$, $Z(i) > 0$, $R(i) > 0$, and $Y(i)$ such that for $i \in \mathcal{N}$,*

$$\begin{bmatrix} \tilde{\Phi}_{11}(i) & 0 & X(i)\mathrm{H}_2 & 0 & \tilde{\Phi}_{15}(i) & \delta X(i)C^T(i)\\ \star & -\beta^{d_2+1}Z(i) & 0 & 0 & X(i)A_d^T(i) & \delta X(i)C_d^T(i)\\ \star & \star & \beta R(i)-I & 0 & 0 & 0\\ \star & \star & \star & -\beta^{\tau+1}R(i) & A_\tau^T(i) & 0\\ \star & \star & \star & \star & -X(i) & 0\\ \star & \star & \star & \star & \star & -\delta X(i) \end{bmatrix} < 0, \tag{10.25}$$

where

$$\begin{cases} \tilde{\Phi}_{11}(i) \triangleq -(2+\beta)X(i) + \beta(d_2 - d_1 + 1)Z(i) + \mathrm{H}_1^{-1}, \\ \tilde{\Phi}_{15}(i) \triangleq X(i)A^T(i) + Y^T(i)B_u^T(i). \end{cases}$$

Then the closed-loop stabilization system in (10.4) is mean-square exponentially stable for any switching signal with average dwell time satisfying $T_a > T_a^* = \text{ceil}\left(-\frac{\ln\mu}{\ln\beta}\right)$, *where* $\mu \geq 1$ *satisfies*

$$X(i) \leq \mu X(j), \quad Z(i) \leq \mu Z(j), \quad R(i) \leq \mu R(j), \quad \forall i,j \in \mathcal{N}. \tag{10.26}$$

In this case, a robustly stabilizing state feedback controller can be chosen by

$$u(k) = Y(i)X^{-1}(i)x(k). \tag{10.27}$$

Proof. By performing a congruence transformation on (10.7) with matrix diag $\{X(i), X(i), I, I, X(i), X(i)\}$ (where $X(i) = P^{-1}(i)$), it follows that

$$\begin{bmatrix} \hat{\Phi}_{11}(i) & 0 & X(i)\mathrm{H}_2 & 0 & X(i)A^T(i) & \delta X(i)C^T(i) \\ \star & -\beta^{d_2+1}Z(i) & 0 & 0 & X(i)A_d^T(i) & \delta X(i)C_d^T(i) \\ \star & \star & \beta R(i) - I & 0 & 0 & 0 \\ \star & \star & \star & -\beta^{\tau+1}R(i) & A_\tau^T(i) & 0 \\ \star & \star & \star & \star & -X(i) & 0 \\ \star & \star & \star & \star & \star & -\delta X(i) \end{bmatrix} < 0, \tag{10.28}$$

where

$$\begin{cases} \hat{\Phi}_{11}(i) \triangleq -\beta X(i) + \beta\left(d_2 - d_1 + 1\right)Z(i) - X(i)\mathrm{H}_1 X(i), \\ Z(i) \triangleq X(i)Q(i)X(i). \end{cases}$$

However, the following matrix inequality holds:

$$\left(X(i) - \mathrm{H}_1^{-1}\right)\mathrm{H}_1\left(X(i) - \mathrm{H}_1^{-1}\right) \geq 0,$$

thus,

$$X(i)\mathrm{H}_1 X(i) \geq 2X(i) - \mathrm{H}_1^{-1}.$$

Therefore, matrix inequality (10.28) holds if the following LMI holds:

$$\begin{bmatrix} \tilde{\Phi}_{11}(i) & 0 & X(i)\mathrm{H}_2 & 0 & X(i)A^T(i) & \delta X(i)C^T(i) \\ \star & -\beta^{d_2+1}Z(i) & 0 & 0 & X(i)A_d^T(i) & \delta X(i)C_d^T(i) \\ \star & \star & \beta R(i)-I & 0 & 0 & 0 \\ \star & \star & \star & -\beta^{\tau+1}R(i) & A_\tau^T(i) & 0 \\ \star & \star & \star & \star & -X(i) & 0 \\ \star & \star & \star & \star & \star & -\delta X(i) \end{bmatrix} < 0, \quad (10.29)$$

where $\tilde{\Phi}_{11}(i)$ is defined in (10.25). From Theorem 10.3.1 and the above derivation we know that the discrete-time switched stochastic time-delay system in (10.1a) with $\omega(k) = 0$ is mean-square exponentially stabilizable, that is, the closed-loop stabilization system in (10.4) is mean-square exponentially stable if the matrix inequality, that is, (10.29) with $A(i)$ replacing by $A(i) + B_u(i)K(i))$ holds.

Furthermore, we define $Y(i) = K(i)X(i)$, we have (10.25), and we know that $K(i) = Y(i)X^{-1}(i)$. The proof is completed. ■

10.4 $\mathcal{H}_\infty$ Control

In this section, we will investigate the weighted $\mathcal{H}_\infty$ performance for system (10.1a)–(10.1c) with $u(k) = 0$. A sufficient condition of the weighted $\mathcal{H}_\infty$ performance will be established, and based on which the $\mathcal{H}_\infty$ controller will be synthesized.

Theorem 10.4.1 *For given constants $\beta > 0$ and $\gamma > 0$, suppose that there exist matrices $P(i) > 0$, $Q(i) > 0$, and $R(i) > 0$ such that for $i \in \mathcal{N}$,*

$$\begin{bmatrix} \check{\Phi}_{11}(i) & 0 & \mathrm{H}_2 & 0 & 0 & A^T(i)P(i) & \delta C^T(i)P(i) \\ \star & -\beta^{d_2+1}Q(i) & 0 & 0 & 0 & A_d^T(i)P(i) & \delta C_d^T(i)P(i) \\ \star & \star & \beta R(i)-I & 0 & 0 & 0 & 0 \\ \star & \star & \star & -\beta^{\tau+1}R(i) & 0 & A_\tau^T(i)P(i) & 0 \\ \star & \star & \star & \star & -\gamma^2 I & B_\omega^T(i)P(i) & \delta D_\omega^T(i)P(i) \\ \star & \star & \star & \star & \star & -P(i) & 0 \\ \star & \star & \star & \star & \star & \star & -\delta P(i) \end{bmatrix} < 0, \quad (10.30)$$

where

$$\check{\Phi}_{11}(i) \triangleq -\beta P(i) + \beta(d_2 - d_1 + 1)Q(i) - \mathrm{H}_1 + L^T(i)L(i).$$

Then the system in (10.1a)–(10.1c) with $u(k) = 0$ is mean-square exponentially stable with a weighted $\mathcal{H}_\infty$ performance level γ for any switching signal with average dwell time satisfying $T_a > T_a^* = \text{ceil}\left(-\frac{\ln\mu}{\ln\beta}\right)$, where $\mu \geq 1$ satisfies (10.8).

Proof. The proof of mean-square exponential stability can be referred to the proof of Theorem 10.3.1. Now, we will establish the weighted $\mathcal{H}_\infty$ performance defined in (10.6). To this end, introduce the following index:

$$\mathcal{J} \triangleq \mathbf{E}\left\{\Delta V(x,\alpha) + (1-\beta)V(x,\alpha) + z^T(k)z(k) - \gamma^2\omega^T(k)\omega(k)\right\},$$

where the Lyapunov function $V(x_k, \alpha_k)$ is given in (10.11). By employing the same techniques used as those in the proof of Theorem 10.3.1, for $k \in [k_l, k_{l+1})$, we have

$$\mathbf{E}\left\{\Delta V(x,\alpha) + (1-\beta)V(x,\alpha) + z^T(k)z(k) - \gamma^2\omega^T(k)\omega(k)\right\} \leq \mathbf{E}\left\{\chi^T(k)\Pi(\alpha)\chi(k)\right\},$$

where $\chi(k) \triangleq \begin{bmatrix} \zeta(k) \\ \omega(k) \end{bmatrix}$ and

$$\Pi(\alpha) \triangleq \begin{bmatrix} \breve{\Phi}_{11}(\alpha) & 0 & \mathrm{H}_2 & 0 & 0 \\ \star & -\beta^{d_2+1}Q(\alpha) & 0 & 0 & 0 \\ \star & \star & \beta R(\alpha) - I & 0 & 0 \\ \star & \star & \star & -\beta^{\tau+1}R(\alpha) & 0 \\ \star & \star & \star & \star & -\gamma^2 I \end{bmatrix}$$
$$+ \begin{bmatrix} A^T(\alpha) \\ A_d^T(\alpha) \\ 0 \\ A_\tau^T(\alpha) \\ B_\omega^T(\alpha) \end{bmatrix} P(\alpha) \begin{bmatrix} A^T(\alpha) \\ A_d^T(\alpha) \\ 0 \\ A_\tau^T(\alpha) \\ B_\omega^T(\alpha) \end{bmatrix}^T + \delta \begin{bmatrix} C^T(\alpha) \\ C_d^T(\alpha) \\ 0 \\ 0 \\ D_\omega^T(\alpha) \end{bmatrix} P(\alpha) \begin{bmatrix} C^T(\alpha) \\ C_d^T(\alpha) \\ 0 \\ 0 \\ D_\omega^T(\alpha) \end{bmatrix}^T.$$

By Schur complement, LMI (10.30) is equal to $\Pi(\alpha_k) < 0$, thus $\mathcal{J} < 0$. Let $\Gamma(k) \triangleq z^T(k)z(k) - \gamma^2\omega^T(k)\omega(k)$, then we have

$$\mathbf{E}\{\Delta V(x(k),\alpha(k))\} < \mathbf{E}\{-(1-\beta)V(x(k),\alpha(k)) - \Gamma(k)\}. \tag{10.31}$$

Therefore, for $k \in [k_l, k_{l+1})$, it holds from (10.31) that

$$\mathbf{E}\{V(x(k),\alpha(k))\} < \beta^{k-k_l}\mathbf{E}\left\{V(x(k_l),\alpha(k_l))\right\} - \mathbf{E}\left\{\sum_{s=k_l}^{k-1}\beta^{k-1-s}\Gamma(s)\right\}. \tag{10.32}$$

Considering (10.8) and (10.11), it follows that

$$\mathbf{E}\left\{V(x(k_l),\alpha(k_l))\right\} \leq \mu\mathbf{E}\left\{V(x(k_l),\alpha(k_{l-1}))\right\}. \tag{10.33}$$

Thus by (10.32)–(10.33) we have

$$\begin{aligned}
\mathbf{E}\left\{V(x(k),\alpha(k)\right\} &< \beta^{k-k_l}\mathbf{E}\left\{V(x(k_l),\alpha(k_l))\right\} - \mathbf{E}\left\{\sum_{s=k_l}^{k-1}\beta^{k-1-s}\Gamma(s)\right\},\\
\mathbf{E}\left\{V(x(k_l),\alpha(k_l))\right\} &< \beta^{k_l-k_{l-1}}\mu\mathbf{E}\left\{V(x(k_{l-1}),\alpha(k_{l-1}))\right\}\\
&\quad -\mu\mathbf{E}\left\{\sum_{s=k_{l-1}}^{k_l-1}\beta^{k_l-1-s}\Gamma(s)\right\},\\
\mathbf{E}\left\{V(x(k_{l-1}),\alpha(k_{l-1}))\right\} &< \beta^{k_{l-1}-k_{l-2}}\mu\mathbf{E}\left\{V(x(k_{l-2}),\alpha(k_{l-2}))\right\}\\
&\quad -\mu\mathbf{E}\left\{\sum_{s=k_{l-2}}^{k_{l-1}-1}\beta^{k_{l-1}-1-s}\Gamma(s)\right\},\\
&\vdots\\
\mathbf{E}\left\{V(x(k_1),\alpha(k_1))\right\} &< \beta^{k_1-k_0}\mu\mathbf{E}\left\{V(x(k_0),\alpha(k_0))\right\} - \mu\mathbf{E}\left\{\sum_{s=k_0}^{k_1-1}\beta^{k_1-1-s}\Gamma(s)\right\}.
\end{aligned}$$

Therefore, it follows from the above inequalities and the relationship $\vartheta = N_\alpha(k_0,k) \leq (k-k_0)/T_a$ that

$$\begin{aligned}
\mathbf{E}\left\{V(x(k),\alpha(k))\right\} &< \beta^{k-k_l}\mathbf{E}\left\{V(x(k_l),\alpha(k_l))\right\} - \mathbf{E}\left\{\sum_{s=k_l}^{k-1}\beta^{k-1-s}\Gamma(s)\right\}\\
&< \beta^{k-k_0}\mu^{N_\alpha(k_0,k)}\mathbf{E}\left\{V(x(k_0),\alpha(k_0))\right\}\\
&\quad -\beta^{k-k_1}\mu^{N_\alpha(k_0,k)}\mathbf{E}\left\{\sum_{s=k_0}^{k_1-1}\beta^{k_1-1-s}\Gamma(s)\right\}\\
&\quad -\beta^{k-k_2}\mu^{N_\alpha(k_1,k)}\mathbf{E}\left\{\sum_{s=k_1}^{k_2-1}\beta^{k_2-1-s}\Gamma(s)\right\}\\
&\quad -\cdots\\
&\quad -\beta^{k-k_{l-1}}\mu^{2}\mathbf{E}\left\{\sum_{s=k_{l-2}}^{k_{l-1}-1}\beta^{k_{l-1}-1-s}\Gamma(s)\right\}\\
&\quad -\beta^{k-k_l}\mu\mathbf{E}\left\{\sum_{s=k_{l-1}}^{k_l-1}\beta^{k_l-1-s}\Gamma(s)\right\}
\end{aligned}$$

$$
\begin{aligned}
&-\mathbf{E}\left\{\sum_{s=k_l}^{k-1}\beta^{k-1-s}\Gamma(s)\right\}\\
&=\beta^{k-k_0}\mu^{N_\alpha(k_0,k)}\mathbf{E}\left\{V(x(k_0),\alpha(k_0))\right\}\\
&-\mathbf{E}\left\{\sum_{s=k_0}^{k-1}\beta^{k-1-s}\mu^{N_\alpha(s,k)}\Gamma(s)\right\}.
\end{aligned}\tag{10.34}
$$

Under zero initial condition, that is, $x(\theta)=\phi(\theta)=0,(-\max\{\tau,d_2\}<\theta\leq 0)$, (10.34) implies

$$
\mathbf{E}\left\{\sum_{s=k_0}^{k-1}\beta^{k-1-s}\mu^{N_\alpha(s,k)}z^T(s)z(s)\right\}<\gamma^2\mathbf{E}\left\{\sum_{s=k_0}^{k-1}\beta^{k-1-s}\mu^{N_\alpha(s,k)}\omega^T(s)\omega(s)\right\}.
$$

Multiplying both sides of the above inequality by $\mu^{-N_\alpha(0,k)}$ yields

$$
\mathbf{E}\left\{\sum_{s=k_0}^{k-1}\beta^{k-1-s}\mu^{-N_\alpha(0,s)}z^T(s)z(s)\right\}<\gamma^2\mathbf{E}\left\{\sum_{s=k_0}^{k-1}\beta^{k-1-s}\mu^{-N_\alpha(0,s)}\omega^T(s)\omega(s)\right\}.\tag{10.35}
$$

Notice that $N_\alpha(0,s)\leq s/T_a$ and $T_a>-\frac{\ln\mu}{\ln\beta}$, so we have $N_\alpha(0,s)\leq -s\frac{\ln\beta}{\ln\mu}$. Thus (10.35) implies

$$
\begin{aligned}
\mathbf{E}\left\{\sum_{s=k_0}^{k-1}\beta^{k-1-s}\mu^{s\frac{\ln\beta}{\ln\mu}}z^T(s)z(s)\right\}&=\mathbf{E}\left\{\sum_{s=k_0}^{k-1}\beta^{k-1-s}\beta^{s}z^T(s)z(s)\right\}\\
&<\gamma^2\mathbf{E}\left\{\sum_{s=k_0}^{k-1}\beta^{k-1-s}\mu^{-N_\alpha(0,s)}\omega^T(s)\omega(s)\right\}\\
&<\gamma^2\mathbf{E}\left\{\sum_{s=k_0}^{k-1}\beta^{k-1-s}\omega^T(s)\omega(s)\right\}.
\end{aligned}
$$

which yields

$$
\mathbf{E}\left\{\sum_{s=k_0}^{\infty}\beta^{s}z^T(s)z(s)\right\}<\mathbf{E}\left\{\sum_{s=k_0}^{\infty}\omega^T(s)\omega(s)\right\}.
$$

By Definition 10.2.2, we know that system (10.1a)–(10.1c) with $u(k)=0$ is mean-square exponentially stable with a weighted $\mathcal{H}_\infty$ performance level γ under α_k. This completes the proof. ■

Remark 10.4 *Note that Theorem 10.4.1 gives a weighted $\mathcal{H}_\infty$ performance for the discrete-time switched stochastic time-delay system in (10.1a)–(10.1c) with $u(k)=0$. The* *term 'weighted' refers to the weighting function β^s in the left-hand side of (10.6). This is also the characteristic of the mean-square exponential stability result to the switched stochastic hybrid system by using the average dwell time approach combining with the piecewise Lyapunov function technique. When setting $\beta = 1$, from the analysis in Remark 10.3, there is no switching. Thus, the result in Theorem 10.4.1 becomes a mean-square asymptotic stability condition with an $\mathcal{H}_\infty$ performance for the deterministic system.* ♦

Now, we are in a position to present a solution to the $\mathcal{H}_\infty$ control problem for the discrete-time switched stochastic time-delay system in (10.1a)–(10.1c).

Theorem 10.4.2 *For given constants $\beta > 0$ and $\gamma > 0$, suppose that there exist matrices $X(i) > 0$, $Z(i) > 0$, $R(i) > 0$, and $Y(i)$ such that for $i \in \mathcal{N}$,*

$$\begin{bmatrix} \tilde{\Phi}_{11}(i) & 0 & X(i)\mathrm{H}_2 & 0 & 0 & \tilde{\Phi}_{15}(i) & \tilde{\Phi}_{16}(i) & \tilde{\Phi}_{17}(i) \\ \star & \tilde{\Phi}_{22}(i) & 0 & 0 & 0 & \tilde{\Phi}_{25}(i) & \tilde{\Phi}_{26}(i) & 0 \\ \star & \star & \tilde{\Phi}_{33}(i) & 0 & 0 & 0 & 0 & 0 \\ \star & \star & \star & \tilde{\Phi}_{44}(i) & 0 & A_\tau^T(i) & 0 & 0 \\ \star & \star & \star & \star & -\gamma^2 I & B_\omega^T(i) & \delta D_\omega^T(i) & 0 \\ \star & \star & \star & \star & \star & -X(i) & 0 & 0 \\ \star & \star & \star & \star & \star & \star & -\delta X(i) & 0 \\ \star & \star & \star & \star & \star & \star & \star & -I \end{bmatrix} < 0, \quad (10.36)$$

where

$$\begin{cases} \tilde{\Phi}_{22}(i) \triangleq -\beta^{d_2+1}Z(i), & \tilde{\Phi}_{16}(i) \triangleq \delta X(i)C^T(i), \\ \tilde{\Phi}_{44}(i) \triangleq -\beta^{\tau+1}R(i), & \tilde{\Phi}_{26}(i) \triangleq \delta X(i)C_d^T(i), \\ \tilde{\Phi}_{25}(i) \triangleq X(i)A_d^T(i), & \tilde{\Phi}_{17}(i) \triangleq X(i)L^T(i), \qquad \tilde{\Phi}_{33}(i) \triangleq \beta R(i) - I. \end{cases}$$

Then the closed-loop system in (10.5a)–(10.5b) is mean-square exponentially stable with a weighted $\mathcal{H}_\infty$ performance level γ for any switching signal with average dwell time satisfying $T_a > T_a^* = \mathrm{ceil}\left(-\frac{\ln \mu}{\ln \beta}\right)$, where $\mu \geq 1$ satisfies

$$X(i) \leq \mu X(j), \quad Z(i) \leq \mu Z(j), \quad R(i) \leq \mu R(j), \quad \forall i, j \in \mathcal{N}. \quad (10.37)$$

In this case, an $\mathcal{H}_\infty$ state feedback controller can be chosen by

$$u(k) = Y(i)X^{-1}(i)x(k). \quad (10.38)$$

Proof. The result can be carried out by employing the same techniques used as those of Theorems 10.3.2 and 10.4.1. ∎

10.5 Illustrative Example

Example 10.5.1 (Stabilization problem) Consider the switched stochastic hybrid system in (10.1a) with $N = 2$ and

$$A(1) = \begin{bmatrix} -0.2 & 0 \\ 0 & -0.1 \end{bmatrix}, \ A_d(1) = \begin{bmatrix} -0.1 & -0.2 \\ 0 & -0.15 \end{bmatrix},$$

$$A(2) = \begin{bmatrix} -0.2 & 0 \\ 0 & -0.1 \end{bmatrix}, \ A_d(2) = \begin{bmatrix} -0.2 & -1.1 \\ 0 & -0.22 \end{bmatrix},$$

$$A_\tau(1) = \begin{bmatrix} -0.1 & 0 \\ 0.1 & -0.3 \end{bmatrix}, \ A_\tau(2) = \begin{bmatrix} -0.1 & 0 \\ 0.2 & -0.36 \end{bmatrix},$$

$$C(1) = \begin{bmatrix} -0.1 & 0 \\ 0.3 & 0.12 \end{bmatrix}, \ C_d(1) = \begin{bmatrix} 0.11 & 0.1 \\ 0.3 & 0.02 \end{bmatrix}, \ B_u(1) = \begin{bmatrix} 0.2 & 0.2 \\ 0.1 & 0.2 \end{bmatrix},$$

$$C(2) = \begin{bmatrix} -0.2 & 0.13 \\ 0.1 & 0.2 \end{bmatrix}, \ C_d(2) = \begin{bmatrix} 0.01 & 0.2 \\ 0.2 & 0.32 \end{bmatrix}, \ B_u(2) = \begin{bmatrix} 0.3 & 0.2 \\ 0.1 & 0.3 \end{bmatrix},$$

and $d(k) = 2 + \frac{1+(-1)^k}{2}$, $\beta = 0.7$. A straightforward calculation gives $d_2 = 3$, $\tau = 1$, and $\delta = 0.314$. Give $\beta = 0.7$ and set $\mu = 1.6$, thus $T_a > T_a^* = \text{ceil}(-\frac{\ln \mu}{\ln \beta}) = 2$. Checking the conditions in (10.7) by using LMI Toolbox, a set of feasible solutions is found. Therefore, the switched stochastic hybrid system (10.1a) with the above parametric matrices is mean-square exponentially stable for $T_a > T_a^* = 2$. Moreover, taking $T_a = 3 > T_a^* = 2$ and according to (10.9) and (10.10), we obtained $a = 2.4410$, $b = 5.6846$, $\eta = \sqrt{\frac{b}{a}} = 2.3842$, and $\rho = \sqrt{\beta \mu^{\frac{1}{T_a}}} = 0.9048$, thus, an estimate of the mean-square of the state decay is given by

$$\mathbf{E}\{\|x(k)\|\} \leq 2.3842 \, \|x(k_0)\|_C \, 0.9048^{(k-k_0)}.$$

Now, we further simulate the stabilization problem. As analyzed above the open-loop system is mean-square exponentially stable when $T_a \geq T_a^* = 2$. Here, to show the effectiveness, we will design a stabilization controller in (10.3) such that the closed-loop system in (10.4) is mean-square exponentially stable for $T_a = 1$ (in this case, the allowable minimum of μ is $\mu_{\min} = 1.0314$). Solving LMI conditions in Theorem 10.3.2, and considering (10.27), we have

$$K(1) = \begin{bmatrix} -0.2039 & 1.0211 \\ 1.0218 & -0.0944 \end{bmatrix}, \quad K(2) = \begin{bmatrix} -2.2237 & 1.1017 \\ 11.1193 & -0.5039 \end{bmatrix}.$$

Therefore, the controller in the form of (10.3) with above control gains can stabilize the open-loop system when $T_a = 1$.

Example 10.5.2 ($\mathcal{H}_\infty$ control problem) Consider the switched stochastic hybrid system in (10.1a)–(10.1c) with $N = 2$ and the following parameters:

$$A(1) = \begin{bmatrix} -0.2 & 0 \\ 0 & -0.1 \end{bmatrix}, \quad A_d(1) = \begin{bmatrix} -0.1 & -0.2 \\ 0 & -0.15 \end{bmatrix},$$

$$A(2) = \begin{bmatrix} -0.2 & 0 \\ 0.2 & -1.1 \end{bmatrix}, \quad A_d(2) = \begin{bmatrix} -0.2 & -0.15 \\ 0.1 & -0.2 \end{bmatrix},$$

$$A_\tau(1) = \begin{bmatrix} -0.1 & 0 \\ 0.1 & -0.3 \end{bmatrix}, \quad B_u(1) = \begin{bmatrix} 0.02 & 0.2 \\ 0.1 & 0.02 \end{bmatrix},$$

$$A_\tau(2) = \begin{bmatrix} -0.2 & 0 \\ 0.1 & -0.3 \end{bmatrix}, \quad B_u(2) = \begin{bmatrix} 1.00 & 1.20 \\ 0.10 & 0.20 \end{bmatrix},$$

$$B_\omega(1) = \begin{bmatrix} 0.3 & 0 \\ 0 & 0.1 \end{bmatrix}, \quad C(1) = \begin{bmatrix} -0.01 & 0 \\ 0.03 & 0.02 \end{bmatrix},$$

$$B_\omega(2) = \begin{bmatrix} 0.1 & 0.1 \\ 0 & 0.5 \end{bmatrix}, \quad C(2) = \begin{bmatrix} 0.02 & 0.01 \\ -0.03 & 0.02 \end{bmatrix},$$

$$C_d(1) = \begin{bmatrix} 0.1 & 0 \\ 0 & 0.02 \end{bmatrix}, \quad D_\omega(1) = \begin{bmatrix} 0.01 & 0.02 \\ 0.03 & -0.01 \end{bmatrix},$$

$$C_d(2) = \begin{bmatrix} 0.01 & 0 \\ 0.1 & 0.1 \end{bmatrix}, \quad D_\omega(2) = \begin{bmatrix} 0 & 0.02 \\ 0.03 & -0.02 \end{bmatrix},$$

$$L(1) = \begin{bmatrix} 0 & 0.002 \end{bmatrix}, \quad L(2) = \begin{bmatrix} 0.001 & 0 \end{bmatrix}.$$

Let $\beta = 0.7$ and

$$f(k) = \begin{bmatrix} \tan(-x_1(k)) - 0.01x_1(k) + 0.01x_2(k) \\ 0.01x_1(k) - \tan(x_2(k)) + 0.01x_2(k) \end{bmatrix},$$

then it is easy to verify that there exist

$$\Pi_1 = \begin{bmatrix} -0.01 & 0.01 \\ -0.02 & 0.01 \end{bmatrix}, \quad \Pi_2 = \begin{bmatrix} 0.01 & 0.01 \\ 0.01 & 0.01 \end{bmatrix},$$

such that (10.2) is satisfied. It can be shown that the switched stochastic hybrid system with the above two subsystems is not stable for a switching signal given in Figure 10.1 (which is generated randomly; here, ‘1’ and ‘2’ represent the first and second subsystems, respectively). Here, our aim is to design a state feedback controller such that the resulting closed-loop system is mean-square exponentially stable with a weighted $\mathcal{H}_\infty$ performance level $\gamma > 0$ for $T_a > T_a^*$.

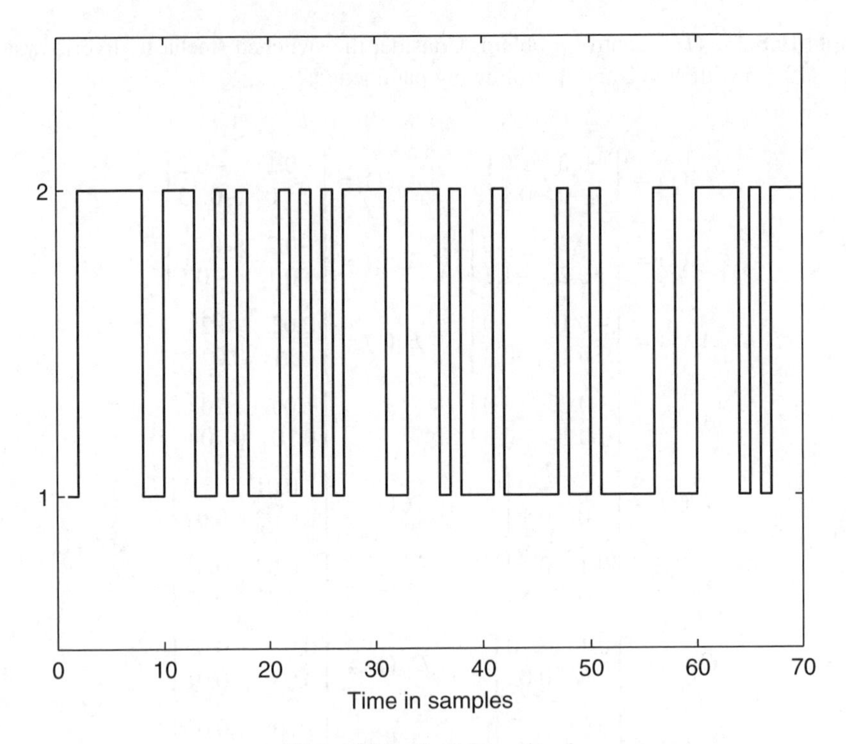

Figure 10.1 Switching signal

Here, for example, we set $T_a^* = \text{ceil}\left(-\frac{\ln\mu}{\ln\beta}\right) = 2$ (in this case, the allowable minimum of μ is $\mu_{\min} = 1.6$). Letting $d_2 = 3$, $\tau = 1$, and $\delta = 0.314$. Solving (10.36)–(10.37) in Theorem 10.4.2, we have $\gamma = 2.1204$ and

$$K(1) = \begin{bmatrix} 1.4425 & -2.6676 \\ -0.7147 & 8.3398 \end{bmatrix},$$

$$K(2) = \begin{bmatrix} 3.6274 & -16.2692 \\ -2.4995 & 13.5442 \end{bmatrix}.$$

Suppose the disturbance input $\omega(k)$ is

$$\omega(k) = \begin{bmatrix} 0.4e^{-2k}\tan(k) & 0.5e^{-k}\sin(2k) \end{bmatrix}^T.$$

Figure 10.2 shows the Brownian path, and Figure 10.3 gives the states of the closed-loop system.

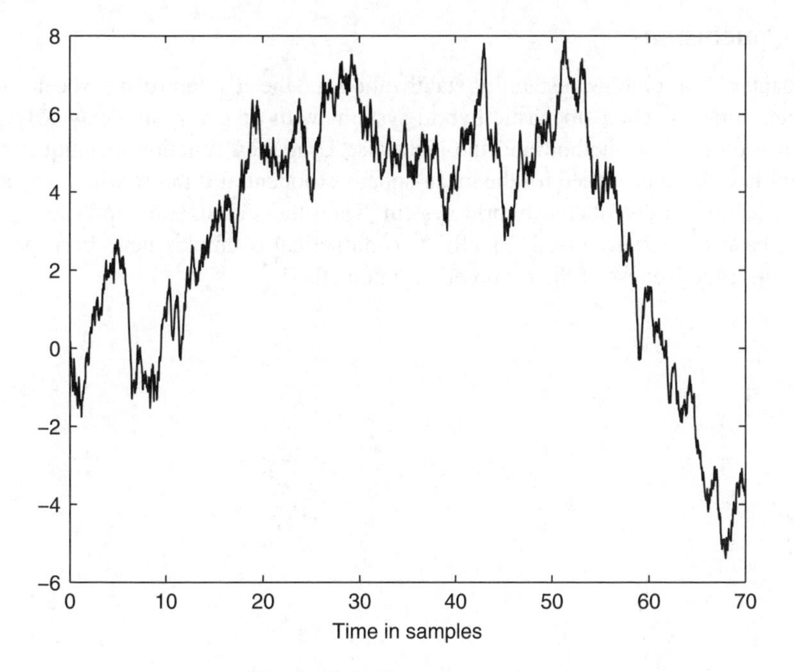

Figure 10.2 Brownian motion

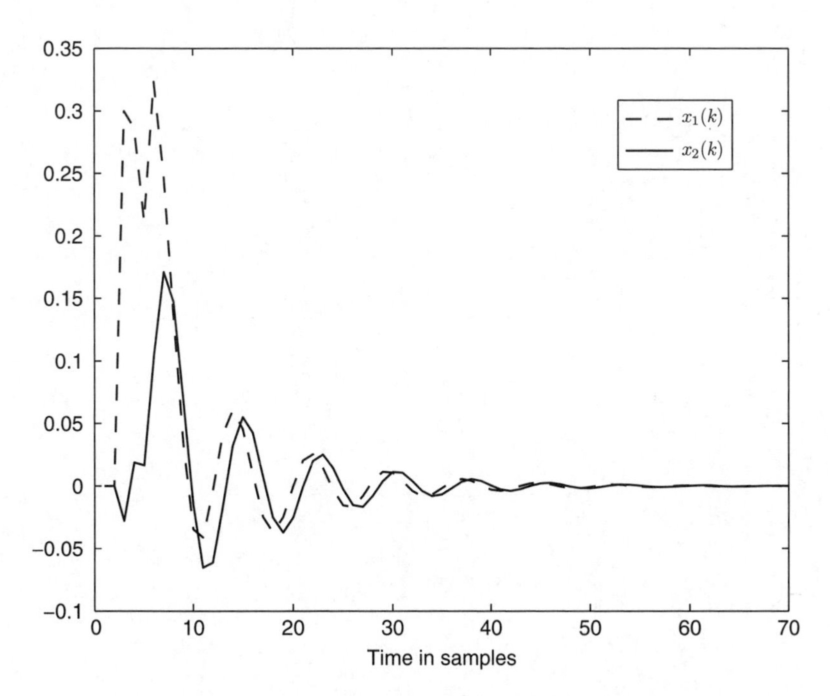

Figure 10.3 States of the closed-loop system

10.6 Conclusion

In this chapter, the problems of stability, stabilization and the $\mathcal{H}_\infty$ control have been considered for discrete-time switched stochastic hybrid systems with time-varying delays. By applying the average dwell time method and the piecewise Lyapunov function technique, sufficient conditions have been proposed for the mean-square exponential stability with a weighted $\mathcal{H}_\infty$ performance for the considered hybrid system. Then the stabilization and the $\mathcal{H}_\infty$ control problems have also been solved. Finally, two numerical examples have been provided to illustrate the effectiveness of the proposed design methods.

11

State Estimation and SMC of Switched Stochastic Hybrid Systems

11.1 Introduction

In Chapter 9, we studied the problems of stability, stabilization, and $\mathcal{H}_\infty$ control for the continuous-time switched stochastic hybrid system. In this chapter, we are interested in investigating the SMC design problem for such hybrid systems, and some results developed in Chapter 9 will be used. First, by designing an integral switching function, we obtain the sliding mode dynamics, which is a switched stochastic hybrid system with the same order as the original systems. Based on the stability analysis result in Chapter 9, a sufficient condition for the existence of the sliding mode is proposed in terms of LMIs, and an explicit parametrization of the desired switching function is also given. Then, a discontinuous SMC law for reaching motion is synthesized, such that the state trajectories of the SMC system can be driven onto a prescribed sliding surface and maintained there for all subsequent time. Moreover, considering that some system state components may not be available in practical applications, we further consider the state estimation problem by designing an observer. Sufficient conditions are also established for the existence and the solvability of the desired observer, and then the observer-based SMC law is synthesized.

11.2 System Description and Preliminaries

Consider the switched stochastic hybrid systems which are established on the probability space $(\Omega, \mathcal{F}, \mathcal{P})$ and are described by

$$\begin{aligned} dx(t) = {} & [A(\alpha(t))x(t) + B(\alpha(t))\,(u(t) + F(\alpha(t))f(x,t))]\,dt \\ & + D(\alpha(t))x(t)d\varpi(t), \end{aligned} \tag{11.1}$$

Sliding Mode Control of Uncertain Parameter-Switching Hybrid Systems, First Edition. Ligang Wu, Peng Shi and Xiaojie Su.

where $x(t) \in \mathbf{R}^n$ is the system state vector; $u(t) \in \mathbf{R}^m$ is the control input; $\varpi(t)$ is a one-dimensional Brownian motion satisfying $\mathbf{E}\{d\varpi(t)\} = 0$, and $\mathbf{E}\{d\varpi^2(t)\} = dt$. $\{(A(\alpha(t)), B(\alpha(t)), D(\alpha(t)), F(\alpha(t))) : \alpha(t) \in \mathcal{N}\}$ is a family of matrices parameterized by an index set $\mathcal{N} = \{1, 2, \ldots, N\}$ and $\alpha(t) : \mathbf{R} \to \mathcal{N}$ is switching signal (denoted by α for simplicity), which is defined as in Chapter 5. For each possible value $\alpha = i$, $i \in \mathcal{N}$, we denote the system matrices associated with mode i by $A(i) = A(\alpha)$, $B(i) = B(\alpha)$, $D(i) = D(\alpha)$, and $F(i) = F(\alpha)$, where $A(i)$, $B(i)$, $D(i)$, and $F(i)$ are constant matrices. The pairs $(A(i), B(i))$ are controllable for $i \in \mathcal{N}$, and matrices $B(i)$ are assumed to be of full column rank. For scalars $\phi(\alpha) > 0$, $i \in \mathcal{N}$, the unknown nonlinear function $f(x, t)$ satisfies

$$\|F(\alpha)f(x,t)\| \le \phi(\alpha), \quad \alpha \in \mathcal{N}. \tag{11.2}$$

The autonomous system of (11.1) can be formulated as

$$dx(t) = A(\alpha)x(t)dt + D(\alpha)x(t)d\varpi(t). \tag{11.3}$$

Definition 11.2.1 *The switched stochastic hybrid system in (11.3) is said to be mean-square exponentially stable under α if its solution $x(t)$ satisfies*

$$\mathbf{E}\{\|x(t)\|\} \le \eta \left\|x(t_0)\right\| e^{-\lambda(t-t_0)}, \quad \forall t \ge t_0,$$

where $\eta \ge 1$ and $\lambda > 0$ are two real constants.

According to Theorem 9.3.2, we have the following result for the mean-square exponential stability of system (11.3).

Theorem 11.2.2 *Given a scalar $\beta > 0$, suppose that there exist matrices $P(i) > 0$ such that for $i \in \mathcal{N}$,*

$$\begin{bmatrix} P(i)A(i) + A^T(i)P(i) + \beta P(i) & D^T(i)P(i) \\ \star & -P(i) \end{bmatrix} < 0.$$

Then the switched stochastic hybrid system in (11.3) is mean-square exponentially stable for any switching signal with average dwell time satisfying $T_a > \frac{\ln \mu}{\beta}$ with $\mu \ge 1$ and satisfying

$$P(i) \le \mu P(j), \quad \forall i, j \in \mathcal{N}.$$

Moreover, an estimate of the mean-square of the state decay is given by

$$\mathbf{E}\{\|x(t)\|\} \le \eta \|x(0)\| e^{-\lambda t},$$

where

$$\begin{cases} \lambda \triangleq \frac{1}{2}\left(\beta - \frac{\ln \mu}{T_a}\right) > 0, \quad \eta \triangleq \sqrt{\frac{b}{a}} \geq 1, \\ a \triangleq \min_{i\in\mathcal{N}} \lambda_{\min}(P(i)), \quad b \triangleq \max_{i\in\mathcal{N}} \lambda_{\max}(P(i)). \end{cases}$$

11.3 Main Results

11.3.1 Sliding Mode Dynamics Analysis

We design the following integral switching function:

$$s(t) = G(i)x(t) - \int_0^t G(i)\Big[A(i) + B(i)K(i)\Big]x(\theta)d\theta, \tag{11.4}$$

where $G(i) \in \mathbf{R}^{m\times n}$ and $K(i) \in \mathbf{R}^{m\times n}$ are real matrices to be designed. In particular, the matrices $G(i)$ are to be chosen such that $G(i)B(i)$ are nonsingular and $G(i)D(i) = 0$ for all $i \in \mathcal{N}$.

Then, the solution of $x(t)$ can be given by

$$x(t) = x(0) + \int_0^t \Big[A(i)x(\theta) + B(i)\left(u(\theta) + F(i)f(x(\theta),\theta)\right)\Big]d\theta$$
$$+ \int_0^t D(i)x(\theta)d\varpi(\theta). \tag{11.5}$$

It follows from (11.4) and (11.5) that

$$s(t) = G(i)x(0)$$
$$+ \int_0^t G(i)\Big[-B(i)K(i)x(\theta) + B(i)\left(u(\theta) + F(i)f(x(\theta),\theta)\right)\Big]d\theta.$$

As is well known that when the system state trajectories reach onto the sliding surface, it follows that $s(t) = 0$ and $\dot{s}(t) = 0$. Therefore, by $\dot{s}(t) = 0$, we get the equivalent control as

$$u_{eq}(t) = K(i)x(t) - F(i)f(x(t),t). \tag{11.6}$$

By substituting (11.6) into (11.1), the sliding mode dynamics can be obtained as

$$dx(t) = (A(i) + B(i)K(i))\,x(t)dt + D(i)x(t)d\varpi(t). \tag{11.7}$$

We will analyze the stability of the sliding mode dynamics in (11.7) based on Theorem 11.2.2, by which an explicit parametrization of the desired switching function designed in (11.4) is given.

Theorem 11.3.1 *For a given constant $\beta > 0$, suppose that there exist matrices $\mathcal{P}(i) > 0$ and $\mathcal{K}(i)$ such that for $i \in \mathcal{N}$,*

$$\begin{bmatrix} A(i)\mathcal{P}(i) + \mathcal{P}(i)A^T(i) + B(i)\mathcal{K}(i) + \mathcal{K}^T(i)B^T(i) + \beta\mathcal{P}(i) & \mathcal{P}(i)D^T(i) \\ \star & -\mathcal{P}(i) \end{bmatrix} < 0. \tag{11.8}$$

Then the sliding mode dynamics in (11.7) is mean-square exponentially stable for any switching signal with average dwell time satisfying $T_a > \frac{\ln\mu}{\beta}$ with $\mu \geq 1$ and satisfying

$$\mathcal{P}(i) \leq \mu\mathcal{P}(j), \quad \forall i,j \in \mathcal{N}. \tag{11.9}$$

Moreover, if the conditions above are feasible, the matrices $K(i)$ in (11.4) can be solved by

$$K(i) = \mathcal{K}(i)\mathcal{P}^{-1}(i). \tag{11.10}$$

Proof. From Theorem 11.2.2, if there exist matrices $P(i) > 0$ such that the following conditions hold for $i \in \mathcal{N}$:

$$\begin{bmatrix} P(i)\left(A(i) + B(i)K(i)\right) + \left(A(i) + B(i)K(i)\right)^T P(i) + \beta P(i) & D^T(i)P(i) \\ \star & -P(i) \end{bmatrix} < 0, \tag{11.11}$$

then the sliding mode dynamics in (11.7) is mean-square exponentially stable for any switching signal with average dwell time satisfying $T_a > \frac{\ln\mu}{\beta}$ with $\mu \geq 1$ and satisfying

$$P(i) \leq \mu P(j), \quad \forall i,j \in \mathcal{N}. \tag{11.12}$$

Letting $\mathcal{P}(i) \triangleq P^{-1}(i)$, and performing a congruence transformation on (11.11) with $\text{diag}\{\mathcal{P}(i), \mathcal{P}(i)\}$, we have that for $i \in \mathcal{N}$,

$$\begin{bmatrix} \left(A(i) + B(i)K(i)\right)\mathcal{P}(i) + \mathcal{P}(i)\left(A(i) + B(i)K(i)\right)^T + \beta\mathcal{P}(i) & \mathcal{P}(i)D^T(i) \\ \star & -\mathcal{P}(i) \end{bmatrix} < 0. \tag{11.13}$$

Let $\mathcal{K}(i) \triangleq K(i)\mathcal{P}(i)$, and we have (11.8) from (11.13). Furthermore, considering (11.12) and noting $\mathcal{P}(i) \triangleq P^{-1}(i)$ yields (11.9). This completes the proof. ■

11.3.2 SMC Law Design

In this section, we will synthesize an SMC law to drive the system state trajectories onto the predefined sliding surface $s(t) = 0$.

Theorem 11.3.2 *Consider the switched stochastic hybrid system (11.1). Suppose that the switching function is designed as (11.4) with $K(i)$ being solved by (11.10) in Theorem 11.3.1. Then the state trajectories of system (11.1) can be driven onto the sliding surface $s(t) = 0$ in a finite time by the following SMC law:*

$$u(t) = K(i)x(t) - (\delta + \gamma(i)) (G(i)B(i))^{-1} \operatorname{sign}(s(t)), \tag{11.14}$$

with

$$\gamma(i) \triangleq \phi(i) \|G(i)B(i)\|,$$

where $\delta > 0$ is a real constant and $G(i)$ are adjustable parameters to be chosen such that $G(i)B(i)$ are nonsingular for $i \in \mathcal{N}$.

Proof. Choose a Lyapunov function of the following form:

$$V(t) = \frac{1}{2}s^T(t)s(t).$$

By (11.4), we have

$$\dot{s}(t) = G(i)B(i)\left[-K(i)x(t) + u(t) + F(i)f(x(t), t)\right].$$

Thus, taking the derivative of $V(t)$ and considering the above equation, we have

$$\begin{aligned} \dot{V}(t) &= s^T(t)\dot{s}(t) \\ &= s^T(t)G(i)B(i)\left[-K(i)x(t) + u(t) + F(i)f(x(t), t)\right]. \end{aligned} \tag{11.15}$$

Substituting (11.14) into (11.15) and noting $\|s(t)\| \leq |s(t)|$, we have

$$\dot{V}(t) \leq -\delta \|s(t)\| \leq -\sqrt{2}\delta V^{\frac{1}{2}}(t). \tag{11.16}$$

It can be shown from (11.16) that there exists an instant $t^* = \sqrt{2V(0)}/\delta$ such that $V(t) = 0$ (equivalently, $s(t) = 0$) when $t \geq t^*$. Thus, it is concluded that the system trajectories can be driven onto the predefined sliding surface in a finite time. ∎

11.4 Observer-Based SMC Design

In this section, we will study the SMC problem under the assumption that some of the system state components are not available. We will first utilize a state observer to generate an estimate of the unmeasured states, and then synthesize an SMC law based on the state estimates. To begin with, we give the following measured output:

$$y(t) = C(i)x(t), \tag{11.17}$$

where $y(t) \in \mathbf{R}^p$ is the measured output. We design the following sliding mode observer to estimate the states of the switched stochastic hybrid system in (11.1):

$$\dot{\hat{x}}(t) = A(i)\hat{x}(t) + B(i)\left(u(t) + v(t)\right) + L(i)\left(y(t) - C(i)\hat{x}(t)\right), \tag{11.18}$$

where $\hat{x}(t) \in \mathbf{R}^n$ represents the estimate of the system states $x(t)$; $L(i) \in \mathbf{R}^{n\times p}$ are the observer gains to be designed; and the control term $v(t)$ is chosen to eliminate the effect of the nonlinear function $f(x, t)$.

Let $e(t) \triangleq x(t) - \hat{x}(t)$ denote the estimation error. According to (11.1) and (11.17)–(11.18), the estimation error dynamics is obtained as

$$\begin{aligned} de(t) = {} & \left[\left(A(i) - L(i)C(i)\right)e(t) - B(i)\left(v(t) - F(i)f(x, t)\right)\right]dt \\ & + D(i)e(t)d\varpi(t) + D(i)\hat{x}(t)d\varpi(t). \end{aligned} \tag{11.19}$$

Remark 11.1 *Notice from (11.19) that the estimation error dynamics corresponds to a switched stochastic hybrid system, and is dependent on the observer feedback matrix $L(i)$ and state estimates $\hat{x}(t)$. This means that the stability analysis of the error dynamics (11.19) is not independent of the observer dynamics (11.18).* ♦

Define the following switching functions in the state estimation space and in the state estimation error space, respectively,

$$s_x(t) = B^T(i)X(i)\left(\hat{x}(t) + \int_0^t B(i)\mathcal{G}(i)\hat{x}(\theta)d\theta\right), \tag{11.20a}$$

$$s_e(t) = B^T(i)X(i)e(t), \tag{11.20b}$$

where $\mathcal{G}(i) \in \mathbf{R}^{m\times n}$ are adjustable matrices which are chosen such that $(A(i) - B(i)\mathcal{G}(i))$ are Hurwitz for $i \in \mathcal{N}$. In addition, $X(i) > 0$ are matrices to be designed such that there always exist appropriately dimensioned matrices $N(i)$ for $B^T(i)X(i) = N(i)C(i)$, thus

$$\begin{aligned} s_e(t) &= B^T(i)X(i)e(t) \\ &= N(i)C(i)\left(x(t) - \hat{x}(t)\right) = N(i)\left(y(t) - C(i)\hat{x}(t)\right). \end{aligned} \tag{11.21}$$

The state-estimate-based SMC laws are designed as

$$u(t) = -(\varrho + \kappa + \chi(t,i) + \phi(i))\,\mathrm{sign}\left(s_x(t)\right), \tag{11.22a}$$

$$v(t) = (\kappa + \phi(i))\,\mathrm{sign}\left(s_e(t)\right), \tag{11.22b}$$

where $\varrho > 0$ and $\kappa > 0$ are two real constants, and

$$\chi(t,i) \triangleq \left\|\left(B^T(i)X(i)B(i)\right)^{-1}\right\| \left[\left(\left\|B^T(i)X(i)\left(A(i) + B(i)\mathcal{G}(i)\right)\right\| \right.\right.$$

$$\left.\left. + \left\|B^T(i)X(i)L(i)C(i)\right\| \right) \|\hat{x}(t)\| + \left\|B^T(i)X(i)L(i)\right\| \|y(t)\| \right].$$

We will show in the following that the sliding motion will be driven onto the specified sliding surface $s_x(t) = 0$ in a finite time.

Theorem 11.4.1 *The state trajectories of systems (11.18) can be driven onto the sliding surface $s_x(t) = 0$ in a finite time by the observer-based SMC law in (11.22a)–(11.22b).*

Proof. Select the following Lyapunov function:

$$\bar{V}(t) = \frac{1}{2} s_x^T(t)\left(B^T(i)X(i)B(i)\right)^{-1} s_x(t). \tag{11.23}$$

Noting $\|s_x(t)\| \leq |s_x(t)|$ and $s_x^T(t)\mathrm{sign}\left(s_e(t)\right) \leq |s_x(t)|$, we have

$$\begin{aligned}
\dot{\bar{V}}(t) &= s_x^T(t)\left(B^T(i)X(i)B(i)\right)^{-1} \dot{s}_x(t) \\
&= s_x^T(t)\left(B^T(i)X(i)B(i)\right)^{-1} B^T(i)X(i)\left(\dot{\hat{x}}(t) + B(i)\mathcal{G}(i)\hat{x}(t)\right) \\
&= s_x^T(t)\left(B^T(i)X(i)B(i)\right)^{-1} B^T(i)X(i)\left\{\left(A(i) + B(i)\mathcal{G}(i)\right)\hat{x}(t)\right. \\
&\quad \left. + B(i)\left(u(t) + v(t)\right) + L(i)\left(y(t) - C(i)\hat{x}(t)\right)\right\}.
\end{aligned} \tag{11.24}$$

Substituting SMC law (11.22a)–(11.22b) into (11.24), we have

$$\begin{aligned}
\dot{\bar{V}}(t) \leq{}& \|s_x(t)\| \left\|\left(B^T(i)X(i)B(i)\right)^{-1}\right\| \left\{ \left\|B^T(i)X(i)\left(A(i) + B(i)\mathcal{G}(i)\right)\hat{x}(t)\right\| \right. \\
& \left. + \left\|B^T(i)X(i)L(i)C(i)\hat{x}(t)\right\| + \left\|B^T(i)X(i)L(i)y(t)\right\| \right\} - \varrho\|s_x(t)\| \\
& - (\kappa + \chi(t,i) + \phi(i))\left|s_x(t)\right| + (\kappa + \phi(i))\left|s_x(t)\right|
\end{aligned}$$

$$
\begin{aligned}
&\leq \|s_x(t)\| \left\| \left(B^T(i)X(i)B(i)\right)^{-1} \right\| \left\{ \left\|B^T(i)X(i)\left(A(i)+B(i)\mathcal{G}(i)\right)\right\| \|\hat{x}(t)\| \right. \\
&\quad \left. + \left\|B^T(i)X(i)L(i)C(i)\right\| \|\tilde{x}(t)\| + \left\|B^T(i)X(i)L(i)\right\| \|y(t)\| \right\} \\
&\quad - \varrho \left\|s_x(t)\right\| - \chi(t,i) \left\|s_x(t)\right\| \\
&= -\varrho \left\|s_x(t)\right\| \leq -\sqrt{2}\tilde{\varrho}\bar{V}^{1/2}(t) < 0, \quad \text{for } \left\|s_x(t)\right\| \neq 0, \qquad (11.25)
\end{aligned}
$$

where $\tilde{\varrho} \triangleq \varrho\sqrt{\lambda_{\min}\left(B^T(i)X(i)B(i)\right)} > 0$. It can be shown from (11.25) that there exists an instant $t^\star = \sqrt{2\bar{V}(0)}/\tilde{\varrho}$ such that $\bar{V}(t) = 0$ (equivalently, $s_x(t) = 0$) when $t \geq t^\star$. Thus, we can say that the system state trajectories can be driven onto the predefined sliding surface in a finite time. This completes the proof. ∎

According to the SMC theory, it follows from $\dot{s}_x(t) = 0$ that the following equivalent control law can be obtained:

$$
\begin{aligned}
u_{eq}(t) = &-\left(B^T(i)X(i)B(i)\right)^{-1} B^T(i)X(i) \\
&\times \left[\left(A(i) + B(i)\mathcal{G}(i) \right) \hat{x}(t) + L(i) \left(y(t) - C(i)\hat{x}(t) \right) \right].
\end{aligned}
$$

Substituting $u_{eq}(t)$ above into (11.18) yields the sliding mode dynamics in the state estimation space, which can be formulated as

$$
\begin{aligned}
\dot{\hat{x}}(t) = &\left[I - B(i)\left(B^T(i)X(i)B(i)\right)^{-1} B^T(i)X(i) \right] \\
&\times \left[A(i)\hat{x}(t) + L(i)C(i)e(t) \right] - B(i)\mathcal{G}(i)\hat{x}(t). \qquad (11.26)
\end{aligned}
$$

In Theorem 11.4.1, our intention is to design an SMC law based on the estimated system states, such that the system state trajectories can be driven onto the predefined sliding surface $s_x(t) = 0$ in a finite time, and the sliding mode dynamics in the state estimation space then results – see (11.26). In the following, we will propose a sufficient stability condition for overall closed-loop system composed of the estimation error dynamics (11.19) and the sliding mode dynamics in the state estimation space (11.26).

Theorem 11.4.2 *Consider the switched stochastic hybrid system in (11.1) with (11.17). Its unmeasured states are estimated by the observer (11.18). The switching functions in the state estimation space and in the state estimation error space are chosen as (11.20a)–(11.20b),* *and the observer-based SMC law is synthesized by (11.22a)–(11.22b). If there exist matrices $X(i) > 0$, $N(i) > 0$ and $\mathcal{L}(i)$ such that for $i \in \mathcal{N}$,*

$$\begin{bmatrix} \Pi_{11}(i) & \Pi_{12}(i) & \sqrt{2}X(i)B(i) & 0 \\ \star & \Pi_{22}(i) & 0 & C^T(i)\mathcal{L}^T(i) \\ \star & \star & -B^T(i)X(i)B(i) & 0 \\ \star & \star & \star & -X(i) \end{bmatrix} < 0, \tag{11.27}$$

$$B^T(i)X(i) - N(i)C(i) = 0, \tag{11.28}$$

where

$$\begin{cases} \Pi_{11}(i) \triangleq X(i)A(i) + A^T(i)X(i) - X(i)B(i)\mathcal{G}(i) - \mathcal{G}^T(i)B^T(i)X(i) \\ \qquad\qquad + A^T(i)X(i)A(i) + D^T(i)X(i)D(i), \\ \Pi_{12}(i) \triangleq \mathcal{L}(i)C(i) + D^T(i)X(i)D(i), \\ \Pi_{22}(i) \triangleq X(i)A(i) + A^T(i)X(i) - \mathcal{L}(i)C(i) - C^T(i)\mathcal{L}^T(i) + D^T(i)X(i)D(i), \end{cases}$$

then the overall closed-loop switched stochastic hybrid system is globally asymptotically stable. Moreover, the observer gain is given by

$$L(i) = X^{-1}(i)\mathcal{L}(i), \quad i \in \mathcal{N}. \tag{11.29}$$

Proof. Select the following Lyapunov functions:

$$\begin{cases} \tilde{V}(\hat{x}, e) \triangleq \tilde{V}(\hat{x}) + \tilde{V}(e), \\ \tilde{V}(\hat{x}) \triangleq \dfrac{1}{2}\hat{x}^T(t)X(i)\hat{x}(t), \\ \tilde{V}(e) \triangleq \dfrac{1}{2}e^T(t)X(i)e(t). \end{cases} \tag{11.30}$$

Then, as with the solution of systems (11.19) and (11.26), we have

$$\begin{aligned} \mathcal{L}\tilde{V}(\hat{x}) = \hat{x}^T(t)X(i)\Big\{ & \Big[I - B(i)\left(B^T(i)X(i)B(i)\right)^{-1}B^T(i)X(i)\Big] \\ & \times \Big[A(i)\hat{x}(t) + L(i)C(i)e(t)\Big] - B(i)\mathcal{G}(i)\hat{x}(t)\Big\}, \end{aligned} \tag{11.31}$$

$$\mathcal{L}\tilde{V}(e) = e^T(t)X(i)\left[\left(A(i) - L(i)C(i)\right)e(t) - B(i)\left(v(t) - F(i)f(x,t)\right)\right]$$
$$+ \frac{1}{2}x^T(t)D^T(i)X(i)D(i)x(t). \tag{11.32}$$

Thus, we have

$$\begin{aligned}\mathcal{L}\tilde{V}(\hat{x}, e) = &\frac{1}{2}\hat{x}^T(t)\left(X(i)A(i) + A^T(i)X(i)\right)\hat{x}(t) + \hat{x}^T(t)X(i)L(i)C(i)e(t)\\ &- \hat{x}^T(t)X(i)B(i)\left(B^T(i)X(i)B(i)\right)^{-1}B^T(i)X(i)A(i)\hat{x}(t)\\ &- \hat{x}^T(t)X(i)B(i)\left(B^T(i)X(i)B(i)\right)^{-1}B^T(i)X(i)L(i)C(i)e(t)\\ &- \hat{x}^T(t)X(i)B(i)\mathcal{G}(i)\hat{x}(t) + e^T(t)X\left(A(i) - L(i)C(i)\right)e(t)\\ &- e^T(t)X(i)B(i)\left(v(t) - F(i)f(x,t)\right) + \frac{1}{2}x^T(t)D^T(i)X(i)D(i)x(t).\end{aligned} \tag{11.33}$$

Notice (11.22) and $\|s_e(t)\| \leq |s_e(t)|$. Thus,

$$\begin{aligned}- e^T(t)X(i)B(i)\left(v(t) - F(i)f(x,t)\right) &= -s_e^T(t)\left(\kappa + \phi(i)\right)\text{sign}\left(s_e(t)\right)\\ &\quad + s_e^T(t)F(i)f(x,t)\\ &\leq -\left(\kappa + \phi(i)\right)\left|s_e^T(t)\right| + \phi(i)\left\|s_e^T(t)\right\|\\ &\leq -\kappa\left\|s_e^T(t)\right\| < 0.\end{aligned} \tag{11.34}$$

However, the following inequalities hold:

$$\begin{aligned}&- \hat{x}^T(t)X(i)B(i)\left(B^T(i)X(i)B(i)\right)^{-1}B^T(i)X(i)A(i)\hat{x}(t)\\ \leq &\frac{1}{2}\left\{\hat{x}^T(t)X(i)B(i)\left(B^T(i)X(i)B(i)\right)^{-1}B^T(i)X(i)\hat{x}(t) + \hat{x}^T(t)A^T(i)X(i)A(i)\hat{x}(t)\right\}\\ &- \hat{x}^T(t)X(i)B(i)\left(B^T(i)X(i)B(i)\right)^{-1}B^T(i)X(i)L(i)C(i)e(t)\\ \leq &\frac{1}{2}\left\{\hat{x}^T(t)X(i)B(i)\left(B^T(i)X(i)B(i)\right)^{-1}B^T(i)X(i)\hat{x}(t)\right.\\ &\left.+ e^T(t)C^T(i)L^T(i)X(i)L(i)C(i)e(t)\right\}.\end{aligned}$$

Considering (11.31)–(11.34), we have

$$\mathcal{L}\tilde{V}(\hat{x}, e) \leq \frac{1}{2}\zeta^T(t)\Omega(i)\zeta(t), \tag{11.35}$$

where $\zeta(t) \triangleq \begin{bmatrix} \hat{x}(t) \\ e(t) \end{bmatrix}$ and $\Omega(i) \triangleq \begin{bmatrix} \Omega_{11}(i) & \Omega_{12}(i) \\ \star & \Omega_{22}(i) \end{bmatrix}$ with

$$\begin{cases} \Omega_{11}(i) \triangleq X(i)\left(A(i) - B(i)\mathcal{G}(i)\right) + \left(A(i) - B(i)\mathcal{G}(i)\right)^T X(i) + A^T(i)X(i)A(i) \\ \qquad + D^T(i)X(i)D(i) + 2X(i)B(i)\left(B^T(i)X(i)B(i)\right)^{-1} B^T(i)X(i) \\ \Omega_{12}(i) \triangleq X(i)L(i)C(i) + D^T(i)X(i)D(i), \\ \Omega_{22}(i) \triangleq X(i)A(i) - X(i)L(i)C(i) + \left(X(i)A(i) - X(i)L(i)C(i)\right)^T \\ \qquad + C^T(i)L^T(i)X(i)L(i)C(i) + D^T(i)X(i)D(i). \end{cases}$$

Let $\mathcal{L}(i) \triangleq X(i)L(i)$ and by Schur complement, (11.27) implies $\Omega(i) < 0$. Thus,

$$\mathcal{L}\tilde{V}(\hat{x}, e) < 0.$$

We know that the overall closed-loop switched stochastic hybrid system composed of the estimation error dynamics (11.19) and the sliding mode dynamics in the state estimation space (11.26) is globally asymptotically stable. This completes the proof. ■

Note that the conditions in Theorem 11.4.2 are not all expressed in LMI form due to (11.28), thus they can not be solved directly by an LMI procedure. In fact, (11.28) can be equivalently converted to

$$\operatorname{trace}\left(\left(B^T(i)X(i) - N(i)C(i)\right)^T \left(B^T(i)X(i) - N(i)C(i)\right) \right) = 0.$$

We consider the following matrix inequalities for scalar $\hbar > 0$,

$$\left(B^T(i)X(i) - N(i)C(i)\right)^T \left(B^T(i)X(i) - N(i)C(i)\right) \leq \hbar I, \quad i \in \mathcal{N}. \tag{11.36}$$

By Schur complement, (11.36) is equivalent to

$$\begin{bmatrix} -\hbar I & \left(B^T(i)X(i) - N(i)C(i)\right)^T \\ \star & -I \end{bmatrix} \leq 0, \quad i \in \mathcal{N}. \tag{11.37}$$

Therefore, when $\hbar > 0$ is chosen as a sufficiently small scalar, (11.28) can be solved through LMI (11.37).

11.5 Illustrative Example

Example 11.5.1 (SMC problem) Consider system (11.1) with $N = 2$ and the following parameters:

$$A(1) = \begin{bmatrix} -1.0 & 0.6 & -2.4 \\ 2.0 & -0.5 & -0.8 \\ 0.1 & 2.2 & 0.5 \end{bmatrix}, \; D(1) = \begin{bmatrix} 0.3 & 0.1 & 0.1 \\ 0.1 & 0.3 & 0.3 \\ 0.2 & 0.1 & 0.1 \end{bmatrix}, \; B(1) = \begin{bmatrix} 1.2 \\ 0.8 \\ 0.5 \end{bmatrix},$$

$$A(2) = \begin{bmatrix} 1.0 & 0.8 & 1.0 \\ 0.0 & 0.5 & -0.6 \\ 0.3 & 0.4 & -0.5 \end{bmatrix}, \; D(2) = \begin{bmatrix} 0.2 & 0.1 & 0.2 \\ 0.1 & 0.3 & 0.1 \\ 0.2 & 0.2 & 0.2 \end{bmatrix}, \; B(2) = \begin{bmatrix} 0.5 \\ 1.2 \\ 0.4 \end{bmatrix},$$

$$F(1) = 1.6, \; F(2) = 2.0, \; f(t) = 0.5 \exp(-t) \sin(t).$$

Set $\beta = 0.5$. It can be shown that the system in (11.1) with $u(t) = 0$ and the above parametric matrices is unstable for a switching signal given in Figure 11.1 (which is generated randomly; here, '1' and '2' represent the first and second subsystems, respectively). Thus, our aim is to design the SMC law $u(t)$ in (11.14) such that the closed-loop system is mean-square exponentially stable for $T_a > T_a^* = 0.1$ (in this case, the allowable minimum of μ is $\mu_{\min} = 1.0513$). To check the stability of the sliding mode dynamics in (11.7) with $T_a > T_a^* = 0.1$

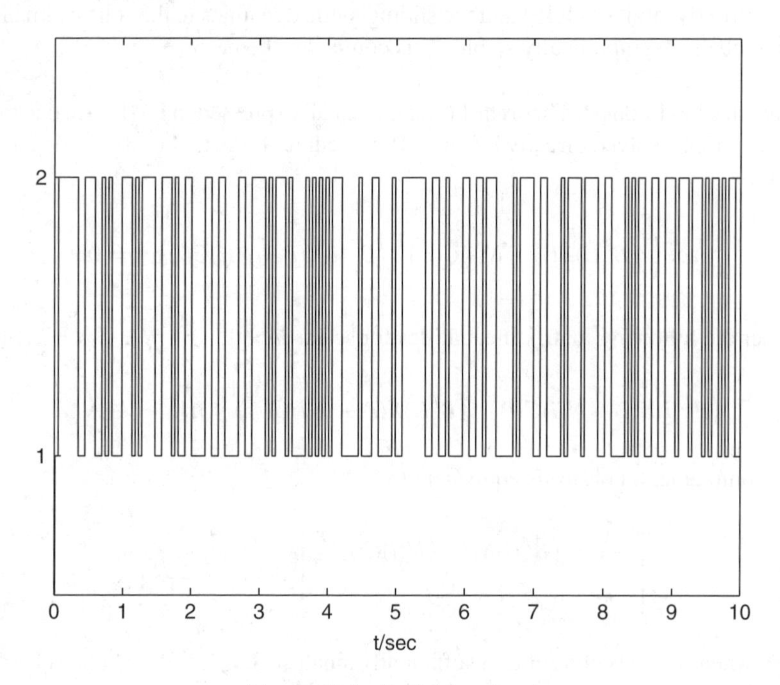

Figure 11.1 Switching signal

(that is, set $\mu = 1.0513$), we solve the conditions (11.8)–(11.9) in Theorem 11.3.1, and by (11.10), we obtain

$$K(1) = \begin{bmatrix} -0.5946 & -2.8781 & -2.0463 \end{bmatrix},$$

$$K(2) = \begin{bmatrix} -3.1291 & -0.6310 & -1.6249 \end{bmatrix}.$$

We choose

$$G(1) = \begin{bmatrix} 5 & 1 & -8 \end{bmatrix}, \quad G(2) = \begin{bmatrix} 4 & 2 & -5 \end{bmatrix}.$$

Thus, the switching function defined in (11.4) is given by

$$s(t) = \begin{cases} s(t,1) = \begin{bmatrix} 5 & 1 & -8 \end{bmatrix} x(t) \\ \qquad - \int_0^t \begin{bmatrix} -5.4648 & -23.1586 & -22.5297 \end{bmatrix} x(\theta) d\theta, & i = 1, \\ s(t,2) = \begin{bmatrix} 4 & 2 & -5 \end{bmatrix} x(t) \\ \qquad - \int_0^t \begin{bmatrix} -5.0097 & 0.6855 & 1.4002 \end{bmatrix} x(\theta) d\theta, & i = 2. \end{cases}$$

and the SMC law designed in (11.14) can be computed as

$$u(t) = \begin{cases} u(t,1) = \begin{bmatrix} -0.5946 & -2.8781 & -2.0463 \end{bmatrix} x(t) \\ \qquad - 0.3571\,(\delta + 2.24)\,\text{sign}\,(s(t,1)), & i = 1, \\ u(t,2) = \begin{bmatrix} -3.1291 & -0.6310 & -1.6249 \end{bmatrix} x(t) \\ \qquad - 0.4167\,(\delta + 2.40)\,\text{sign}\,(s(t,2)), & i = 2, \end{cases}$$

where $\delta > 0$ is an adjustable constant.

To prevent the control signals from chattering, we replace $\text{sign}\,(s(t))$ in the SMC law with $s(t)/\left(0.01 + \|s(t)\|\right)$. Set $\delta = 0.5$ and suppose that the initial condition is $x(0) = \begin{bmatrix} -1.0 & 0.5 & 1.0 \end{bmatrix}^T$. By using the discretization approach [96], we simulate standard Brownian motion. Some initial parameters are given as follows: the simulation time $t \in [0, T^*]$ with $T^* = 10$, the normally distributed variance $\delta t = \frac{T^*}{N^*}$ with $N^* = 2^{11}$, step size $\Delta t = \rho \delta t$ with $\rho = 2$, and the number of discretized Brownian paths $p = 10$. The simulation results are given in Figures 11.2–11.6. Specifically, Figures 11.2–11.4 are the simulation results along an individual discretized Brownian path, with Figure 11.2 showing the states of the closed-loop system under the designed SMC law. The switching function and the SMC input are given in Figures 11.3 and 11.4, respectively. Figures 11.5–11.6 are the simulation results on $x(t)$ and $s(t)$ along 10 individual paths (dotted lines) and the average over 10 paths (solid line), respectively.

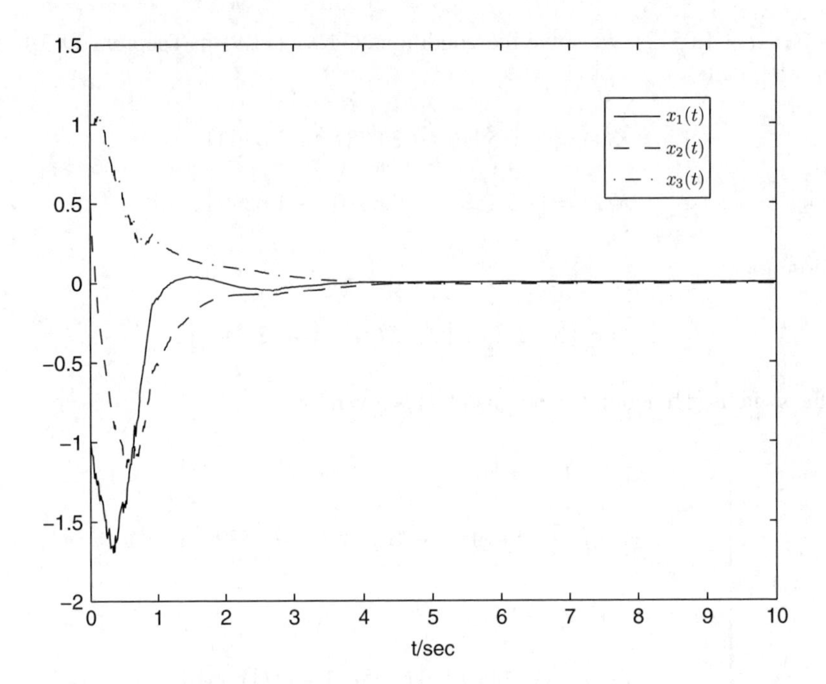

Figure 11.2 States of the closed-loop system

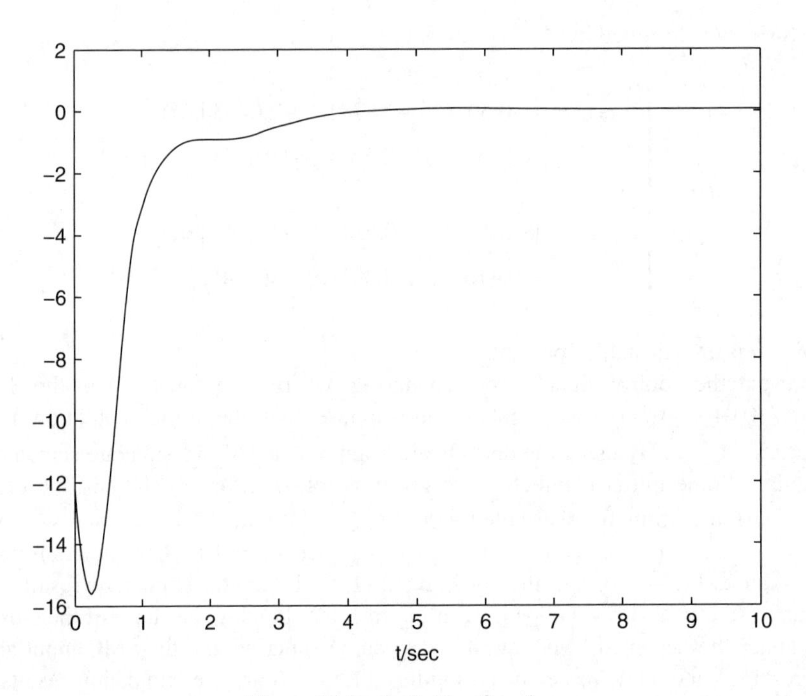

Figure 11.3 Switching function

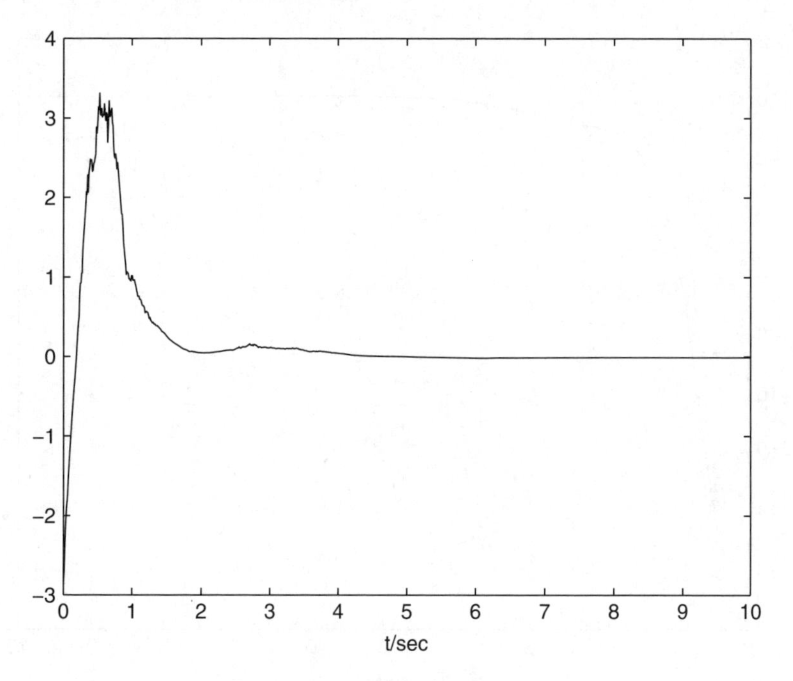

Figure 11.4 Control input

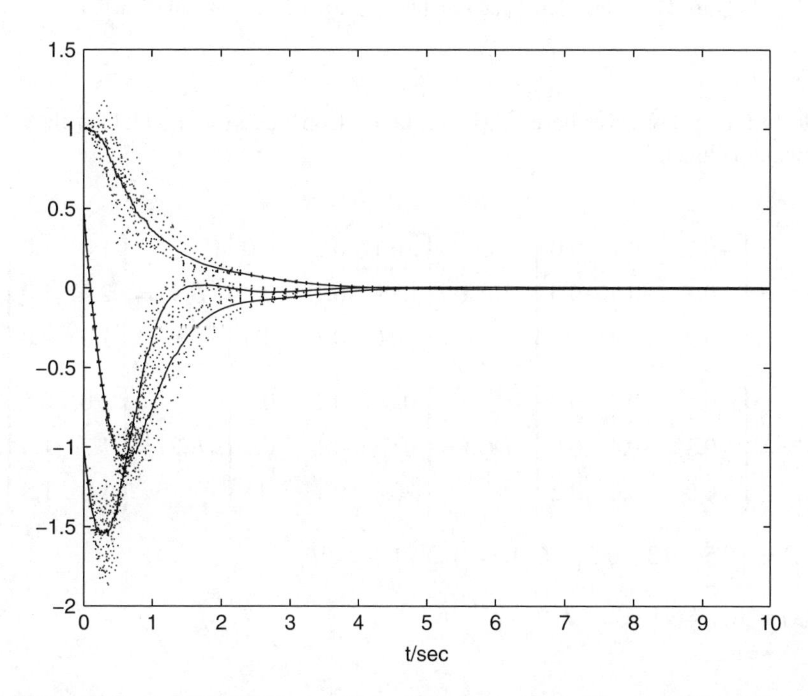

Figure 11.5 Individual paths and the average of the states of the closed-loop system

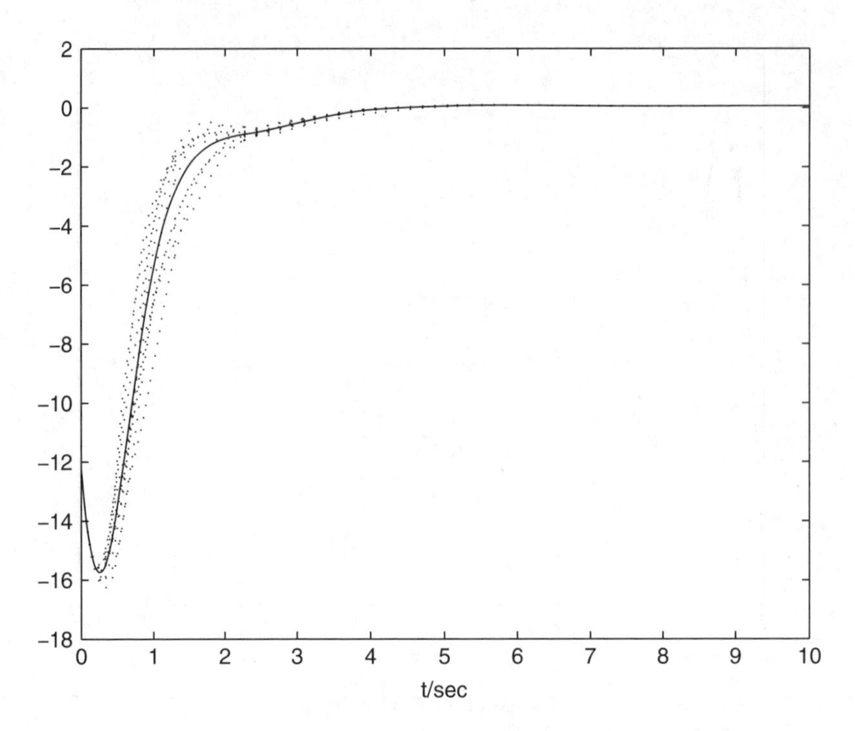

Figure 11.6 Individual paths and the average of the switching function

Example 11.5.2 (Observer-based SMC problem) Consider system (11.1) with $N = 2$ and the following parameters:

$$A(1) = \begin{bmatrix} -0.7 & 0.2 & 0.0 \\ 0.3 & -0.4 & 0.0 \\ 0.0 & 0.4 & 0.2 \end{bmatrix}, \; D(1) = \begin{bmatrix} 0.1 & 0.2 & 0.0 \\ 0.03 & 0.1 & 0.2 \\ 0.0 & 0.1 & 0.05 \end{bmatrix}, \; B(1) = \begin{bmatrix} 1.0 & 1.0 \\ 2.0 & 1.0 \\ 1.0 & 2.0 \end{bmatrix},$$

$$A(2) = \begin{bmatrix} -0.5 & 0.2 & 0.0 \\ 0.3 & -0.4 & 0.0 \\ 0.0 & 0.2 & 0.2 \end{bmatrix}, \; D(2) = \begin{bmatrix} 0.02 & 0.2 & 0.0 \\ 0.03 & 0.1 & 0.1 \\ 0.0 & 0.1 & 0.15 \end{bmatrix}, \; B(2) = \begin{bmatrix} 2.0 & 0.5 \\ 1.3 & 1.0 \\ 1.0 & 1.5 \end{bmatrix},$$

$$C(1) = \begin{bmatrix} 0.5 & 0.3 & 0.5 \end{bmatrix}, \; C(2) = \begin{bmatrix} 0.3 & 0.4 & 0.7 \end{bmatrix},$$

$$F(1) = 1.0, \; F(2) = 2.0, \; f(t) = e^{-t}\sin(t).$$

In this example, we will consider the SMC design in the case where some of the system state components are not available. We design a sliding mode observer in the form of (11.18) to estimate the system states, and then synthesize the observer-based SMC laws in (11.22a)–(11.22b) for the reaching motion. First, we select matrices $\mathcal{G}(1)$ and $\mathcal{G}(2)$ as follows:

$$\mathcal{G}(1) = \mathcal{G}(2) = \begin{bmatrix} 0.5 & 2.5 & -2.0 \\ 0.3 & -1.5 & 4.0 \end{bmatrix},$$

which guarantee that both $A(1) - B(1)\mathcal{G}(1)$ and $A(2) - B(2)\mathcal{G}(2)$ are Hurwitz. Then, solving (11.27) and (11.37), and by (11.29), we obtain

$$L(1) = \begin{bmatrix} 0.3177 \\ 1.0438 \\ 0.7120 \end{bmatrix}, \; L(2) = \begin{bmatrix} -0.6163 \\ -0.0363 \\ 1.4330 \end{bmatrix}.$$

According to (11.20a), we have

$$s_x(t) = \begin{cases} s_x(t,1) = \begin{bmatrix} 0.3118 & 0.2078 & 0.3001 \\ 0.3495 & 0.2001 & 0.3700 \end{bmatrix} \hat{x}(t) \\ \quad + \int_0^t \begin{bmatrix} 0.8496 & 0.8891 & 2.4239 \\ 0.9467 & 0.8649 & 2.9188 \end{bmatrix} \hat{x}(\theta)d\theta, \quad i = 1, \\ s_x(t,2) = \begin{bmatrix} 0.0909 & 0.1146 & 0.1573 \\ 0.0671 & 0.0916 & 0.1427 \end{bmatrix} \hat{x}(t) \\ \quad + \int_0^t \begin{bmatrix} 0.3628 & 0.6261 & 0.6079 \\ 0.2997 & 0.4812 & 0.5648 \end{bmatrix} \hat{x}(\theta)d\theta, \quad i = 2, \end{cases}$$

and by (11.21), we have

$$s_e(t) = \begin{cases} s_e(t,1) = \begin{bmatrix} 0.6242 \\ 0.7115 \end{bmatrix} (y(t) - C(i)\hat{x}(t)), \quad i = 1, \\ s_e(t,2) = \begin{bmatrix} 0.2476 \\ 0.2117 \end{bmatrix} (y(t) - C(i)\hat{x}(t)), \quad i = 2. \end{cases}$$

Thus, the state estimate-based SMC laws designed in (11.22a)–(11.22b) are computed with $\phi(1) = 1$, $\phi(2) = 2$ and

$$\begin{cases} \chi(t,1) = 140.8915\|\hat{x}(t)\| + 5.3635\|y(t)\|, \\ \chi(t,2) = 120.2777\|\hat{x}(t)\| + 2.2355\|y(t)\|. \end{cases}$$

11.6 Conclusion

The problem of the SMC of a continuous-time switched stochastic hybrid system has been investigated in this chapter. An integral switching function has been designed, and a sufficient condition for the existence of sliding mode has been established in terms of LMIs, and an explicit parametrization of the desired switching function has also been given. Then, a discontinuous SMC law for reaching motion has been synthesized to drive the system state trajectories onto the predefined sliding surface in a finite time. Moreover, we have further studied the observer design and observer-based SMC problems for the case that some system state components are not accessible. Sufficient conditions have also been proposed for the existence of the desired sliding mode, and the observer-based SMC law has been designed for the reaching motion. Two numerical examples have been provided to illustrate the effectiveness of the proposed design scheme.

12

SMC with Dissipativity of Switched Stochastic Hybrid Systems

12.1 Introduction

Dissipativity theory has played a critical part in analysis and control design of linear and nonlinear systems, especially for high-order systems, since from the practical application point of view, many systems need to be dissipative for achieving effective noise attenuation. It has been recognized that for more abstract systems one can still associate with them an energy-like function (called the storage function) and an input-power-like function (called the supply rate). Dissipativity is then characterized by storage functions and supply rates, which represent the energy stored inside the system and energy supplied from outside the system, respectively. Roughly speaking, dissipative systems are those for which the increase in stored energy is never larger than the amount of energy supplied by the environment, that is, dissipative systems can only dissipate but also not generate energy. The dissipative systems theory is closely related to the dynamic properties of a process and, in particular, to its stability properties.

In this chapter, we will study dissipativity analysis and SMC design for switched stochastic hybrid systems. A more general supply rate is proposed, and a strict $(\mathcal{Z}, \mathcal{Y}, \mathcal{X})$-dissipativity is defined, which includes $\mathcal{H}_\infty$, positive realness, and passivity as its special cases. The main idea is to introduce the strict $(\mathcal{Z}, \mathcal{Y}, \mathcal{X})$-dissipativity into the analysis of sliding mode dynamics so as to improve the transient performance of the SMC system. The objective is to conduct dissipativity analysis and investigate the dissipativity-based SMC design scheme, with a view to contributing to the development of SMC design and the dissipativity analysis methods for the switched stochastic hybrid system. Specifically, an integral sliding surface is designed such that the sliding mode exists with the same order as the original system.

Then, by using the average dwell time approach and the piecewise Lyapunov function technique, a sufficient condition is established in terms of LMIs, which guarantees the sliding mode dynamics to be mean-square exponentially stable with a strict dissipativity performance.

Sliding Mode Control of Uncertain Parameter-Switching Hybrid Systems, First Edition. Ligang Wu, Peng Shi and Xiaojie Su.

In addition, a solution to the dissipativity synthesis is provided by designing a discontinuous SMC law such that the system state trajectories can be driven onto the predefined sliding surface in a finite time and maintained there for all subsequent time.

12.2 Problem Formulation and Preliminaries

12.2.1 System Description

Consider the continuous-time switched stochastic hybrid systems, which are established on the probability space $(\Omega, \mathcal{F}, \mathcal{P})$, and are described by

$$\begin{aligned} dx(t) &= \{A(\alpha(t))x(t) + B(\alpha(t))[u(t) + f(x(t), t, \alpha(t))] \\ &\quad + E(\alpha(t))\omega(t)\}\, dt + F(\alpha(t))x(t)d\varpi(t), \qquad (12.1a) \\ z(t) &= C(\alpha(t))x(t) + D(\alpha(t))\omega(t), \qquad (12.1b) \end{aligned}$$

where $x(t) \in \mathbf{R}^n$ is the system state vector; $u(t) \in \mathbf{R}^m$ is the control input; $z(t) \in \mathbf{R}^q$ is the controlled output; $\omega(t) \in \mathbf{R}^p$ which belongs to $\mathcal{L}_2[0, \infty)$, is either a disturbance input or a reference signal; $\varpi(t)$ is a one-dimensional Brownian motion satisfying $\mathbf{E}\{d\varpi(t)\} = 0$; and $\mathbf{E}\left\{d\varpi^2(t)\right\} = dt$. In addition, $f(x(t), t, \alpha(t)) \in \mathbf{R}^m$ is an unknown nonlinear function satisfying

$$\|f(x(t), t, \alpha(t))\| \leq \phi(\alpha(t))\, \|x(t)\|,$$

where $\phi(\alpha(t)) > 0$, and $\phi(\alpha(t))\, \|x(t)\|$ is an upper bound of the norm of the nonlinear function.

Here, $\left\{(A(\alpha(t)), B(\alpha(t)), C(\alpha(t)), D(\alpha(t)), E(\alpha(t)), F(\alpha(t))) : \ \alpha(t) \in \mathcal{N}\right\}$ in system (12.1a)–(12.1b) is a family of matrices parameterized by an index set $\mathcal{N} = \{1, 2, \ldots, N\}$ and $\alpha(t) : \mathbf{R} \to \mathcal{N}$ is a piecewise constant function of time t called a switching signal (denoted by α for simplicity), which is defined as in Chapter 5. For each possible value $\alpha = i\ (i \in \mathcal{N})$, we will denote the system matrices associated with mode i by $A(i) = A(\alpha)$, $B(i) = B(\alpha)$, $C(i) = C(\alpha)$, $D(i) = D(\alpha)$, $E(i) = E(\alpha)$, and $F(i) = F(\alpha)$, where $A(i)$, $B(i)$, $C(i)$, $D(i)$, $E(i)$, and $F(i)$ are constant matrices.

Assumption 12.1 *For each $\alpha \in \mathcal{N}$, the pair $(A(\alpha), B(\alpha))$ in system (12.1a) is controllable and the matrix $B(\alpha)$ has full column rank.*

First, we consider the following switched stochastic hybrid systems:

$$\begin{aligned} dx(t) &= [A(\alpha)x(t) + E(\alpha)\omega(t)]dt + F(\alpha)x(t)d\varpi(t), \qquad (12.2a) \\ z(t) &= C(\alpha)x(t) + D(\alpha)\omega(t), \qquad (12.2b) \end{aligned}$$

where $x(t) \in X \in \mathbf{R}^n$ is the state vector; $\omega(t) \in \Omega \in \mathbf{R}^p$ is the input; and $z(t) \in Z \in \mathbf{R}^q$ is the controlled output.

Definition 12.2.1 *The switched stochastic hybrid system (12.2a) with $\omega(t) = 0$ is said to be mean-square exponentially stable under α if its solution $x(t)$ satisfies*

$$\mathbf{E}\{\|x(t)\|\} \leq \eta \|x(t_0)\| e^{-\lambda(t-t_0)}, \quad \forall t \geq t_0,$$

where $\eta \geq 1$ and $\lambda > 0$ are real constants.

12.2.2 Dissipativity

In this section, we will give a brief introduction to dissipative systems. Dissipative systems can be regarded as a generalization of passive systems with more general internal and supplied energies [269]. A system is called 'dissipative' if there is 'power dissipation' in the system. Dissipative systems are those that cannot store more energy than that supplied by the environment and/or by other systems connected to them, that is, dissipative systems can only dissipate but not generate energy [169].

According to [97], associated with the switched stochastic hybrid system (12.2a)–(12.2b) is a real-valued function $\Phi(\omega, z)$ called the *supply rate*, which is formally defined as follows.

Definition 12.2.2 (Supply Rate) *The supply rate is a real-valued function: $\Phi(\omega, z)$: $\Omega \times Z \rightarrow \mathbf{R}$, which is assumed to be locally Lebesgue integrable independently of the input and the initial conditions, that is, for any $\omega \in \Omega$, $z \in Z$, and $t^* \geq 0$, it holds that $\int_0^{t^*} |\Phi(\omega(t), z(t))| dt < +\infty$.*

The classical form of dissipativity in [97] is obviously applicable to the switched stochastic hybrid system in (12.1a)–(12.1b).

Definition 12.2.3 (Dissipative system) *The switched stochastic hybrid system (12.2a)–(12.2b) with supply rate $\Phi(\omega, z)$ is said to be dissipative if there exists a nonnegative function $V(x) : X \rightarrow \mathbf{R}$, called the storage function, such that the following dissipation inequality holds:*

$$\mathbf{E}\left\{V(x(t^*)) - V(x(0))\right\} \leq \mathbf{E}\left\{\int_0^{t^*} \Phi(\omega(t), z(t)) dt\right\}, \tag{12.3}$$

for all initial condition $x_0 \in X$, input $\omega \in \Omega$, and $t^* \geq 0$ (or said differently: for all admissible inputs $\omega(t)$ that drive the state from $x(0)$ to $x(t^*)$ on the interval $[0, t^*]$, where $x(t^*)$ is the state variable at time $t = t^*$).

Remark 12.1 *Inequality (12.3) is called the dissipation inequality and it formalizes the property that the increase in stored energy is never greater than the amount of energy supplied by the environment. Passive systems are a special class of dissipative systems that have a bilinear supply rate, that is, $\Phi(\omega, z) = z^T \omega$. If a system with a constant* positive *feedforward of $\mathcal{X}$ is passive, then the process is dissipative with respect to the supply rate $\Phi(\omega, z) = z^T \omega + \omega^T \mathcal{X} \omega$, where $\mathcal{X} = \mathcal{X}^T \in \mathbf{R}^{p \times p}$. Similarly, if a system with a constant* negative *feedback of $\mathcal{Z}$ is passive, then the process is dissipative with respect to the supply rate $\Phi(\omega, z) = z^T \mathcal{Z} z + z^T \omega$, where $\mathcal{Z} = \mathcal{Z}^T \in \mathbf{R}^{p \times p}$.* ♦

Motivated by the above facts, a more general supply rate is proposed in the following definition.

Definition 12.2.4 *Given matrices $\mathcal{Z} \in \mathbf{R}^{q\times q}$, $\mathcal{X} \in \mathbf{R}^{p\times p}$, and $\mathcal{Y} \in \mathbf{R}^{q\times p}$ with $\mathcal{Z}$ and $\mathcal{X}$ being symmetric, the switched stochastic hybrid system (12.2a)–(12.2b) is called $(\mathcal{Z}, \mathcal{Y}, \mathcal{X})$-dissipative if for some real function $\gamma(\cdot)$ with $\gamma(0) = 0$,*

$$\mathbf{E}\left\{\int_0^{t^*} \begin{bmatrix} z(t) \\ \omega(t) \end{bmatrix}^T \begin{bmatrix} \mathcal{Z} & \mathcal{Y} \\ \star & \mathcal{X} \end{bmatrix} \begin{bmatrix} z(t) \\ \omega(t) \end{bmatrix} dt \right\} + \gamma(x_0) \geq 0, \quad \forall t^* \geq 0. \tag{12.4}$$

Furthermore, if for some scalar $\delta > 0$ and $\forall t^* \geq 0$,

$$\mathbf{E}\left\{\int_0^{t^*} \begin{bmatrix} z(t) \\ \omega(t) \end{bmatrix}^T \begin{bmatrix} \mathcal{Z} & \mathcal{Y} \\ \star & \mathcal{X} \end{bmatrix} \begin{bmatrix} z(t) \\ \omega(t) \end{bmatrix} dt \right\} + \gamma(x_0) \geq \delta \int_0^{t^*} \omega^T(t)\omega(t)dt, \tag{12.5}$$

then the switched stochastic hybrid system (12.2a)–(12.2b) is called strictly $(\mathcal{Z}, \mathcal{Y}, \mathcal{X})$-$\delta$-dissipative.

Remark 12.2 *In Definition 12.2.4, we assume that $\mathcal{Z} \leq 0$, thus the performance defined in Definition 12.2.4 includes $\mathcal{H}_\infty$, positive realness, and passivity as special cases. Specifically,*

Case 1. When $\mathcal{Z} = -I$, $\mathcal{Y} = 0$, and $\mathcal{X} - \delta I = \gamma^2 I$, (12.5) reduces to an $\mathcal{H}_\infty$ performance requirement.

Case 2. When $\mathcal{Z} = 0$, $\mathcal{Y} = I$, $\mathcal{X} - \delta I = 0$, and $\gamma(x_0) = 0$, (12.5) corresponds to an extended strict positive real problem.

Case 3. When $\mathcal{Z} = -\theta I$, $\mathcal{Y} = (1-\theta)I$, and $\mathcal{X} - \delta I = \theta\gamma^2 I$, $\theta \in [0, 1]$, (12.5) corresponds to the mixed $\mathcal{H}_\infty$ and positive real performances.

Case 4. When $\mathcal{Z} = 0$, $\mathcal{Y} = I$, and $\mathcal{X} - \delta I = \kappa I$, (12.5) corresponds to a passivity problem. ♦

12.3 Dissipativity Analysis

In this section, we will apply the average dwell time method combined with the piecewise Lyapunov function technique to investigate the dissipativity and the mean-square exponential stability for the switched stochastic hybrid system in (12.2a)–(12.2b).

Theorem 12.3.1 *Given matrices $0 \geq \mathcal{Z} \in \mathbf{R}^{q\times q}$, $\mathcal{X} \in \mathbf{R}^{p\times p}$, and $\mathcal{Y} \in \mathbf{R}^{q\times p}$, with $\mathcal{Z}$ and $\mathcal{X}$ being symmetric, and scalars $\beta > 0$, $\delta > 0$, suppose that there exist matrices $0 < P(i) \in \mathbf{R}^{n\times n}$ such that for $i \in \mathcal{N}$,*

$$\Pi(i) \triangleq \begin{bmatrix} \Pi_{11}(i) & \Pi_{12}(i) \\ \star & \Pi_{22}(i) \end{bmatrix} < 0, \tag{12.6}$$

where

$$\begin{cases} \Pi_{11}(i) \triangleq P(i)A(i) + A^T(i)P(i) + \beta P(i) + F^T(i)P(i)F(i) - C^T(i)\mathcal{Z}C(i), \\ \Pi_{12}(i) \triangleq P(i)E(i) - C^T(i)\mathcal{Y}^T - C^T(i)\mathcal{Z}D(i), \\ \Pi_{22}(i) \triangleq -\mathcal{X} + \delta I - \mathcal{Y}D(i) - D^T(i)\mathcal{Y}^T - D^T(i)\mathcal{Z}D(i), \end{cases}$$

then the switched stochastic hybrid system in (12.2a)–(12.2b) is strictly $(\mathcal{Z}, \mathcal{Y}, \mathcal{X})$*-*$\delta$*-dissipative in the sense of Definition 12.2.4 for any switching signal with the average dwell time satisfying* $T_a > \frac{\ln \mu}{\beta}$ *(where* $\mu \geq 1$*) and satisfying*

$$P(i) \leq \mu P(j), \quad \forall i, j \in \mathcal{N}. \tag{12.7}$$

Proof. Choose the following Lyapunov function:

$$V(x, \alpha) \triangleq x^T(t)P(\alpha)x(t), \tag{12.8}$$

where $P(\alpha) > 0$, $\alpha \in \mathcal{N}$ are to be determined. Then, as with the solution of system (12.2a)–(12.2b), by Itô's formula, we obtain the stochastic differential as

$$\begin{aligned} dV(x, \alpha) &= \mathcal{L}V(x, \alpha)dt + 2x^T(t)P(\alpha)F(\alpha)x(t)d\varpi(t), \\ \mathcal{L}V(x, \alpha) &= 2x^T(t)P(\alpha)\left[A(\alpha)x(t) + E(\alpha)\omega(t)\right] + x^T(t)F^T(\alpha)P(\alpha)F(\alpha)x(t) \\ &= x^T(t)\left[P(\alpha)A(\alpha) + A^T(\alpha)P(\alpha) + F^T(\alpha)P(\alpha)F(\alpha)\right]x(t) \\ &\quad + 2x^T(t)P(\alpha)E(\alpha)\omega(t). \end{aligned} \tag{12.9}$$

To show the strict dissipativity of system (12.2a)–(12.2b), we consider (12.9), and for any nonzero $\omega(t) \in \mathcal{L}_2[0, \infty)$, it follows that

$$\begin{aligned} \Gamma(x, \alpha) &\triangleq \mathcal{L}V(x, \alpha) + \beta V(x, \alpha) - z^T(t)\mathcal{Z}z(t) - 2\omega^T(t)\mathcal{Y}z(t) - \omega^T(t)(\mathcal{X} - \delta I)\omega(t) \\ &\triangleq \begin{bmatrix} x(t) \\ \omega(t) \end{bmatrix}^T \Pi(\alpha) \begin{bmatrix} x(t) \\ \omega(t) \end{bmatrix}, \end{aligned}$$

where $\Pi(\alpha)$ is defined in (12.6). By (12.6) we have $\Gamma(x, \alpha) < 0$, that is,

$$\mathcal{L}V(x, \alpha) < -\beta V(x, \alpha) + \Psi(t),$$

where

$$\Psi(t) \triangleq z^T(t)\mathcal{Z}z(t) + 2\omega^T(t)\mathcal{Y}z(t) + \omega^T(t)(\mathcal{X} - \delta I)\omega(t).$$

Thus, we have

$$\begin{aligned} dV(x, \alpha) &= \mathcal{L}V(x, \alpha)dt + 2x^T(t)P(\alpha)F(\alpha)x(t)d\varpi(t) \\ &< -\beta V(x, \alpha)dt + \Psi(t)dt + 2x^T(t)P(\alpha)F(\alpha)x(t)d\varpi(t). \end{aligned}$$

Observe that

$$\begin{aligned} d\left[e^{\beta t}V(x, \alpha)\right] &= \beta e^{\beta t}V(x, \alpha)dt + e^{\beta t}dV(x, \alpha) \\ &< e^{\beta t}\Psi(t)dt + 2e^{\beta t}x^T(t)P(\alpha)F(\alpha)x(t)d\varpi(t). \end{aligned} \tag{12.10}$$

Integrating both sides of (12.10) from $t^\star > 0$ to t and then taking expectations results in

$$\mathbf{E}\left\{e^{\beta t}V(x,\alpha)\right\} - \mathbf{E}\left\{e^{\beta t^\star}V(x(t^\star),\alpha(t^\star))\right\} < \mathbf{E}\left\{\int_{t^\star}^{t} e^{\beta\tau}\Psi(\tau)d\tau\right\},$$

which is equivalent to

$$\mathbf{E}\left\{V(x,\alpha)\right\} < e^{-\beta(t-t^\star)}\mathbf{E}\left\{V(x(t^\star),\alpha(t^\star))\right\} + \mathbf{E}\left\{\int_{t^\star}^{t} e^{-\beta(t-\tau)}\Psi(s)d\tau\right\}. \quad (12.11)$$

Now, for an arbitrary piecewise constant switching signal α, and for any $t > 0$, we let $0 = t_0 < t_1 < \cdots < t_k < \cdots$ $(k = 0, 1, \ldots)$ denote the switching points of α over the interval $(0, t)$. As mentioned earlier, the i_kth subsystem is activated when $t \in [t_k, t_{k+1})$.

According to (12.11) and letting $t^\star = t_k$, we have

$$\mathbf{E}\left\{V(x,\alpha)\right\} < e^{-\beta(t-t_k)}\mathbf{E}\left\{V(x(t_k),\alpha(t_k))\right\} + \mathbf{E}\left\{\int_{t_k}^{t} e^{-\beta(t-\tau)}\Psi(\tau)d\tau\right\}. \quad (12.12)$$

Using (12.7) and (12.8), at switching instant t_k, we have

$$\mathbf{E}\left\{V(x(t_k),\alpha(t_k))\right\} \le \mu\mathbf{E}\left\{V(x(t_k^-),\alpha(t_k^-))\right\}. \quad (12.13)$$

Therefore, it follows from (12.12)–(12.13) and the relationship $\vartheta = N_\alpha(0,t) \le (t-0)/T_a$ that

$$\begin{aligned}
\mathbf{E}\left\{V(x,\alpha)\right\} &< \mu e^{-\beta(t-t_k)}\mathbf{E}\left\{V(x(t_k^-),\alpha(t_k^-))\right\} + \mathbf{E}\left\{\int_{t_k}^{t} e^{-\beta(t-\tau)}\Psi(\tau)d\tau\right\} \\
&\le \mu^\vartheta e^{-\beta t}\mathbf{E}\left\{V(x(0),\alpha(0))\right\} + \mu^\vartheta\mathbf{E}\left\{\int_{0}^{t_1} e^{-\beta(t-\tau)}\Psi(\tau)d\tau\right\} \\
&\quad + \mu^{\vartheta-1}\mathbf{E}\left\{\int_{t_1}^{t_2} e^{-\beta(t-\tau)}\Psi(\tau)d\tau\right\} + \cdots \\
&\quad + \mu^0\mathbf{E}\left\{\int_{t_k}^{t} e^{-\beta(t-\tau)}\Psi(\tau)d\tau\right\} \\
&= \mathbf{E}\left\{\int_{0}^{t} e^{-\beta(t-\tau)+N_\alpha(\tau,t)\ln\mu}\Psi(\tau)d\tau\right\} \\
&\quad + e^{-\beta t+N_\alpha(0,t)\ln\mu}V(x(0),\alpha(0)).
\end{aligned} \quad (12.14)$$

Under the zero initial condition, that is, $x(0) = 0$, (12.14) implies

$$\mathbf{E}\left\{V(x,\alpha)\right\} \le \mathbf{E}\left\{\int_{0}^{t} e^{-\beta(t-\tau)+N_\alpha(\tau,t)\ln\mu}\Psi(\tau)d\tau\right\}. \quad (12.15)$$

Multiplying both sides of (12.15) by $e^{-N_\alpha(0,t)\ln\mu}$ yields

$$\begin{aligned}\mathbf{E}\left\{e^{-N_\alpha(0,t)\ln\mu}V(x,\alpha)\right\} &\leq \mathbf{E}\left\{\int_0^t e^{-\beta(t-\tau)-N_\alpha(0,\tau)\ln\mu}\Psi(\tau)d\tau\right\} \\ &\leq \mathbf{E}\left\{\int_0^t e^{-\beta(t-\tau)}\Psi(\tau)d\tau\right\} \\ &\leq \mathbf{E}\left\{\int_0^t \Psi(\tau)d\tau\right\}. \end{aligned} \quad (12.16)$$

Noting $N_\alpha(0,t) \leq t/T_a$ and $T_a > T_a^* = \ln\mu/\beta$, we have $N_\alpha(0,t)\ln\mu \leq \beta t$. Thus, (12.16) implies

$$\mathbf{E}\left\{e^{-\beta t}V(x,\alpha)\right\} \leq \mathbf{E}\left\{\int_0^t \Psi(\tau)d\tau\right\}.$$

It is true that for arbitrary $t^* \geq 0$,

$$0 \leq \mathbf{E}\left\{e^{-\beta t^*}V(x,\alpha)\right\} \leq \mathbf{E}\left\{\int_0^{t^*} \Psi(t)dt\right\},$$

which satisfies (12.5). Hence, the proof is completed. ■

From the proof of Theorem 12.3.1, we also have the following result.

Theorem 12.3.2 *Given a scalar $\beta > 0$, suppose that there exist matrices $P(i) > 0$ such that for $i \in \mathcal{N}$,*

$$\Pi_{11}(i) \triangleq P(i)A(i) + A^T(i)P(i) + \beta P(i) + F^T(i)P(i)F(i) < 0. \quad (12.17)$$

Then the switched stochastic hybrid system (12.2a)–(12.2b) with $\omega(t) = 0$ is mean-square exponentially stable for any switching signal with the average dwell time satisfying $T_a > \frac{\ln\mu}{\beta}$ (where $\mu \geq 1$) and satisfying (12.7). Moreover, an estimate of the mean square of the state decay is given by

$$\mathbf{E}\{\|x(t)\|\} \leq \eta\,\|x(0)\|\,e^{-\lambda t}, \quad (12.18)$$

where

$$\begin{cases}\lambda = \dfrac{1}{2}\left(\beta - \dfrac{\ln\mu}{T_a}\right) > 0, \quad \eta = \sqrt{\dfrac{b}{a}} \geq 1, \\ a = \min\limits_{\forall i\in\mathcal{N}} \lambda_{\min}(P(i)), \quad b = \max\limits_{\forall i\in\mathcal{N}} \lambda_{\max}(P(i)). \end{cases} \quad (12.19)$$

Proof. Choose the Lyapunov function as (12.8). Inequality (12.17) implies

$$P(\alpha)A(\alpha) + A^T(\alpha)P(\alpha) + F^T(\alpha)P(\alpha)F(\alpha) < -\beta P(\alpha).$$

Considering (12.9) for $\omega(t) = 0$, we have

$$\mathcal{L}V(x,\alpha) < -\beta x^T(t)P(\alpha)x(t) = -\beta V(x,\alpha).$$

Thus,

$$dV(x,\alpha) < -\beta V(x,\alpha)dt + 2x^T(t)P(\alpha)F(\alpha)x(t)d\varpi(t).$$

Observe that

$$\begin{aligned} d\left[e^{\beta t}V(x,\alpha)\right] &= \beta e^{\beta t}V(x,\alpha)dt + e^{\beta t}dV(x,\alpha) \\ &< e^{\beta t}\left[\beta V(x,\alpha)dt - \beta V(x,\alpha)dt + 2x^T(t)P(\alpha)F(\alpha)x(t)d\varpi(t)\right] \\ &= 2e^{\beta t}x^T(t)P(\alpha)F(\alpha)x(t)d\varpi(t). \end{aligned} \tag{12.20}$$

Integrating both sides of (12.20) from $T > 0$ to t and taking expectations, with some mathematical operations, we have

$$\mathbf{E}\left\{V(x,\alpha)\right\} < e^{-\beta(t-T)}\mathbf{E}\left\{V(x(T),\alpha(T))\right\}. \tag{12.21}$$

As the analysis made in the proof of Theorem 12.3.1, we let $0 = t_0 < t_1 < \cdots < t_k < \cdots$ $(k = 1, 2, \ldots)$ denote the switching points of α over the interval $(0, t)$, and suppose that the i_kth subsystem is activated when $t \in [t_k, t_{k+1})$. Letting $T = t_k$ in (12.21) gives

$$\mathbf{E}\left\{V(x,\alpha)\right\} < e^{-\beta(t-t_k)}\mathbf{E}\left\{V(x(t_k),\alpha(t_k))\right\}. \tag{12.22}$$

Therefore, it follows from (12.13) and (12.22), and the relationship $\vartheta = N_\alpha(0,t) \leq (t-0)/T_a$ that

$$\begin{aligned} \mathbf{E}\left\{V(x,\alpha)\right\} &< e^{-\beta(t-t_k)}\mu\mathbf{E}\left\{V(x(t_k^-),\alpha(t_k^-))\right\} \\ &\leq \cdots \\ &\leq e^{-\beta(t-0)}\mu^{\vartheta}\mathbf{E}\left\{V(x(0),\alpha(0))\right\} \\ &\leq e^{-(\beta-\ln\mu/T_a)t}\mathbf{E}\left\{V(x(0),\alpha(0))\right\} \\ &= e^{-(\beta-\ln\mu/T_a)t}V(x(0),\alpha(0)). \end{aligned} \tag{12.23}$$

Note from (12.8) that

$$\mathbf{E}\left\{V(x,\alpha)\right\} \geq a\mathbf{E}\left\{\|x(t)\|^2\right\}, \quad V(x(0),\alpha(0)) \leq b\,\|x(0)\|^2, \tag{12.24}$$

where a and b are defined in (12.19). Combining (12.23) and (12.24) together yields

$$\begin{aligned}\mathbf{E}\left\{\|x(t)\|^2\right\} &\leq \frac{1}{a}\mathbf{E}\left\{V(x,\alpha)\right\}\\ &\leq \frac{b}{a}e^{-(\beta-\ln\mu/T_a)t}\|x(0)\|^2,\end{aligned}$$

which implies (12.18). By Definition 12.2.1 with $t_0 = 0$, system (12.2a)–(12.2b) with $u(t) = 0$ is mean-square exponentially stable. ■

12.4 Sliding Mode Control

12.4.1 Sliding Mode Dynamics

We design the following integral switching function:

$$s(t) = G(i)x(t) - \int_0^t G(i)\left[A(i) + B(i)K(i)\right]x(\theta)d\theta, \tag{12.25}$$

where $K(i) \in \mathbf{R}^{m\times n}$ are real matrices to be designed, and matrices $G(i)$ are to be chosen such that $G(i)B(i)$ are nonsingular and $G(i)F(i) = 0$ for $i \in \mathcal{N}$.

The solution of $x(t)$ can be given by

$$x(t) = x(0) + \int_0^t \left[A(i)x(\theta) + B(i)\left(u(\theta) + f(x(\theta),\theta,i)\right) + E(i)\omega(\theta)\right]d\theta + \int_0^t F(i)x(\theta)d\varpi(\theta). \tag{12.26}$$

It follows from (12.25) and (12.26) that

$$s(t) = G(i)x(0) + \int_0^t G(i)\left[-B(i)K(i)x(\theta) + B(i)\left(u(\theta) + f(x(\theta),\theta,i)\right) + E(i)\omega(\theta)\right]d\theta.$$

It is well known that when the system state trajectories reach onto the sliding surface, it follows that $s(t) = 0$ and $\dot{s}(t) = 0$. Therefore, by $\dot{s}(t) = 0$ we get the equivalent control as

$$u_{eq}(t) = K(i)x(t) - f(x(t),t,i) - \left(G(i)B(i)\right)^{-1}G(i)E(i)\omega(t). \tag{12.27}$$

By substituting (12.27) into (12.1a), the sliding mode dynamics can be obtained as

$$dx(t) = \left\{\left[A(i) + B(i)K(i)\right]x(t) + \left[I - B(i)\left(G(i)B(i)\right)^{-1}G(i)\right]E(i)\omega(t)\right\}dt + F(i)x(t)d\varpi(t). \tag{12.28}$$

For notational simplicity, we define

$$\begin{cases} \tilde{E}(i) \triangleq \left[I - B(i)(G(i)B(i))^{-1} G(i)\right] E(i), \\ \tilde{A}(i) \triangleq A(i) + B(i)K(i). \end{cases}$$

Thus, the sliding mode dynamics in (12.28) combined with the controlled output equation in (12.1b) can be formulated as

$$dx(t) = \left[\tilde{A}(i)x(t) + \tilde{E}(i)\omega(t)\right] dt + F(i)x(t)d\varpi(t), \tag{12.29a}$$

$$z(t) = C(i)x(t) + D(i)\omega(t). \tag{12.29b}$$

In this chapter, we choose $G(i) = B^T(i)X(i)$, where $0 < X(i) \in \mathbf{R}^{n\times n}$ is to be designed later. Thus,

$$\tilde{E}(i) \triangleq \left[I - B(i)\left(B^T(i)X(i)B(i)\right)^{-1} B^T(i)X(i)\right] E(i). \tag{12.30}$$

Note that the positive definiteness of matrix $X(i)$ guarantees the nonsingularity of $G(i)B(i)$ due to Assumption 12.1.

The above analysis gives the first step of the SMC for the switched stochastic hybrid system (12.1a)–(12.1b). Specifically, we design an integral-type sliding surface as given in (12.25) so that the dynamics restricted to the sliding surface (i.e. the sliding mode dynamics) has the form of (12.29a)–(12.29b). The remaining problems to be addressed in this chapter can be stated as follows.

1. Dissipativity analysis. Given all the system matrices in (12.1a)–(12.1b), determine the matrices $G(i)$ and $K(i)$ in the switching function (12.25) such that the sliding mode dynamics in (12.29a)–(12.29b) is mean-square exponentially stable and strictly $(\mathcal{Z}, \mathcal{Y}, \mathcal{X})$-$\delta$-dissipative in the sense of Definitions 12.2.1 and 12.2.4, respectively.
2. SMC law design. Synthesize an SMC law to globally drive the system state trajectories onto the predefined sliding surface $s(t) = 0$ in a finite time and maintain them there for all subsequent time. This is the second step of SMC.

12.4.2 Sliding Mode Dynamics Analysis

First, we give the following result for the dissipativity of the sliding mode dynamics in (12.29a)–(12.29b).

Theorem 12.4.1 *Given matrices $0 \geq \mathcal{Z} \in \mathbf{R}^{q\times q}$, $\mathcal{X} \in \mathbf{R}^{p\times p}$, and $\mathcal{Y} \in \mathbf{R}^{q\times p}$, with $\mathcal{Z}$ and $\mathcal{X}$ being symmetric, and scalars $\beta > 0$, $\delta > 0$, suppose that there exist matrices $X(i) > 0$ such that for $i \in \mathcal{N}$,*

$$\tilde{\Pi}(i) \triangleq \begin{bmatrix} \tilde{\Pi}_{11}(i) & \tilde{\Pi}_{12}(i) & X(i)B(i) \\ \star & \tilde{\Pi}_{22}(i) & 0 \\ \star & \star & -B^T(i)X(i)B(i) \end{bmatrix} < 0, \tag{12.31}$$

where

$$\begin{cases} \tilde{\Pi}_{11}(i) \triangleq X(i)\tilde{A}(i) + \tilde{A}^T(i)X(i) + \beta X(i) + F^T(i)X(i)F(i) - C^T(i)\mathcal{Z}C(i), \\ \tilde{\Pi}_{12}(i) \triangleq X(i)E(i) - C^T(i)\mathcal{Y}^T - C^T(i)\mathcal{Z}D(i), \\ \tilde{\Pi}_{22}(i) \triangleq -\mathcal{X} + \delta I - \mathcal{Y}D(i) - D^T(i)\mathcal{Y}^T + E^T(i)X(i)E(i) - D^T(i)\mathcal{Z}D(i). \end{cases}$$

Then the sliding mode dynamics in (12.29a)–(12.29b) is mean-square exponentially stable and strictly $(\mathcal{Z}, \mathcal{Y}, \mathcal{X})$-$\delta$-dissipative for any switching signal with the average dwell time satisfying $T_a > \frac{\ln\mu}{\beta}$ (where $\mu \geq 1$) and satisfying

$$X(i) \leq \mu X(j), \quad \forall i,j \in \mathcal{N}.$$

Proof. The result can be obtained by employing the same techniques as used in the proof of Theorem 12.3.1 and noticing (12.30) and

$$\begin{aligned} &-2x^T(t)X(i)B(i)\left(B^T(i)X(i)B(i)\right)^{-1}B^T(i)X(i)E(i)\omega(t) \\ &\leq x^T(t)X(i)B(i)\left(B^T(i)X(i)B(i)\right)^{-1}B^T(i)X(i)x(t) + \omega^T(t)E^T(i)X(i)E(i)\omega(t). \end{aligned}$$

Thus, the detailed proof is omitted for brevity. ■

In the following, based on the result in Theorem 12.4.1, we are in a position to present a solution to the dissipativity synthesis problem for the sliding mode dynamics in (12.29a)–(12.29b).

Theorem 12.4.2 *Given matrices $0 \geq \mathcal{Z} \in \mathbf{R}^{q\times q}$, $\mathcal{X} \in \mathbf{R}^{p\times p}$, and $\mathcal{Y} \in \mathbf{R}^{q\times p}$, with $\mathcal{Z}$ and $\mathcal{X}$ being symmetric, and scalars $\beta > 0$, $\delta > 0$, suppose that there exist matrices $X(i) > 0$, $\mathscr{X}(i) > 0$, and $\mathscr{Y}(i)$ such that for $i \in \mathcal{N}$,*

$$\begin{bmatrix} \check{\Pi}_{11}(i) & \check{\Pi}_{12}(i) & B(i) & \mathscr{X}(i)F^T(i) & \mathscr{X}(i)C^T(i)\mathcal{Z} \\ \star & \tilde{\Pi}_{22}(i) & 0 & 0 & 0 \\ \star & \star & -B^T(i)X(i)B(i) & 0 & 0 \\ \star & \star & \star & -\mathscr{X}(i) & 0 \\ \star & \star & \star & \star & \mathcal{Z} \end{bmatrix} < 0, \tag{12.32a}$$

$$B^T(i)X(i)F(i) = 0, \tag{12.32b}$$

$$X(i)\mathscr{X}(i) = I, \tag{12.32c}$$

where $\tilde{\Pi}_{22}(i)$ is defined in Theorem 12.4.1 and

$$\begin{cases} \check{\Pi}_{11}(i) \triangleq A(i)\mathscr{X}(i) + \mathscr{X}(i)A^T(i) + B(i)\mathscr{Y}(i) + \mathscr{Y}^T(i)B^T(i) + \beta\mathscr{X}(i), \\ \check{\Pi}_{12}(i) \triangleq E(i) - \mathscr{X}(i)C^T(i)\mathcal{Y}^T - \mathscr{X}(i)C^T(i)\mathcal{Z}D(i). \end{cases}$$

Then the sliding mode dynamics in (12.29a)–(12.29b) is mean-square exponentially stable and strictly $(\mathcal{Z}, \mathcal{Y}, \mathcal{X})$-$\delta$-dissipative for any switching signal with the average dwell time satisfying $T_a > \frac{\ln \mu}{\beta}$ (where $\mu \geq 1$) and satisfying

$$X(i) \leq \mu X(j), \quad \forall i,j \in \mathcal{N}. \tag{12.33}$$

Moreover, if the above conditions are feasible, then the matrix variable $K(i)$ in (12.25) can be computed by

$$K(i) = \mathscr{Y}(i)\mathscr{X}^{-1}(i) = \mathscr{Y}(i)X(i). \tag{12.34}$$

Proof. Let $\mathscr{X}(i) \triangleq X^{-1}(i)$ and $\mathscr{Y}(i) \triangleq K(i)\mathscr{X}(i)$. Then by performing a congruence transformation on (12.31) with $\text{diag}\{\mathscr{X}(i), I, I\}$ and by Schur complement, the result can be obtained. ■

Remark 12.3 *Note that there exist two matrix equalities of (12.32b) and (12.32c) in Theorem 12.4.2, which cannot be solved directly by applying the LMI Toolbox in the Matlab environment. In the following, we will propose some algorithms to solve them. First, to solve (12.32b), for a scalar $\varepsilon > 0$, we consider the following matrix inequalities:*

$$\left(B^T(i)X(i)F(i)\right)^T \left(B^T(i)X(i)F(i)\right) \leq \varepsilon I, \quad i \in \mathcal{N}. \tag{12.35}$$

By Schur complement, (12.35) is equivalent to

$$\begin{bmatrix} -\varepsilon I & \left(B^T(i)X(i)F(i)\right)^T \\ \star & -I \end{bmatrix} \leq 0, \quad i \in \mathcal{N}. \tag{12.36}$$

Therefore, when $\varepsilon > 0$ is chosen as a sufficiently small scalar, matrix equality (12.32b) can be solved through LMI (12.36). Next, we use the CCL method to solve (12.32c) by formulating it into a sequential optimization problem subject to LMI constraints. ♦

Now, combining the methods for solving (12.32b)–(12.32c) together, we introduce the following minimization problem involving LMI conditions instead of the original nonconvex feasibility problem formulated in Theorem 12.4.2.

Problem SMA (Sliding mode analysis):

$$\min \text{trace}\left(\sum_{i \in \mathcal{N}} X(i)\mathscr{X}(i)\right)$$

subject to (12.32a), (12.33), (12.36) and

$$\begin{bmatrix} X(i) & I \\ I & \mathscr{X}(i) \end{bmatrix} \geq 0, \quad i \in \mathcal{N}. \tag{12.37}$$

Remark 12.4 *By CCL method [66], if the solution of the above minimization problem is Nn, that is,* min trace$(\sum_{i\in\mathcal{N}} X(i)\mathcal{X}(i)) = Nn$, *then the conditions in Theorem 12.4.2 are solvable. Although it is still not possible to always find the global optimal solution, the proposed minimization problem is easier to solve than the original nonconvex feasibility problem.* ♦

12.4.3 SMC Law Design

In this section, we will synthesize a discontinuous SMC law, by which the state trajectories of the switched stochastic hybrid system (12.1a)–(12.1b) can be driven onto the pre-specified sliding surface $s(t) = 0$ in a finite time and then are maintained there for all subsequent time.

Theorem 12.4.3 *Consider the switched stochastic hybrid system (12.1a)–(12.1b). Suppose that the switching function is designed as (12.25) with $K(i)$ being solved by (12.34), and $G(i)$ is chosen as $G(i) = B^T(i)X(i)$ with $X(i) > 0$ being solved in Theorem 12.4.2. Then the state trajectories of system (12.1a)–(12.1b) can be driven onto the sliding surface $s(t) = 0$ in a finite time by the following SMC law:*

$$u(t) = K(i)x(t) - \rho(t,i)\text{sign}\,(s(t)), \tag{12.38}$$

where

$$\rho(t,i) = \varrho + \phi(i)\,\|x(t)\| + \left\|\left(B^T(i)X(i)B(i)\right)^{-1} B^T(i)X(i)E(i)\right\| \|\omega(t)\|,$$

with ϱ being a positive constant.

Proof. Choose a Lyapunov function of the following form:

$$W(t) = \frac{1}{2}s^T(t)\left(B^T(i)X(i)B(i)\right)^{-1} s(t).$$

According to (12.27), we have

$$\begin{aligned}\dot{s}(t) &= -B^T(i)X(i)B(i)K(i)x(t) + B^T(i)X(i)E(i)\omega(t)\\ &\quad + B^T(i)X(i)B(i)u(t) + B^T(i)X(i)B(i)f(x(t),t,i).\end{aligned}$$

Thus, taking the derivative of $W(t)$ and considering the above equation, we have

$$\begin{aligned}\dot{W}(t) &= s^T(t)\left(B^T(i)X(i)B(i)\right)^{-1}\dot{s}(t)\\ &= -\,s^T(t)K(i)x(t) + s^T(t)\left(B^T(i)X(i)B(i)\right)^{-1} B^T(i)X(i)E(i)\omega(t)\\ &\quad + s^T(t)\left(u(t) + f(x(t),t,i)\right).\end{aligned} \tag{12.39}$$

Substituting (12.38) into (12.39) and noting that $\|s(t)\| \leq |s(t)|$, we have

$$\dot{W}(t) \leq -\varrho \|s(t)\| \leq -\tilde{\varrho} W^{\frac{1}{2}}(t), \tag{12.40}$$

where

$$\tilde{\varrho} \triangleq \varrho \sqrt{2/\lambda_{\max}\left[\left(B^T(i)X(i)B(i)\right)^{-1}\right]} > 0.$$

It can be shown from (12.40) that there exists an instant $t^* = 2\sqrt{W(0)}/\tilde{\varrho}$ such that $W(t) = 0$ (equivalently, $s(t) = 0$) when $t \geq t^*$. Thus, we can say that the system state trajectories can be driven onto the predefined sliding surface in a finite time. ■

12.5 Illustrative Example

Example 12.5.1 Consider system (12.1a)–(12.1b) with $N = 2$ and the following parameters

$$A(1) = \begin{bmatrix} -0.7 & 0.6 & -1.9 \\ 2.2 & -0.5 & -0.8 \\ 0.1 & 1.7 & 0.9 \end{bmatrix}, \; F(1) = \begin{bmatrix} 0.3 & 0.1 & 0.1 \\ 0.1 & 0.3 & 0.3 \\ 0.2 & 0.1 & 0.1 \end{bmatrix}, \; B(1) = \begin{bmatrix} 1.2 \\ 0.8 \\ 0.5 \end{bmatrix},$$

$$A(2) = \begin{bmatrix} 1.0 & -0.6 & 1.0 \\ -0.4 & 0.7 & -0.6 \\ 0.3 & 0.3 & -0.8 \end{bmatrix}, \; F(2) = \begin{bmatrix} 0.2 & 0.1 & 0.2 \\ 0.1 & 0.3 & 0.1 \\ 0.2 & 0.2 & 0.2 \end{bmatrix}, \; B(2) = \begin{bmatrix} 0.5 \\ 1.2 \\ 0.8 \end{bmatrix},$$

$$E(1) = \begin{bmatrix} 0.2 \\ 0.1 \\ 0.2 \end{bmatrix}, \; E(2) = \begin{bmatrix} 0.3 \\ 0.2 \\ 0.2 \end{bmatrix}, \; D(1) = 0.2, \; D(2) = 0.4,$$

$$C(1) = [1.5 \quad 1.0 \quad 0.8], \; C(2) = [0.5 \quad 1.2 \quad 0.6].$$

Suppose $\beta = 0.5$ and

$$f(x(t), t, 1) = f(x(t), t, 2) = 0.5 \exp(-t) \sin\left(\sqrt{x_1^2(t) + x_2^2(t) + x_3^2(t)}\right),$$

(thus $\phi(1)$ and $\phi(2)$ can be chosen as $\phi(1) = \phi(2) = 0.5$), and the exogenous input $\omega(t)$ is given by $\omega(t) = 1/(1 + t^2)$.

It is found that the system in (12.1a)–(12.1b) with $u(t) = 0$ and $\omega(t) = 0$ and above parametric matrices is unstable for a switching signal given in Figure 12.1 (which is generated randomly; here, '1' and '2' represent the first and second subsystems, respectively). Therefore, our

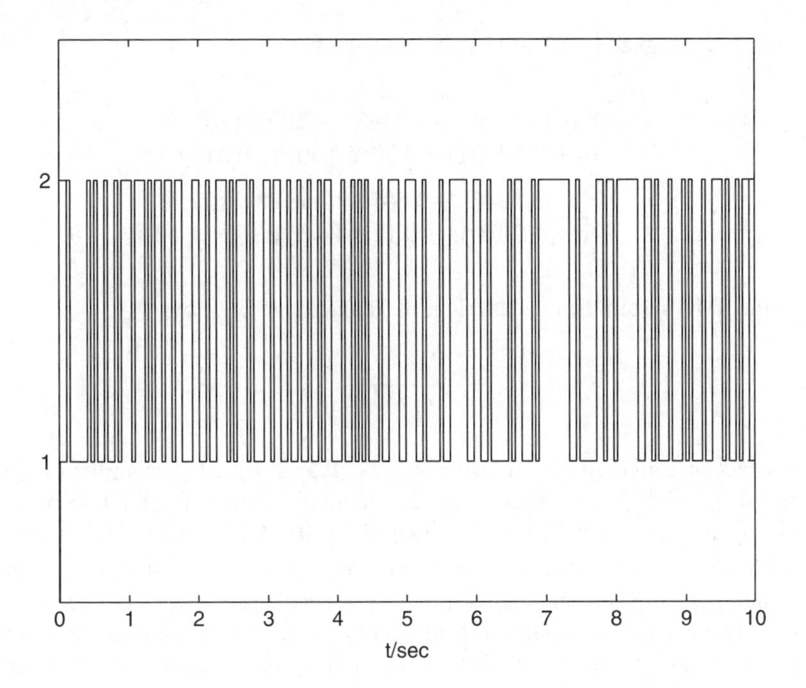

Figure 12.1 Switching signal

aim is to design the SMC law $u(t)$ in (12.38) such that the resulting closed-loop system is mean-square exponentially stable and strictly $(\mathcal{Z}, \mathcal{Y}, \mathcal{X})$-$\delta$-dissipative for $T_a > T_a^* = 0.1$ (in this case, the allowable minimum of μ is $\mu_{\min} = 1.0513$). Firstly, we need to check the stability and the strict $(\mathcal{Z}, \mathcal{Y}, \mathcal{X})$-$\delta$-dissipativity of the sliding mode dynamics in (12.29a)–(12.29b) with $T_a > T_a^* = 0.1$ (that is, set $\mu = 1.0513$). To this end, we choose $\mathcal{Z}$, $\mathcal{Y}$, $\mathcal{X}$, and δ as $\mathcal{Z} = -1.0$, $\mathcal{Y} = 1.5$, $\mathcal{X} = 1.7$, and $\delta = 0.1$, respectively. Solve the conditions (12.32a)–(12.33) in Theorem 12.4.2 according to Remark 12.3 and applying the CCL method, and by (12.34), we have

$$K(1) = 10^3 \times [-2.1557 \quad 1.8405 \quad -2.0972],$$
$$K(2) = 10^3 \times [3.6247 \quad -3.8879 \quad -6.1889].$$

Thus, the switching function defined in (12.25) is given by

$$s(t) = \begin{cases} s(t,1) = & [2.5025 \quad 0.5052 \quad -4.0126]\,x(t) \\ & -\int_0^t 10^3 \times [-4.9132 \quad -5.0005 \quad -11.3543]\,x(\theta)d\theta, \quad i = 1, \\ s(t,2) = & [-1.5017 \quad 1.5107 \quad 2.5025]\,x(t) \\ & -\int_0^t 10^3 \times [10.3551 \quad -11.3575 \quad -22.6221]\,x(\theta)d\theta, \quad i = 2, \end{cases}$$

and the SMC law designed in (12.38) can be computed as

$$u(t) = \begin{cases} u(t,1) = & 10^3 \times [-2.1557 \quad 1.8405 \quad -2.0972]\, x(t) \\ & -(\varrho + 0.5\,\|x(t)\| + 0.3083\,\|\omega(t)\|)\,\mathrm{sign}\,(s(t,1)), \qquad i = 1, \\ u(t,2) = & 10^3 \times [3.6247 \quad -3.8879 \quad -6.1889]\, x(t) \\ & -(\varrho + 0.5\,\|x(t)\| + 0.1216\,\|\omega(t)\|)\,\mathrm{sign}\,(s(t,2)), \qquad i = 2. \end{cases}$$

To prevent the SMC system from chattering, we replace sign $(s(t))$ by

$$\frac{s(t)}{\|s(t)\| + 0.01}.$$

Set $\varrho = 0.5$ and the initial condition $x(0) = [-1.0 \quad 0.5 \quad 1.0]^T$. By using the discretization approach [96], we simulate standard Brownian motion. Some initial parameters are given as follows: the simulation time $t \in [0, T^*]$ with $T^* = 10$, the normally distributed variance $\delta t = \frac{T^*}{N^*}$ with $N^* = 2^{11}$, the step size $\Delta t = \rho \delta t$ with $\rho = 2$, and the number of discretized Brownian paths $p = 10$. The simulation results are presented in Figures 12.2–12.5. Specifically, Figures 12.2–12.3 display the simulation results along an individual discretized Brownian path, with Figure 12.2 showing the states of the closed-loop system, and the switching function is given in Figure 12.3. Figures 12.4–12.5 are the simulation results on $x(t)$ and $s(t)$ along 10 individual paths (dotted lines) and the average over 10 paths (solid line), respectively.

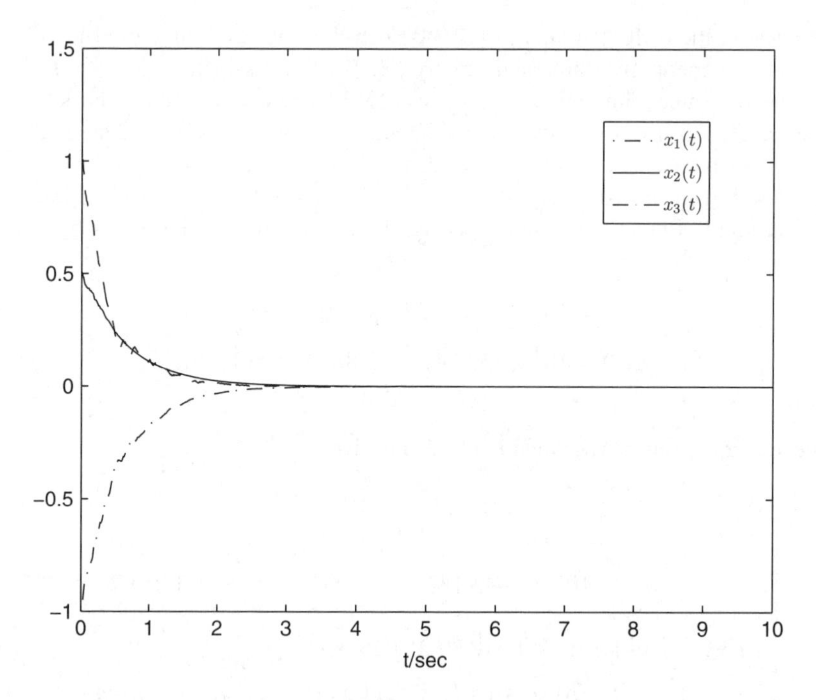

Figure 12.2 States of the closed-loop system

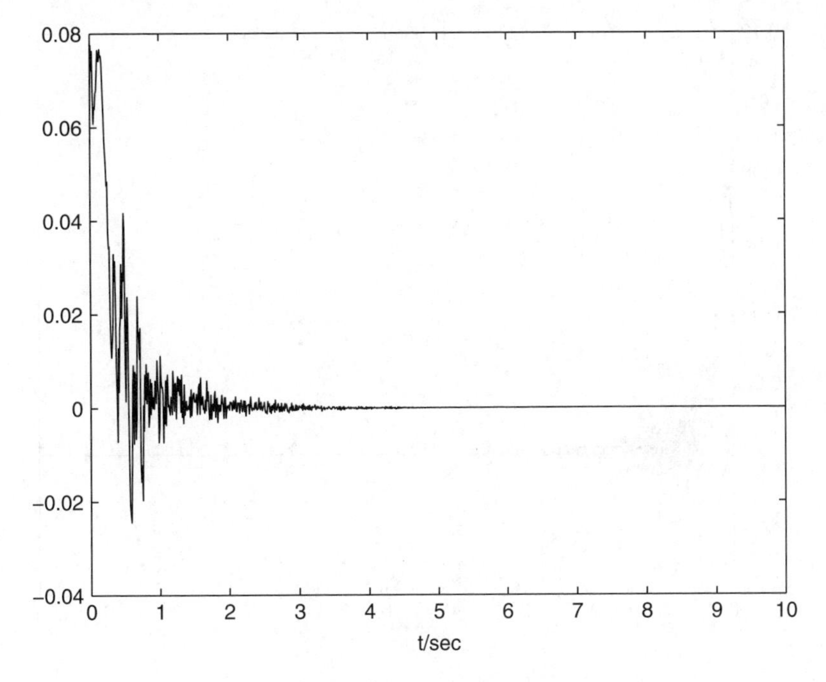

Figure 12.3 Switching function

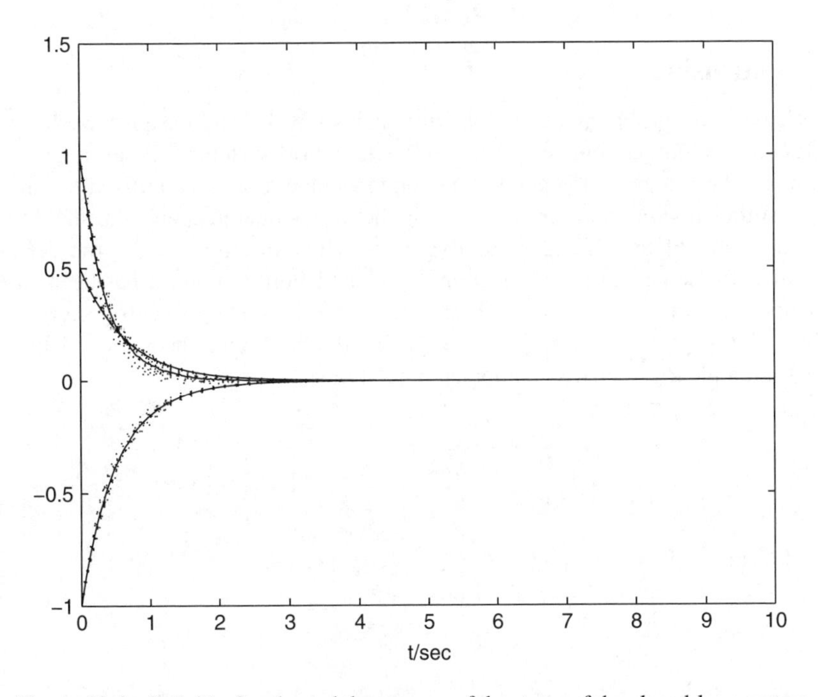

Figure 12.4 Individual paths and the average of the states of the closed-loop system

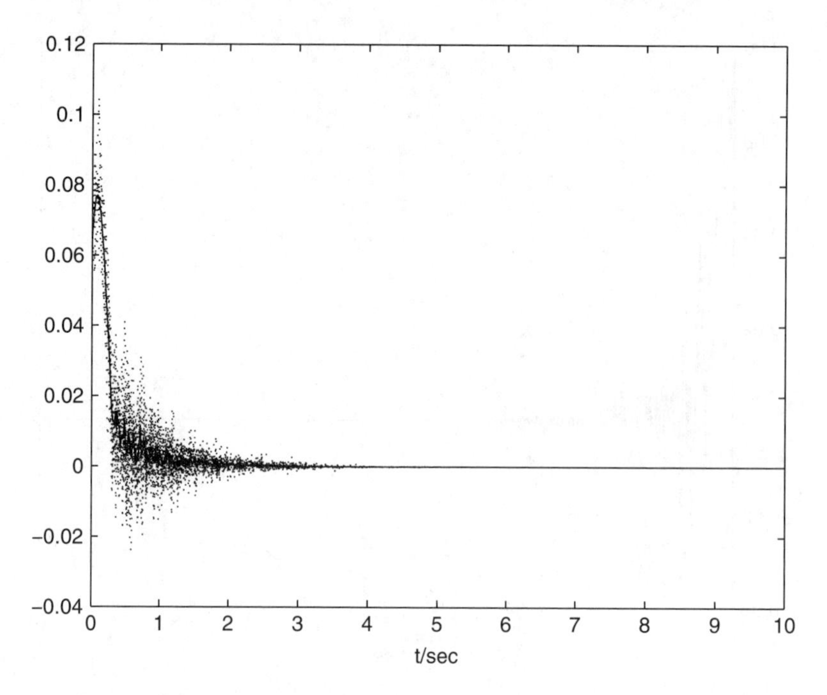

Figure 12.5 Individual paths and the average of the switching function

12.6 Conclusion

In this chapter, the problems of dissipativity analysis and SMC design have been studied for a class of continuous-time switched stochastic hybrid systems. The average dwell time approach and the piecewise Lyapunov function technique have been utilized to establish the LMI-type sufficient condition for guaranteeing the mean-square exponential stability and the strict dissipativity of the sliding mode dynamics. This was followed by the derivation of the condition for achieving the dissipativity synthesis. Furthermore, it has been shown that a discontinuous SMC law can be synthesized to drive the system state trajectories onto the predefined sliding surface in a finite time. Finally, the developed theory was validated by a numerical example together with computer simulations.

References

[1] K. Abidi, J. Xu, and X. Yu, On the discrete-time integral sliding-mode control, *IEEE Transactions on Automatic Control*, Vol. 52, No. 4, pp. 709–715, 2007.
[2] V. Acary, B. Brogliato, and Y. V. Orlov, Chattering-free digital sliding-mode control with state observer and disturbance rejection, *IEEE Transactions on Automatic Control*, Vol. 57, No. 5, pp. 1087–1101, 2012.
[3] A. Alessandri, M. Baglietto, and G. Battistelli, A maximum-likelihood Kalman filter for switching discrete-time linear systems, *Automatica*, Vol. 46, No. 11, 1870–1876, 2010.
[4] L. I. Allerhand and U. Shaked, Robust stability and stabilization of linear switched systems with dwell time, *IEEE Transactions on Automatic Control*, Vol. 56, No. 2, pp. 381–386, 2011.
[5] A. C. Antoulas, *Approximation of Large-Scale Dynamic Systems*, SIAM Press, 2005.
[6] Z. Artstein and J. Ronen, On stabilization of switched linear systems, *Systems and Control Letters*, Vol. 57, No. 11, pp. 919–926, 2008.
[7] G. A. Jr. Baker, P. Graves-Morris, *Padé Approximants*, New York, Cambridge University Press, 1996.
[8] S. Baromand and B. Labibi, Covariance control for stochastic uncertain multivariable systems via sliding mode control strategy, *IET Control Theory and Applications*, Vol. 6, No. 3, pp. 349–356, 2012.
[9] G. Bartolini and P. Pydynowski, An improved, chattering free, VSC scheme for uncertain dynamical systems, *IEEE Transactions on Automatic Control*, Vol. 41, No. 8, pp. 1220–1226, 1996.
[10] G. Bartolini, A. Ferrara, and E. Usai, Chattering avoidance by second-order sliding mode control, *IEEE Transactions on Automatic Control*, Vol. 43, No. 2, pp. 241–246, 1998.
[11] A. Bartoszewicz, Chattering attenuation in sliding mode control systems, *Control and Cybernetics*, Vol. 29, No. 2, pp. 585–594, 2000.
[12] M. Basin, J. Rodriguez-Gonzalez, L. Fridman, and P. Acosta, Integral sliding mode design for robust filtering and control of linear stochastic time-delay systems, *International Journal of Robust and Nonlinear Control*, Vol. 15, No. 9, pp. 407–421, 2005.
[13] M. Basin, A. Ferreira, and L. Fridman, Sliding mode identification and control for linear uncertain stochastic systems, *International Journal of Systems Science*, Vol. 38, No. 11, pp. 861–869, 2007.
[14] G. Battistelli, On stabilization of switching linear systems, *Automatica*, Vol. 49, No. 5, pp. 1162–1173, 2013.
[15] D. E. C. Belkhiat, N. Messai, and N. Manamanni, Design of a robust fault detection based observer for linear switched systems with external disturbances, *Nonlinear Analysis – Hybrid Systems*, Vol. 5, No. 2, pp. 206–219, 2011.
[16] A. Bemporad, G. Ferrari-Trecate, and M. Morari, Observability and controllability of piecewise affine and hybrid systems, *IEEE Transactions on Automatic Control*, Vol. 45, No. 10, pp. 1864–1876, 2000.
[17] S. C. Bengea and R. A. DeCarlo, Optimal control of switching systems, *Automatica*, Vol. 41, No. 1, pp. 11–27, 2005.
[18] A. Birouche, B. Mourllion, and M. Basset, Model order-reduction for discrete-time switched linear systems, *International Journal of Systems Science*, Vol. 43, No. 9, pp. 1753–1763, 2012.
[19] I. Boiko and L. Fridman, Analysis of chattering in continuous sliding-mode controllers, *IEEE Transactions on Automatic Control*, Vol. 50, No. 9, pp. 1442–1446, 2005.

Sliding Mode Control of Uncertain Parameter-Switching Hybrid Systems, First Edition. Ligang Wu, Peng Shi and Xiaojie Su.

[20] E.-K. Boukas and H. Yang, Exponential stabilizability of stochastic systems with Markovian jumping parameters, *Automatica*, Vol. 35, No. 8, pp. 1437–1441, 1999.

[21] E.-K. Boukas, P. Shi, and S. K. Nguang, Robust $\mathcal{H}_\infty$ control for linear Markovian jump systems with unknown nonlinearities, *Journal of Mathematical Analysis and Applications*, Vol. 282, No. 1, pp. 241–255, 2003.

[22] E.-K. Boukas, *Stochastic Hybrid Systems: Analysis and Design*. Boston: Birkhauser, 2005.

[23] E.-K. Boukas, Stabilization of stochastic singular nonlinear hybrid systems, *Nonlinear Analysis*, Vol. 64, No. 2, pp. 217–228, 2006.

[24] E.-K. Boukas, $\mathcal{H}_\infty$ control of discrete-time Markov jump systems with bounded transition probabilities, *Optimal Control Applications and Methods*, Vol. 30, No. 5, pp. 477–494, 2009.

[25] S. Boyd, L. El Ghaoui, E. Feron, and V. Balakrishnan, *Linear Matrix Inequalities in Systems and Control Theory*, Philadelphia, PA, SIAM, 1994.

[26] M. Branicky, Multiple Lyapunov functions and other analysis tools for switched and hybrid systems, *IEEE Transactions on Automatic Control*, Vol. 43, No. 4, pp. 475–482, 1998.

[27] W. Cao and J. Xu, Nonlinear integral-type sliding surface for both matched and unmatched uncertain systems, *IEEE Transactions on Automatic Control*, Vol. 49, No. 8, pp. 1355–1360, 2004.

[28] Y. Cao and J. Lam, Stochastic stabilizability and $\mathcal{H}_\infty$ control for discrete-time jump linear systems with time delay, *Journal of The Franklin Institute*, Vol. 336, No. 8, pp. 1263–1281, 1999.

[29] Y. Cao and J. Lam, Robust $\mathcal{H}_\infty$ control of uncertain Markovian jump systems with time delay, *IEEE Transactions on Automatic Control*, Vol. 45, No. 1, pp. 77–83, 2000.

[30] F. Castaños and L. Fridman, Analysis and design of integral sliding manifolds for systems with unmatched perturbations, *IEEE Transactions on Automatic Control*, Vol. 51, No. 5, pp. 853–858, 2006.

[31] M. Chan, C. Tao, and T. Lee, Sliding mode controller for linear systems with mismatched time-varying uncertainties, *Journal of The Franklin Institute*, Vol. 337, No. 2, pp. 105–115, 2000.

[32] J.-L. Chang, Dynamic output integral sliding-mode control with disturbance attenuation, *IEEE Transactions on Automatic Control*, Vol. 54, No. 11, pp. 2653–2658, 2009.

[33] K.-Y. Chang and W.-J. Wang, Robust covariance control for perturbed stochastic multivariable system via variable structure control, *Systems and Control Letters*, Vol. 37, No. 5, pp. 323–328, 1999.

[34] B. Chen, J. Huang, and Y. Niu, Sliding mode control for Markovian jumping systems with actuator nonlinearities, *International Journal of Systems Science*, Vol. 43, No. 4, pp. 656–664, 2012.

[35] B. Chen, Y. Niu, and Y. Zou, Adaptive sliding mode control for stochastic Markovian jumping systems with actuator degradation, *Automatica*, Vol. 49, No. 6, pp. 1748–1754, 2013.

[36] M. Chen, C. Chen, and F. Yang, An LTR-observer-based dynamic sliding mode control for chattering reduction, *Automatica*, Vol. 43, No. 6, pp. 1111–1116, 2007.

[37] D. Cheng, Controllability of switched bilinear systems, *IEEE Transactions on Automatic Control*, Vol. 50, No. 4, pp. 511–515, 2005.

[38] D. Cheng, L. Guo, Y. Lin, and Y. Wang, Stabilization of switched linear systems, *IEEE Transactions on Automatic Control*, Vol. 50, No. 5, pp. 661–666, 2005.

[39] C. S. Chiu, Derivative and integral terminal sliding mode control for a class of MIMO nonlinear systems, *Automatica*, Vol. 48, No. 2, pp. 316–326, 2012.

[40] H. H. Choi, On the existence of linear sliding surfaces for a class of uncertain dynamic systems with mismatched uncertainties, *Automatica*, Vol. 35, No. 10, pp. 1707–1715, 1999.

[41] H. H. Choi, Variable structure control of dynamical systems with mismatched norm-bounded uncertainties: an LMI approach, *International Journal of Control*, Vol. 74, No. 13, pp. 1324–1334, 2001.

[42] H. H. Choi, Variable structure output feedback control design for a class of uncertain dynamic systems, *Automatica*, Vol. 38, No. 2, pp. 335–341, 2002.

[43] H. H. Choi, LMI-based sliding surface design for integral sliding mode control of mismatched uncertain systems, *IEEE Transactions on Automatic Control*, Vol. 52, No. 4, pp. 736–742, 2007.

[44] J. C. Y. Chung and C. L. Lin, A transformed Lure problem for sliding mode control and chattering reduction, *IEEE Transactions on Automatic Control*, Vol. 44, No. 3, pp. 563–568, 1999.

[45] P. Colaneri, J. C. Geromel, and A. Astolfi, Stabilization of continuous-time switched nonlinear systems, *Systems and Control Letters*, Vol. 57, No. 1, pp. 95–103, 2008.

[46] J. Daafouz, P. Riedinger, and C. Iung, Stability analysis and control synthesis for switched systems: a switched Lyapunov function approach, *IEEE Transactions on Automatic Control*, Vol. 47, No. 11, pp. 1883–1887, 2002.

[47] J. Daafouz and J. Bernussou, Robust dynamic output feedback control for switched systems, in *Proceedings of the 41st IEEE Conference on Decision and Control*, Las Vegas, Nevada, USA, 10–13 December, 2002, pp. 4389–4394.

[48] J. M. Daly and D. W. L. Wang, Output feedback sliding mode control in the presence of unknown disturbances, *Systems and Control Letters*, Vol. 58, No. 3, pp. 188–193, 2009.

[49] M. R. Davoodi, A. Golabi, H. A. Talebi, and H.R. Momeni. Simultaneous fault detection and control design for switched linear systems based on dynamic observer, *Optimal Control Application and Methods*, Vol. 34, No. 1, pp. 35–52, 2013.

[50] G. S. Deaecto, J. C. Geromel, and J. Daafouz, Trajectory-dependent filter design for discrete-time switched linear systems, *Nonlinear Analysis – Hybrid Systems*, Vol. 4, No. 1, pp. 1–8, 2010.

[51] G. S. Deaecto, J. C. Geromel, and J. Daafouz, Dynamic output feedback $\mathcal{H}_\infty$ control of switched linear systems, *Automatica*, Vol. 47, No. 8, pp. 1713–1720, 2011.

[52] R. A. Decarlo, S. H. Zak, and G. P. Matthews, Variable structure control of nonlinear multivariable systems: a tutorial, *Proceedings of the IEEE*, Vol. 76, No. 3, pp. 212–232, 1988.

[53] R. A. Decarlo, M. S. Branicky, S. Pettersson, and B. Lennartson, Perspectives and results on the stability and stabilizability of hybrid systems, *Proceedings of The IEEE*, Vol. 88, No. 7, pp. 1069–1082, 2000.

[54] C. E. de Souza and M. D. Fragoso, $\mathcal{H}_\infty$ filtering for Markovian jump linear systems, *International Journal of System Sciences*, Vol. 33, No. 11, pp. 909–915, 2002.

[55] C. E. de Souza and M. D. Fragoso, $\mathcal{H}_\infty$ filtering for discrete-time linear systems with Markovian jumping parameters, *International Journal of Robust and Nonlinear Control*, Vol. 13, No. 14, pp. 1299–1316, 2003.

[56] C. E. de Souza, Robust stability and stabilization of uncertain discrete time Markovian jump linear systems, *IEEE Transactions on Automatic Control*, Vol. 51, No. 5, pp. 836–841, 2006.

[57] C. E. de Souza, A. Trofino, and K. A. Barbosa, Mode-independent $\mathcal{H}_\infty$ filters for Markovian jump linear systems, *IEEE Transactions on Automatic Control*, Vol. 51, No. 11, pp. 1837–1841, 2006.

[58] B. Du, J. Lam, Z. Shu, and Z. Wang, A delay-partitioning projection approach to stability analysis of continuous systems with multiple delay components, *IET Control Theory and Applications*, Vol. 3, No. 4, pp. 383–390, 2009.

[59] D. Du, B. Jiang, and P. Shi, Active fault-tolerant control for switched systems with time delay, *International Journal of Adaptive Control and Signal Processing*, Vol. 25, No. 5, pp. 466–480, 2011.

[60] C. Edwards and S. K. Spurgeon, *Sliding Mode Control: Theory and Applications*, Taylor and Francis, London, 1998.

[61] C. Edwards and S. K. Spurgeon, On the limitations of some variable structure output feedback controller designs, *Automatica*, Vol. 36, No. 5, pp. 743–748, 2000.

[62] C. Edwards, A. Akoachere, and S. K. Spurgeon, Sliding-mode output feedback controller design using linear matrix inequalities, *IEEE Transactions on Automatic Control*, Vol. 46, No. 1, pp. 115–119, 2001.

[63] C. Edwards and S. K. Spurgeon, Linear matrix inequality methods for designing sliding mode output feedback controllers, *IEE Proceedings – Control Theory and Applications*, Vol. 150, No. 5, pp. 539–545, 2003.

[64] C. Edwards, S. K. Spurgeon, and R. G. Hebden, On the design of sliding mode output feedback controllers, *International Journal of Control*, Vol. 76, No. 9-10, pp. 893–905, 2003.

[65] M. O. Efe, C. Ünsal, O. Kaynak, and X. Yu, Variable structure control of a class of uncertain systems, *Automatica*, Vol. 40, No. 1, pp. 59–64, 2004.

[66] L. El Ghaoui, F. Oustry, and M. AitRami, A cone complementarity linearization algorithm for static output-feedback and related problems, *IEEE Transactions on Automatic Control*, Vol. 42, No. 8, pp. 1171–1176, 1997.

[67] N. H. El-Farra, P. Mhaskar, and P. D. Christofides, Output feedback control of switched nonlinear systems using multiple Lyapunov functions, *Systems and Control Letters*, Vol. 54, No. 12, pp. 1163–1182, 2005.

[68] R. El-Khazal and R. Decarlo, Output feedback variable structure control design, *Automatica*, Vol. 31, No. 6, pp. 805–816, 1995.

[69] R. El-Khazali, Variable structure robust control of uncertain time-delay systems, *Automatica*, Vol. 34, No. 3, pp. 327–332, 1998.

[70] S. V. Emelyanov, *Variable Structure Control Systems*, Moscow: Nauka (in Russia), 1970.

[71] L. Fang, H. Lin, and P. J. Antsaklis, Stabilization and performance analysis for a class of switched systems, in *Proceedings of the 43rd IEEE Conference on Decision and Control*, Atlantis, Paradise Island, Bahamas, USA, December 14–17, 2004, pp. 3265–3270.

[72] Y. Feng, X. Yu, and F. Han, On nonsingular terminal sliding-mode control of nonlinear systems, *Automatica*, Vol. 49, No. 6, pp. 1715–1722, 2013.

[73] Z. Feng, J. Lam, H. Gao and B. Du, Improved stability and stabilization results for discrete singular delay systems via delay partitioning, in *Proceedings of the 48th Conference on Decision and Control and the 28th Chinese Control Conference*, Shanghai, China, December 16–18, 2009, pp. 7210–7215.

[74] E. Fridman, New Lyapunov–Krasovskii functionals for stability of linear retarded and neutral type systems, *Systems and Control Letters*, Vol. 43, No. 4, pp. 309–319, 2001.

[75] E. Fridman, F. Gouaisbaut, M. Dambrine, and J.-P. Richard, Sliding mode control of systems with time-varying delays via descriptor approach, *International Journal of Systems Science*, Vol. 34, No. 8-9, pp. 553–559, 2003.

[76] H. Gao, J. Lam, S. Xu, and C. Wang, Stabilization and $\mathcal{H}_\infty$ control of two-dimensional Markovian jump systems, *IMA Journal of Mathematical Control and Information*, Vol. 21, No. 4, pp. 377–392, 2004.

[77] W. Gao, *Fundamentals of Variable Structure Control Theory*, Beijing: Press of Science and Technology in China (in Chinese), 1990.

[78] W. Gao and J. C. Hung, Variable structiire control of nonlinear systems: a new approach, *IEEE Transactions on Industrial Electronics*, Vol. 40, No. 1, pp. 45–55, 1993.

[79] J. C. Geromel and P. Colaneri, Stability and stabilization of discrete time switched systems, *International Journal of Control*, Vol. 79, No. 7, pp. 719–728, 2006.

[80] J. C. Geromel, P. Colaneri, and P. Bolzern, Dynamic output feedback control of switched linear systems, *IEEE Transactions on Automatic Control*, Vol. 53, No. 3, pp. 720–733, 2008.

[81] J. C. Geromel, P. Colaneri, and P. Bolzern, Passivity of switched linear systems: analysis and control design, *Systems and Control Letters*, Vol. 61, No. 4, pp. 549–554, 2012.

[82] K. Glover, All optimal Hankel norm approximations of linear multivariable systems and their $\mathcal{L}_\infty$ error bounds, *International of Control*, Vol. AC-39, No. 6, pp. 1115–1193, 1984.

[83] G. Golo and Č. Milosavljević, Robust discrete-time chattering free sliding mode control, *Systems and Control Letters*, Vol. 41, No. 1, pp. 19–28, 2000.

[84] C. A. C. Gonzaga, M. Jungers, and J. Daafouz, Stability analysis and stabilisation of switched nonlinear systems, *International Journal of Control*, Vol. 85, No. 7, pp. 822–829, 2012.

[85] F. Gouaisbaut, M. Dambrine, and J.-P. Richard, Robust control of delay systems: a sliding mode control design via LMI, *Systems and Control Letters*, Vol. 46, No. 4, pp. 219–230, 2002.

[86] F. Gouaisbaut and D. Peaucelle, Delay-dependent stability analysis of linear time delay systems, in *IFAC Workshop on Time Delay System*, L'Aquila, Italy, 10–12 July, 2006.

[87] E. J. Grimme, *Krylov Projection Methods for Model Reduction*, PhD Thesis, ECE Dept, University of Illinois, Urbana-Champaign, 1997.

[88] K. Gu, V. Kharitonov, and J. Chen, *Stability of Time-Delay Systems*. Cambridge, MA, Birkhauser, 2003.

[89] S. Gugercin and A. C. Antoulas, A survey of model reduction by balanced truncation and some new results, *International of Control*, Vol. 77, No. 8, pp. 748–766, 2004.

[90] X. Han, E. Fridman, S. K. Spurgeon, and C. Edwards, On the design of sliding-mode static-output-feedback controllers for systems with state delay, *IEEE Transactions on Industrial Electronics*, Vol. 56, No. 9, pp. 3656–3664, 2009.

[91] X. Han, E. Fridman, and S. K. Spurgeon, Sliding mode control in the presence of input delay: a singular perturbation approach, *Automatica*, Vol. 48, No. 8, pp. 1904–1912, 2012.

[92] J. P. Hespanha and A. S. Morse, Stability of switched systems with average dwell-time, in *Proceedings of the 38th IEEE Conference on Decision and Control*, 1999, pp. 2655–2660, 1999.

[93] J. P. Hespanha, Uniform stability of switched linear systems: extensions of Lasalle's invariance principle, *IEEE Transactions on Automatic Control*, Vol. 49, No. 4, pp. 470–482, 2004.

[94] J. P. Hespanha, H. Unbehauen, Ed., Stabilization through hybrid control, in *Encyclopedia Life Support Systems (EOLSS)*, Vol. Control Systems, Robotics and Automation, Oxford, UK, 2004.

[95] L. Hetel, J. Daafouz, and C. Iung, Stabilization of arbitrary switched linear systems with unknown time-varying delays, *IEEE Transactions on Automatic Control*, Vol. 51, No. 10, pp. 1668–1674, 2006.

[96] D. J. Higham, An algorithmic introduction to numerical simulation of stochastic differential equations, *SIAM Review*, Vol. 43, No. 3, pp. 525–546, 2001.

[97] D. J. Hill and P. J. Moylan, Dissipative dynamical systems: basic input-output and state properties, *Journal of The Franklin Institute*, Vol. 309, No. 5, pp. 327–357, 1980.

[98] D. W. C. Ho and Y. Niu, Robust fuzzy design for nonlinear uncertain stochastic systems via sliding-mode control, *IEEE Transactions on Fuzzy Systems*, Vol. 15, No. 3, pp. 350–358, 2007.

[99] J. Hu, Z. Wang, H. Gao, and L. K. Stergioulas, Robust sliding mode control for discrete stochastic systems with mixed time delays, randomly occurring uncertainties, and randomly occurring nonlinearities, *IEEE Transactions on Industrial Electronics*, Vol. 59, No. 7, pp. 3008–3015, 2012.

[100] T. Hu, L. Ma, and Z. Lin, Stabilization of switched systems via composite quadratic functions, *IEEE Transactions on Automatic Control*, Vol. 53, No. 11, pp. 2571–2585, 2008.

[101] J. Y. Hung, W. Gao, and J. C. Hung, Variable structure control: a survey, *IEEE Transactions on Industrial Electronics*, Vol. 40, No. 1, pp. 2–22, 1993.

[102] H. Ishii and B. A. Francis, Stabilizing a linear system by switching control with dwell time, *IEEE Transactions on Automatic Control*, Vol. 47, No. 12, pp. 1962–1973, 2002.

[103] Y. Itkis, *Control Systems of Variable Structure*. New York: Wiley, 1976.

[104] Y. Ji and H. J. Chizeck, Controllability, stabilizability, and continuous-time Markovian jump linear quadratic control, *IEEE Transactions on Automatic Control*, Vol. 35, No. 7, pp. 777–788, 1990.

[105] Z. Ji, L. Wang, and X. Guo, On controllability of switched linear systems, *IEEE Transactions on Automatic Control*, Vol. 53, No. 3, pp. 796–801, 2008.

[106] M. Kamgarpour and C. Tomlin, On optimal control of non-autonomous switched systems with a fixed mode sequence, *Automatica*, Vol. 48, No. 6, pp. 1177–1181, 2012.

[107] A. Khalid, J.-X. Xu, and X. Yu, On the discrete-time integral sliding-mode control, *IEEE Transactions on Automatic Control*, Vol. 52, No. 4, pp. 709–715, 2007.

[108] K.-S. Kim, Y. Park, and S.-H. Oh, Designing robust sliding hyperplanes for parametric uncertain systems: a Riccati approach, *Automatica*, Vol. 36, No. 7, pp. 1041–1048, 2000.

[109] C. King and R. Shorten, A singularity test for the existence of common quadratic Lyapunov functions for pairs of stable LTI systems, in *Proceedings of the American Control Conference*, Boston, Massachusetts, USA, 30 June–2 July, 2004, pp. 3881–3884.

[110] V. B. Kolmanovskii, S.-I. Niculescu, and J.-P. Richard, On the Lyapunov–Krasovskii functionals for stability analysis of linear delay systems, *International Journal of Control*, Vol. 72, No. 4, pp. 374–384, 1999.

[111] G. Kotsalis, A. Megretski, and M. A. Dahleh, Model reduction of discrete-time Markov jump linear systems, in *Proceedings of the 2006 American Control Conference*, Minneapolis, Minnesota, USA, June 14-16, 2006, pp. 454–459.

[112] C.-M. Kwan, Sliding mode control of linear systems with mismatched uncertainties, *Automatica*, Vol. 31, No. 2, pp. 303–307, 1995.

[113] C.-M. Kwan, Further results on variable output feedback controllers, *IEEE Transactions on Automatic Control*, Vol. 46, No. 9, pp. 1505–1508, 2001.

[114] S. Laghrouche, F. Plestan, and A. Glumineau, Higher order sliding mode control based on integral sliding mode, *Automatica*, Vol. 43, No. 3, pp. 531–537, 2007.

[115] J. Lam, Z. Shu, S. Xu, and E.-K. Boukas, Robust $\mathcal{H}_\infty$ control of descriptor discrete-time Markovian jump systems, *International Journal of Control*, Vol. 80, No. 3, pp. 374–385, 2007.

[116] H. Lee, V. I. Utkin, and A. Malinin, Chattering reduction using multiphase sliding mode control, *International Journal of Control*, Vol. 82, No. 9, pp. 1720–1737, 2009.

[117] F. H. F. Leung, L. K. Wong, and P. K. S. Tam, Algorithm for eliminating chattering in sliding mode control *Electronics Letters*, Vol. 32, No. 6, pp. 599–601, 1996.

[118] A. Levant and L. Alelishvili, Integral high-order sliding modes, *IEEE Transactions on Automatic Control*, Vol. 52, No. 7, pp. 1278–1282, 2007.

[119] J. Li and G.-H. Yang, Simultaneous fault detection and control for switched systems under asynchronous switching, *Proceedings of the Institution of Mechanical Engineers, Part I: Journal of Systems and Control Engineering*, Vol. 227, No. I1, pp. 70–84, 2013.

[120] X. Li, *Sliding Mode Control of Uncertain Time Delay Systems*, PhD Thesis, Purdue University, 2002.

[121] X. Li and C. E. de Souza, Delay-dependent robust stability and stabilisation of uncertain linear delay systems: a linear matrix inequality approach, *IEEE Transactions on Automatic Control*, Vol. 42, No. 8, pp. 1144–1148, 1997.

[122] X. Li and C. E. de Souza, Criteria for robust stability and stabilization of uncertain linear systems with state delay, *Automatica*, Vol. 33, No. 9, pp. 1657–1662, 1997.

[123] X. Li and R. A. Decarlo, Robust sliding mode control of uncertain time delay systems, *International Journal of Control*, Vol. 76, No. 13, pp. 1296–1305, 2003.

[124] Z. Li, C. Wen, and Y. C. Soh, Stabilization of a class of switched systems via designing switching laws, *IEEE Transactions on Automatic Control*, Vol. 46, No. 4, pp. 665–670, 2001.

[125] J. Lian, J. Zhao, and G. M. Dimirovski, Integral sliding mode control for a class of uncertain switched nonlinear systems, *European Journal of Control*, Vol. 16, No. 1, pp. 16–22, 2010.
[126] J. Lian and Y. Ge, Robust $\mathcal{H}_\infty$ output tracking control for switched systems under asynchronous switching, *Nonlinear Analysis – Hybrid Systems*, Vol. 8, pp. 57–68, 2013.
[127] D. Liberzon, J. P. Hespanha, and A. S. Morse, Stability of switched linear systems: a Lie-algebraic condition, *System and Control Letters*, Vol. 37, No. 3, pp. 117–122, 1999.
[128] D. Liberzon and A. S. Morse, Basic problems in stability and design of switched systems, *IEEE Control Systems Magazine*, Vol. 19, No. 5, pp. 59–70, 1999.
[129] D. Liberzon, *Switching in Systems and Control*, Birkhauser, Boston, 2003.
[130] H. Lin and P. J. Antsaklis, Switching stabilizability for continuous-time uncertain switched linear systems, *IEEE Transactions on Automatic Control*, Vol. 52, No. 4, pp. 633–646, 2007.
[131] H. Lin and P. J. Antsaklis, Stability and stabilizability of switched linear systems: a survey of recent results, *IEEE Transactions on Automatic Control*, Vol. 54, No. 2, pp. 308–322, 2009.
[132] C. Liu, C. Li, and C. Li, Controllability and observability of switched linear systems with continuous-time and discrete-time subsystems, *IET Control Theory and Applications*, Vol. 6, No. 6, pp. 855–863, 2012.
[133] Q. Liu, W. Wang, and D. Wang, New results on model reduction for discrete-time switched systems with time delay, *International Journal of Innovative Computing Information and Control*, Vol. 8, No. 5A, pp. 3431–3440, 2012.
[134] S. Ma and E.-K. Boukas, Robust $\mathcal{H}_\infty$ filtering for uncertain discrete Markov jump singular systems with mode-dependent time-delay, *IET Control Theory and Applications*, Vol. 3, No. 3, pp. 351–361, 2009.
[135] S. Ma and E.-K. Boukas, A singular system approach to robust sliding mode control for uncertain Markov jump systems, *Automatica*, Vol. 45, No. 11, pp. 2707–2713, 2009.
[136] M. S. Mahmoud and P. Shi, *Methodologies for Control of Jump Time-Delay Systems*, Kluwer Academic Publishers, Boston, 2003.
[137] M. S. Mahmoud, Delay-dependent $\mathcal{H}_\infty$ filtering of a class of switched discrete-time state delay systems, *Signal Processing*, Vol. 88, No. 11, pp. 2709–2719, 2008.
[138] M. S. Mahmoud, Generalized $\mathcal{H}_2$ control of switched discrete-time systems with unknown delays, *Applied Mathematics and Computation*, Vol. 211, No. 1, pp. 33–44, 2009.
[139] M. S. Mahmoud and P. Shi, Asynchronous $\mathcal{H}_\infty$ filtering of discrete-time switched systems, *Signal Processing*, Vol. 92, No. 10, pp. 2356–2364, 2012.
[140] I. Malloci, J. Daafouz, and C. Iung, Stability and stabilization of two time scale switched systems in discrete time, *IEEE Transactions on Automatic Control*, Vol. 55, No. 6, pp. 1434–1438, 2010.
[141] Z. Man, A. P. Paplinski, and H. R. Wu, A robust MIMO terminal sliding mode control scheme for rigid robotic manipulators, *IEEE Transactions on Automatic Control*, Vol. 39, No. 12, pp. 2464–2469, 1994.
[142] Z. Man and X. Yu, Terminal sliding mode control of MIMO linear systems, *IEEE Transactions on Circuits and Systems I: Fundamental Theory and Applications*, Vol. 44, No. 11, pp. 1065–1070, 1997.
[143] J. L. Mancilla-Aguilar and R. A. Garcia, Some results on the stabilization of switched systems, *Automatica*, Vol. 49, No. 2, pp. 441–447, 2013.
[144] X. Mao and C. Yuan, *Stochastic Differential Equations with Markovian Switching*, Imperial College Press, 2006.
[145] X. Mao, *Stochastic Differential Equations and Applications*, 2nd Edition, Horwood, 2008.
[146] M. Margaliot and D. Liberzon, Lie-algebraic stability conditions for nonlinear switched systems and differential inclusions, *Systems and Control Letters*, Vol. 55, No. 1, pp. 8-16, 2006.
[147] N. Meskin and K. Khorasani, Fault detection and isolation of discrete-time Markovian jump linear systems with application to a network of multi-agent systems having imperfect communication channels, *Automatica*, Vol. 45, No. 9, pp. 2032–2040, 2009.
[148] A. N. Michel, Recent trends in the stability analysis of hybrid dynamical systems, *IEEE Transactions on Circuits and Systems – I: Fundamental Theory and Applications*, Vol. 46, No. 1, pp. 120–134, 1999.
[149] S. Mitra, D. Liberzon, and N. Lynch, Verifying average dwell time of hybrid systems, *ACM Transactions on Embedded Computing Systems*, Vol. 8, No. 1, pp. 1–37, 2008.
[150] N. Monshizadeh, H. L. Trentelman, and M. K. Camlibel, A simultaneous balanced truncation approach to model reduction of switched linear systems, *IEEE Transactions on Automatic Control*, Vol. 57, No. 12, pp. 3118–3131, 2012.
[151] A. S. Morse, Supervisory control of families of linear set-point controllers Part I. Exact matching, *IEEE Transactions on Automatic Control*, Vol. 41, No. 10, pp. 1413–1431, 1996.

[152] M. Nader and K. Khashayar, A geometric approach to fault detection and isolation of continuous-time Markovian jump linear systems, *IEEE Transactions on Automatic Control*, Vol. 55, No. 6, pp. 1343–1357, 2010.

[153] B. Niu and J. Zhao, Robust $\mathcal{H}_\infty$ control for a class of uncertain nonlinear switched systems with average dwell time, *International Journal of Control*, Vol. 86, No. 6, pp. 1107–1117, 2013.

[154] Y. Niu, J. Lam, X. Wang, and D. W. C. Ho, Observer-based sliding mode control for nonlinear state-delayed systems, *International Journal of Systems Science*, Vol. 35, No. 2, pp. 139–150, 2004.

[155] Y. Niu, W. C. H. Daniel, and J. Lam, Robust integral sliding mode control for uncertain stochastic systems with time-varying delay, *Automatica*, Vol. 41, No. 5, pp. 873–880, 2005.

[156] Y. Niu and D. W. C. Ho, Robust observer design for Itô stochastic time-delay systems via sliding mode control, *Systems and Control Letters*, Vol. 55, No. 10, pp. 781–793, 2006.

[157] Y. Niu, D. W. C. Ho, and X. Wang, Sliding mode control for Itô stochastic systems with Markovian switching, *Automatica*, Vol. 43, No. 10, pp. 1784–1790, 2007.

[158] Y. Niu, D. W. C. Ho, and X. Wang, Robust $\mathcal{H}_\infty$ control for nonlinear stochastic systems: a sliding-mode approach, *IEEE Transactions on Automatic Control*, Vol. 53, No. 7, pp. 1695–1701, 2008.

[159] T. Ooba and Y. Funahashi, On a common quadratic Lyapunov function for widely distant systems, *IEEE Transactions on Automatic Control*, Vol. 42, No. 2, pp. 1697–1699, 1997.

[160] Y. Orlov, W. Perruquetti, and J.-P. Richard, Sliding mode control synthesis of uncertain time-delay systems, *Asian Journal of Control*, Vol. 5, No. 4, pp. 568–577, 2003.

[161] Y. Orlov, Finite time stability and robust control synthesis of uncertain switched systems, *SIAM Journal on Control and Optimization*, Vol. 43, No. 4, pp. 1253–1271, 2005.

[162] S. Oucheriah, Exponential stabilization of linear delayed systems using sliding-mode controllers, *IEEE Transactions on Circuits and Systems I: Fundamental Theory and Applications*, Vol. 50, No. 6, pp. 826–830, 2003.

[163] P. Park, D. J. Choi, and S. G. Kong, Output feedback variable structure control for linear systems with uncertainties and disturbances, *Automatica*, Vol. 43, No. 1, pp. 72–79, 2007.

[164] P. Peleties, R. DeCarlo, Asymptotic stability of m-switched systems using Lyapunov-like functions, in *Proceedings of the American Control Conference*, Boston, Massachusetts, USA, 26–28 June, 1991, pp. 1679–1684.

[165] C. D. Persis, R. D. Santis, and A. S. Morse, Switched nonlinear systems with state-dependent dwell time, *Systems and Control Letters*, Vol. 50, No. 4, pp. 291–302, 2003.

[166] Y. Qiao and D. Cheng, On partitioned controllability of switched linear systems, *Automatica*, Vol. 45, No. 1, pp. 225–229, 2009.

[167] J. Qiu, G. Feng, and J. Yang, Robust mixed $\mathcal{H}_2/\mathcal{H}_\infty$ filtering design for discrete-time switched polytopic linear systems, *IET Control Theory and Applications*, Vol. 2, No. 5, pp. 420–430, 2008.

[168] J.-P. Richard, Time-delay systems: an overview of some recent advances and open problems, *Automatica*, Vol. 39, No. 10, pp. 1667–1694, 2003.

[169] O. J. Rojas, J. Bao, and P. L. Lee, On dissipativity, passivity and dynamic operability of nonlinear processes, *Journal of Process Control*, Vol. 18, No. 5, pp. 515–526, 2008.

[170] M. Rubagotti, A. Estrada, F. Castaños, A. Ferrara, and L. Fridman, Integral sliding mode control for nonlinear systems with matched and unmatched perturbations, *IEEE Transactions on Automatic Control*, Vol. 56, No. 11, pp. 2699–2704, 2011.

[171] C. Seatzu, D. Corona, A. Giua, and A. Bemporad, Optimal control of continuous-time switched affine systems, *IEEE Transactions on Automatic Control*, Vol. 51, No. 5, pp. 726–741, 2006.

[172] A. Sellami, D. Arzelier, R. M'hiri, and J. Zrida, A sliding mode control approach for systems subjected to a norm-bounded uncertainty, *International Journal of Robust and Nonlinear Control*, Vol. 17, No. 4, pp. 327–346, 2007.

[173] H. R. Shaker and R. Wisniewski, Model reduction of switched systems based on switching generalized gramians, *International Journal of Innovative Computing Information and Control*, Vol. 8, No. 7B, pp. 5025–5044, 2012.

[174] P. Shi, E.-K. Boukas, and R. K. Agarwal, Control of Markovian jump discrete-time systems with norm bounded uncertainty and unknown delay, *IEEE Transactions on Automatic Control*, Vol. 44, No. 11, pp. 2139–2144, 1999.

[175] P. Shi, E.-K. Boukas, and R. K. Agarwal, Kalman filtering for continuous-time uncertain systems with Markovian jumping parameters, *IEEE Transactions on Automatic Control*, Vol. 44, No. 8, pp. 1592–1597, 1999.

[176] P. Shi, E.-K. Boukas, and Y. Shi, On stochastic stabilization of discrete-time Markovian jump systems with delay in state, *Stochastic Analysis and Applications*, Vol. 21, No. 4, pp. 935–951, 2003.

[177] P. Shi, M. Mahmoud, S. Nguang, and A. Ismail, Robust filtering for jumping systems with mode-dependent delays, *Signal Processing*, Vol. 86, No. 1, pp. 140–152, 2006.

[178] P. Shi, Y. Xia, G. P. Liu, and D. Rees, On designing of sliding-mode control for stochastic jump systems, *IEEE Transactions on Automatic Control*, Vol. 51, No. 1, pp. 97–103, 2006.

[179] X. Shi, D.-W. Ding, and X. Li, Model reduction of discrete-time switched linear systems over finite-frequency ranges, *Nonlinear Dynamics*, Vol. 71, No. 1-2, pp. 361–370, 2013.

[180] R. Shorten and K. Narendra, Necessary and sufficient conditions for the existence of a common quadratic Lyapunov function for two stable second order linear time-invariant systems, in *Proceedings of the American Control Conference*, San Diego, California, USA, June, 1999, pp. 1410–1414.

[181] R. Shorten and K. Narendra, Necessary and sufficient conditions for the existence of a CQLF for a finite number of stable LTI systems, *International Journal of Adaptive Control and Signal Processing*, Vol. 16, No. 10, pp. 709–728, 2002.

[182] R. Shorten, K. Narendra, and O. Mason, A result on common quadratic Lyapunov functions, *IEEE Transactions on Automatic Control*, Vol. 48, No. 1, pp. 110–113, 2003.

[183] J. J. Slotine and S. S. Sastry, Tracking control of nonlinear systems using sliding surface with application to robot manipulator, *International Journal of Control*, Vol. 38, No. 2, pp. 931–938, 1983.

[184] S. K. Spurgeon, Sliding mode observers: a survey *International Journal of Systems Science*, Vol. 39, No. 8, pp. 751–764, 2008.

[185] G. Stikkel, J. Bokor, and Z. Szabo, Necessary and sufficient condition for the controllability of switching linear hybrid systems, *Automatica*, Vol. 40, No. 6, pp. 1093–1097, 2004.

[186] W.-C. Su, S. V. Drakunovs, and Ü. Özgüner, Constructing discontinuity surfaces for variable structure Systems: a Lyapunov approach, *Automatica*, Vol. 32, No. 6, pp. 925–928, 1996.

[187] M. Sun, J. Lam, S. Xu, and Y. Zou, Robust exponential stabilization for Markovian jump systems with mode-dependent input delay, *Automatica*, Vol. 43, No. 10, pp. 1799–1807, 2007.

[188] X. Sun, J. Zhao, and D. J. Hill, Stability and $\mathcal{L}_2$-gain analysis for switched delay systems: a delay-dependent method, *Automatica*, Vol. 42, No. 10, pp. 1769–1774, 2006.

[189] Z. Sun, S. Ge, and T. Lee, Controllability and reachability criteria for switched linear systems, *Automatica*, Vol. 38, No. 5, pp. 775–786, 2002.

[190] Z. Sun and S. Ge, Dynamic output feedback stabilization of a class of switched linear systems, *IEEE Transactions on Circuits and Systems I: Fundamental Theory and Applications*, Vol. 50, No. 8, pp. 1111–1115, 2003.

[191] Z. Sun and S. S. Ge, *Switched Linear Systems: Control and Design*, New York: Springer-Verlag, 2005.

[192] Z. Sun, Stabilization and optimization of switched linear systems, *Automatica*, Vol. 42, No. 5, pp. 783–788, 2006.

[193] R. H. C. Takahashi and P. L. D. Peres, $\mathcal{H}_2$ guaranteed cost-switching surface design for sliding modes with nonmatching disturbances, *IEEE Transactions on Automatic Control*, Vol. 44, No. 11, pp. 2214–2218, 1999.

[194] C. W. Tao, W.-Y. Wang, and M.-L. Chan, Design of sliding mode controllers for bilinear systems with time varying uncertainties, *IEEE Transactions on Systems, Man, and Cybernetics, Part B: Cybernetics*, Vol. 34, No. 1, pp. 639–645, 2004.

[195] E. Uezato, M. Ikeda, Strict LMI conditions for stability, robust stabilization, and $\mathcal{H}_\infty$ control of descriptor systems, in *Proceedings of the 38th Conference on Decision and Control*, Phoenix, Arizona USA, 7–10 December, pp. 4092–4097, 1999.

[196] V. I. Utkin, Variable structure systems with sliding modes, *IEEE Transactions on Automatic Control*, Vol. 22, No. 2, pp. 212–222, 1977.

[197] V. I. Utkin, *Sliding Modes in Control and Optimization*, Springer-Verlag, Berlin, 1992.

[198] V. I. Utkin and J. Shi, Integral sliding mode in systems operating under uncertainty conditions, in *Proceedings of the 35th Conference on Decision and Control*, Kobe, Japan, Dec. 1996, pp. 4591–4596.

[199] V. I. Utkin, J. Guldner, and J. Shi, *Sliding Mode Control in Electromechanical Systems*, Taylor and Francis, London, 1999.

[200] V. I. Utkin and H. Lee, Chattering problem in sliding mode control systems, in *Proceedings of the 2006 International Workshop on Variable Structure Systems*, Alghero, Italy, June 5–7, 2006, pp. 346–350.

[201] S. T. Venkcataraman and S. Gulati, Control of nonlinear systems using terminal sliding modes, in *Proceedings of the 1992 American Control Conference*, Chicago, Illinois, 24–26 June, 1992, pp. 1291–1292.

[202] D. Wang, W. Wang, and P. Shi, $\mathcal{H}_\infty$ filtering of discrete-time switched systems with state delays via switched Lyapunov function approach, *IEEE Transactions on Automatic Control*, Vol. 54, No. 6, pp. 1428–1429, 2009.

[203] D. Wang, W. Wang, and P. Shi, Robust fault detection for switched linear systems with state delays, *IEEE Transactions on Systems, Man, and Cybernetics, Part B: Cybernetics*, Vol. 39, No. 3, pp. 800–805, 2009.

[204] D. Wang, W. Wang, and P. Shi, Delay-dependent model reduction for continuous-time switched state-delayed systems, *International Journal of Adaptive Control and Signal Processing*, Vol. 25, No. 9, pp. 843–854, 2011.

[205] R. Wang, G. Jin, and J. Zhao, Robust fault-tolerant control for a class of switched nonlinear systems in lower triangular form, *Asian Journal of Control*, Vol. 9, No. 1, pp. 68–72, 2007.

[206] Y. Wang, G. Xie, and L. Wang, Controllability of switched time-delay systems under constrained switching, *Journal of Mathematical Analysis and Applications*, Vol. 286, No. 2, pp. 397–421, 2003.

[207] Z. Wang, H. Qiao, and K. J. Burnham, On stabilization of bilinear uncertain time-delay stochastic systems with Markovian jumping parameters, *IEEE Transactions on Automatic Control*, Vol. 47, No. 4, pp. 640–646, 2002.

[208] Z. Wang, J. Lam, and X. Liu, Robust filtering for discrete-time Markovian jump delay systems, *IEEE Signal Processing Letters*, Vol. 11, No. 8, pp. 659–662, 2004.

[209] L. K. Wong, F. H. F. Leung, and P. K. S. Tam, A chattering elimination algorithm for sliding mode control of uncertain non-linear systems, *Mechatronics*, Vol. 8, No. 7, pp. 765–775, 1998.

[210] L. Wu and J. Lam, Sliding mode control of switched hybrid systems with time-varying delay, *International Journal of Adaptive Control and Signal Process*, Vol. 22, No. 10, pp. 909–931, 2008.

[211] L. Wu, P. Shi, H. Gao, and C. Wang, $\mathcal{H}_\infty$ filtering for 2D Markovian jump systems, *Automatica*, Vol. 44, No. 7, pp. 1849–1858, 2008.

[212] L. Wu, C. Wang, and Q. Zeng, Observer-based sliding mode control for a class of uncertain nonlinear neutral delay systems, *Journal of The Franklin Institute*, Vol. 345, No. 3, pp. 233–253, 2008.

[213] L. Wu and D. W. C. Ho, Reduced-order $\mathcal{L}_2$-$\mathcal{L}_\infty$ filtering for a class of nonlinear switched stochastic systems, *IET Control Theory and Applications*, Vol. 3, No. 5, pp. 493–508, 2009.

[214] L. Wu, D. W. C. Ho, and J. Lam, $\mathcal{H}_\infty$ model reduction for continuous-time switched stochastic hybrid systems, *International Journal of Systems Science*, Vol. 40, No. 12, pp. 1241–1251, 2009.

[215] L. Wu and J. Lam, Weighted $\mathcal{H}_\infty$ filtering of switched systems with time-varying delay: average dwell time approach, *Circuits, Systems and Signal Processing*, Vol. 28, No. 6, pp. 1017–1036, 2009.

[216] L. Wu, T. Qi, and Z. Feng, Average dwell time approach to $\mathcal{L}_2$-$\mathcal{L}_\infty$ control of switched delay systems via dynamic output feedback, *IET Control Theory and Applications*, Vol. 3, No. 10, pp. 1425–1436, 2009.

[217] L. Wu and Z. Wang, Guaranteed cost control of switched systems with neutral delay via dynamic output feedback, *International Journal of Systems Science*, Vol. 40, No. 7, pp. 717–728, 2009.

[218] L. Wu and W. X. Zheng, Weighted $\mathcal{H}_\infty$ model reduction for linear switched systems with time-varying delay, *Automatica*, Vol. 45, No. 1, pp. 186–193, 2009.

[219] L. Wu and W. X. Zheng, Passivity-based sliding mode control of uncertain singular time-delay systems, *Automatica*, Vol. 45, No. 9, pp. 2120–2127, 2009.

[220] L. Wu, Z. Feng, and W. X. Zheng, Exponential stability analysis for delayed neural networks with switching parameters: average dwell time approach, *IEEE Transactions on Neural Networks*, Vol. 21, No. 9, pp. 1396–1407, 2010.

[221] L. Wu and D. W. C. Ho, Sliding mode control of singular stochastic hybrid systems, *Automatica*, Vol. 46, No. 4, pp. 779–783, 2010.

[222] L. Wu, D. W. C. Ho, and C. W. Li, Stabilisation and performance synthesis for switched stochastic systems, *IET Control Theory and Applications*, Vol. 4, No. 10, pp. 1877–1888, 2010.

[223] L. Wu, P. Shi, and H. Gao, State estimation and sliding-mode control of Markovian jump singular systems, *IEEE Transactions on Automatic Control*, Vol. 55, No. 5, pp. 1213–1219, 2010.

[224] L. Wu, D. W. C. Ho, and C. W. Li, Sliding mode control of switched hybrid systems with stochastic perturbation, *Systems and Control Letters*, Vol. 60, No. 8, pp. 531–539, 2011.

[225] L. Wu, X. Su, and P. Shi, Sliding mode control with bounded $\mathcal{L}_2$ gain performance of Markovian jump singular time-delay systems, *Automatica*, Vol. 48, No. 8, pp. 1929–1933, 2012.

[226] L. Wu, X. Yao, and W. X. Zheng, Generalized $\mathcal{H}_2$ fault detection for Markovian jumping two-dimensional systems, *Automatica*, Vol. 48, No. 8, pp. 1741–1750, 2012.

[227] L. Wu, W. X. Zheng, and H. Gao, Dissipativity-based sliding mode control of switched stochastic systems, *IEEE Transactions on Automatic Control*, Vol. 58, No. 3, pp. 785–791, 2013.

[228] M. Wu, Y. He, J.-H. She, and G.-P. Liu, Delay-dependent criteria for robust stability of time-varying delay systems, *Automatica*, Vol. 40, No. 8, pp. 1435–1439, 2004.

[229] Y. Wu and X. Yu, Variable structure control design for uncertain dynamic systems with disturbances in input and output channels, *Automatica*, Vol. 35, No. 2, pp. 311–319, 1999.

[230] Z. Wu, H. Su, and J. Chu, Delay-dependent $\mathcal{H}_\infty$ control for singular Markovian jump systems with time delay, *Optimal Control Applications and Methods*, Vol. 30, No. 5, pp. 443–461, 2009.

[231] Z. Wu, H. Su, and J. Chu, $\mathcal{H}_\infty$ filtering for singular Markovian jump systems with time delay, *International Journal of Robust and Nonlinear Control*, Vol. 20, No. 8, pp. 939–957, 2010.

[232] K. Wulff, F. Wirth, and R. Shorten, A control design method for a class of switched linear systems, *Automatica*, Vol. 45, No. 11, pp. 2592–2596, 2009.

[233] Z. Xi and T. Hesketh, Discrete time integral sliding mode control for systems with matched and unmatched uncertainties, *IET Control Theory and Applications*, Vol. 4, No. 5, pp. 889–896, 2010.

[234] Y. Xia and Y. Jia, Robust sliding-mode control for uncertain time-delay systems: an LMI approach, *IEEE Transactions on Automatic Control*, Vol. 48, No. 6, pp. 1086–1091, 2003.

[235] G. Xie, D. Zheng, and L. Wang, Controllability of switched linear systems, *IEEE Transactions on Automatic Control*, Vol. 47, No. 8, pp. 1401–1405, 2002.

[236] G. Xie and L. Wang, Necessary and sufficient conditions for controllability and observability of switched impulsive control systems, *IEEE Transactions on Automatic Control*, Vol. 49, No. 6, pp. 960–966, 2004.

[237] W. Xie, C. Wen, and Z. Li, Input-to-state stabilization of switched nonlinear systems, *IEEE Transactions on Automatic Control*, Vol. 46, No. 7, pp. 1111–1116, 2001.

[238] J. Xiong, J. Lam, H. Gao, and D. W. C. Ho, On robust stabilization of Markovian jump systems with uncertain switching probabilities, *Automatica*, Vol. 41, no. 5, pp. 897–903, 2005.

[239] J. Xiong and J. Lam, Stabilization of discrete-time Markovian jump linear systems via time-delayed controllers, *Automatica*, Vol. 42, No. 5, pp. 747–753, 2006.

[240] S. Xu and T. Chen, Robust $\mathcal{H}_\infty$ control for uncertain stochastic systems with state delay, *IEEE Transactions on Automatic Control*, Vol. 47, No. 12, pp. 2089–2094, 2002.

[241] S. Xu and T. Chen, Reduced-order $\mathcal{H}_\infty$ filtering for stochastic systems, *IEEE Transactions on Signal Processing*, Vol. 50, No. 12, pp. 2998–3007, 2002.

[242] S. Xu, T. Chen, and J. Lam, Robust $\mathcal{H}_\infty$ filtering for uncertain Markovian jump systems with mode-dependent time-delays, *IEEE Transactions on Automatic Control*, Vol. 48, No. 5, pp. 900–907, 2003.

[243] S. Xu and J. Lam, Improved delay-dependent stability criteria for time-delay systems, *IEEE Transactions on Automatic Control*, Vol. 50, No. 3, pp. 384–387, 2005.

[244] S. Xu and J. Lam, *Robust Control and Filtering of Singular Systems*, Berlin: Springer, 2006.

[245] S. Xu, J. Lam, and X. Mao, Delay-dependent $\mathcal{H}_\infty$ control and filtering for uncertain Markovian jump systems with time-varying delays, *IEEE Transactions on Circuits and Systems – I: Fundamental Theory and Applications*, Vol. 54, No. 9, pp. 2070–2077, 2007.

[246] S. Xu and J. Lam, A survey of linear matrix inequality techniques in stability analysis of delay systems, *International Journal of Systems Science*, Vol. 39, No. 12, 1095–1113, 2008.

[247] X. Xu and P. J. Antsaklis, Stabilization of second-order LTI switched systems, *International Journal of Control*, Vol. 73, No. 14, pp. 1261–1279, 2000.

[248] X. Xu and P. J. Antsaklis, Optimal control of switched systems based on parameterization of the switching instants, *IEEE Transactions on Automatic Control*, Vol. 49, No. 1, pp. 2–16, 2004.

[249] X. Xu and G. Zhai, Practical stability and stabilization of hybrid and switched systems, *IEEE Transactions on Automatic Control*, Vol. 50, No. 11, pp. 1897–1903, 2005.

[250] J.-J. Yan, Sliding mode control design for uncertain time-delay systems subjected to a class of nonlinear inputs, *International Journal of Robust and Nonlinear Control*, Vol. 13, No. 6, pp. 519–532, 2003.

[251] W. Yan and J. Lam, An approximate approach to $\mathcal{H}_2$ optimal model reduction, *IEEE Transactions on Automatic Control*, Vol. 44, No. 7, pp. 1341–1358, 1999.

[252] X. Yan, S. K. Spurgeon, and C. Edwards, Sliding mode control for time-varying delayed systems based on a reduced-order observer, *Automatica*, Vol. 46, No. 8, pp. 1354–1362, 2010.

[253] X. Yan, S. K. Spurgeon, and C. Edwards, Static output feedback sliding mode control for time-varying delay systems with time-delayed nonlinear disturbances, *International Journal of Robust and Nonlinear Control*, Vol. 20, No. 7, pp. 777–788, 2010.

[254] X. Yao, L. Wu, W. X. Zheng, and C. Wang, Robust $\mathcal{H}_\infty$ filtering of Markovian jump stochastic systems with uncertain transition probabilities, *International Journal of Systems Science*, Vol. 42, No. 7, pp. 1219–1230, 2011.

[255] X. Yao, L. Wu, and W. X. Zheng, Fault detection filter design for Markovian jump singular systems with intermittent measurements, *IEEE Transactions on Signal Processing*, Vol. 59, No. 7, pp. 3099–3109, 2011.

[256] X. Yao, L. Wu, W. X. Zheng, and C. Wang, Quantized $\mathcal{H}_\infty$ filtering for Markovian jump LPV systems with intermittent measurements, *International Journal of Robust and Nonlinear Control*, Vol. 23, No. 1, pp. 1–14, 2013.

[257] H. Ye, A. Michel, L. Hou, Stability theory for hybrid dynamical systems, *IEEE Transactions on Automatic Control*, Vol. 43, No. 4, pp. 461–474, 1998.

[258] K. D. Young and S. Drakunov, Sliding mode control with chattering reduction, in *Proceedings of the 1992 American Control Conference*, Chicago, Illinois, 24–26 June, 1992, pp. 1291–1292.

[259] K. D. Young, Ed., *Variable Structure Control for Robotics and Aerospace Applications*. New York: Elsevier, 1993.

[260] K. D. Young, V. I. Utkin, and U. Ü. Özgüner, A control engineer's guide to sliding mode control, in *Proceedings of the IEEE International Workshop on Variable Structure Systems*, pages 1–14, Tokyo, Japan, 1996.

[261] J. Yu, F. Liu, X. Yu, C. Wu, and L. Wu, Fault detection of discrete-time switched systems with distributed delays: input-output approach, *International Journal of Systems Science*, Vol. 44, No. 12, pp. 2255–2272, 2013.

[262] X. Yu, C. Wu, F. Liu, and L. Wu, Sliding mode control of discrete-time switched systems with time-delay, *Journal of The Franklin Institute*, Vol. 350, No. 1, pp. 19–33, 2013.

[263] S. H. Zak and S. Hui, On variable structure output feedback controllers for uncertain dynamic systems, *IEEE Transactions on Automatic Control*, Vol. 38, No. 10, pp. 1509–1512, 1993.

[264] G. Zhai, B. Hu, K. Yasuda, and A. N. Michel, Disturbance attenuation properties of time-controlled switched systems, *Journal of The Franklin Institute*, Vol. 338, No. 7, pp. 765–779, 2001.

[265] G. Zhai, B. Hu, K. Yasuda, and A. N. Michel, Qualitative analysis of discrete-time switched systems, in *Proceedings of the American Control Conference*, Anchorage, Alaska, USA, May 8–10, 2002, pp. 1880–1885.

[266] L. Zhang, B. Huang, and J. Lam, $\mathcal{H}_\infty$ model reduction of Markovian jump linear systems, *Systems and Control Letters*, Vol. 50, No. 2, pp. 103–118, 2003.

[267] W. Zhang, A. Abate, J. Hu, and M. P. Vitus, Exponential stabilization of discrete-time switched linear systems, *Automatica*, Vol. 45, No. 11, pp. 2526–2536, 2009.

[268] J. Zhao and D. J. Hill, On stability, $\mathcal{L}_2$-gain and $\mathcal{H}_\infty$ control for switched systems, *Automatica*, Vol. 44, No. 5, pp. 1220–1232, 2008.

[269] J. Zhao and D. J. Hill, Dissipativity theory for switched systems, *IEEE Transactions on Automatic Control*, Vol. 53, No. 4, pp. 941–953, 2008.

[270] S. Zhao and J. Sun, Controllability and observability for time-varying switched impulsive controlled systems, *International Journal of Robust and Nonlinear Control*, Vol. 20, No. 12, pp. 1313–1325, 2010.

[271] Y. Zhao, H. Gao, J. Lam, and B. Du, Stability and stabilization of delay T-S fuzzy systems: a delay partitioning approach, *IEEE Transactions on Fuzzy Systems*, Vol. 17, No. 4, pp. 750–762, 2009.

[272] Y. Zheng, G. Feng, and J. Qiu, Exponential $\mathcal{H}_\infty$ filtering for discrete-time switched state-delay systems under asynchronous switching, *Asian Journal of Control*, Vol. 15, No. 2, pp. 479–488, 2013.

[273] M. Zhong, J. Lam, S. X. Ding, and P. Shi, Robust fault detection of Markovian jump systems, *Circuits, Systems and Signal Processing*, Vol. 23, No. 5, pp. 387–407, 2004.

[274] M. Zhong, H. Ye, P. Shi, and G. Wang, Fault detection for Markovian jump systems, *IEE Proceedings – Control Theory and Applications*, Vol. 152, No. 4, pp. 397–402, 2005.

Index

Sliding Mode Control of Uncertain Parameter-Switching Hybrid Systems, First Edition. Ligang Wu, Peng Shi and Xiaojie Su.